ENGINEERING PROBLEM SOLVING WITH C

FOURTH EDITION

Delores M. Etter

Department of Electrical Engineering
Southern Methodist University
Dallas, TX

International Edition contributions by
Moumita Mitra Manna
Department of Computer Science
Bangabasi College
Kolkata

PEARSON

Boston   Columbus   Indianapolis   New York   San Francisco   Upper Saddle River
Amsterdam   Cape Town   Dubai   London   Madrid   Milan   Munich   Paris   Montréal   Toronto
Delhi   Mexico City   São Paulo   Sydney   Hong Kong   Seoul   Singapore   Taipei   Tokyo

Editorial Director: *Marcia Horton*
Editor in Chief: *Michael Hirsch*
Acquisitions Editor: *Tracy Dunkelberger*
Editorial Assistant: *Emma Snider*
Director of Marketing: *Patrice Jones*
Marketing Manager: *Yez Alayan*
Marketing Coordinator: *Kathryn Ferranti*
Director of Production: *Vince O'Brien*
Managing Editor: *Jeff Holcomb*
Production Project Manager: *Kayla Smith-Tarbox*
Publisher, International Edition: *Angshuman Chakraborty*
Acquisitions Editor, International Edition:
    Somnath Basu
Publishing Assistant, International Edition: *Shokhi Shah*
Print and Media Editor, International Edition:
    Ashwitha Jayakumar

Project Editor, International Edition: *Jayashree Arunachalam*
Publishing Administrator, International Edition:
    Hema Mehta
Operations Supervisor: *Alan Fischer*
Manufacturing Buyer: *Lisa McDowell*
Art Director: *Anthony Gemmellaro*
Text and Cover Designer: *Jodi Notowitz*
Manager, Visual Research: *Karen Sanatar*
Photo Researcher: *AV Katy Holihan*
Manager, Rights and Permissions: *Michael Joyce*
Text Permission Coordinator: *Nicole Coffineau*
Lead Media Project Manager: *Daniel Sandin*
Full-Service Project Management: *Integra*
Composition: *Integra*
Cover Printer: *Moore Langen*

Pearson Education Limited
Edinburgh Gate
Harlow
Essex CM20 2JE
England

and Associated Companies throughout the world

Visit us on the World Wide Web at:
www.pearsoninternationaleditions.com

© Pearson Education Limited 2013

ISBN 10: 0-273-76820-4
ISBN 13: 978-0-273-76820-3

British Library Cataloguing-in-Publication Data
A catalogue record for this book is available from the British Library

10 9 8 7 6 5 4 3 2 1
14 13 12

Typeset in 10/12, Times by Integra
Printed and bound by Courier/Westford in The United States of America

The publisher's policy is to use paper manufactured from sustainable forests.

In memory of my dearest Mother,
Muerladene Janice Van Camp

PREFACE

Engineers use computers to solve a variety of problems ranging from the evaluation of a simple function to solving a system of nonlinear equations. Thus, **C has become the language of choice for many engineers and scientists**, not only because it has powerful commands and data structures, but also because it can easily be used for system-level operations. Since C is a language that a new engineer is likely to encounter in a job, it is a good choice for an introduction to computing for engineers. Therefore, this text was written to introduce engineering problem solving with the following **objectives**:

- to develop a **consistent methodology for solving engineering problems**;

- to present the **fundamental capabilities of C**, the language of choice for many practicing engineers and scientists; and

- to illustrate the problem-solving process with C through a variety of **interesting engineering examples and applications**.

To accomplish these objectives, **Chapter 1 presents a five-step process** that is used consistently in the rest of the text for solving engineering problems. **Chapters 2 through 7 present the fundamental capabilities of C** for solving engineering problems. **Chapter 8 is an introduction to object-oriented programming using C++**. Object-oriented programming is gaining popularity in many fields of engineering and science, and is likely to be seen in the workplace. Throughout all these chapters, we present a large number of examples from many different engineering and scientific disciplines. The solutions to these examples are developed using the five-step process and ANSI C (and ANSI C++ in Chapter 8), which are the standards developed by the American National Standards Institute.

Changes to the Fourth Edition

- The new theme for this edition is Crime Scene Investigation (CSI). Learning about the technology behind crime scene investigation is not only very interesting, but it provides a number of problems for which we can develop C program solutions.

- Section 1.2 has been rewritten to include discussion on current topics such as cloud computing and kernels.

- A new four-color insert has been added to define an important area of crime scene investigation—biometrics. Biometrics is a term used to describe the physical or behavioral characteristics that can be used to identify a person. The insert includes discussion on fingerprints, face recognition, iris recognition, DNA, and speech recognition.

- Each chapter begins with a photo and a related discussion on a technology used in crime scene investigation. Then, within each chapter after Chapter 1, an associated application section has been added so that in addition to learning all the key features of C, you will also learn about forensic anthropology, face recognition and surveillance video, iris recognition, speech analysis and speech recognition, DNA analysis, fingerprint recognition, and hand recognition. In these application sections, we develop a C solution to a problem related to the crime scene technology.

- New Modify! problems have been added to each new application.

- The material in Chapter 8 on C++ has been updated to reflect the new C++ standards.

Prerequisites

No prior experience with the computer is assumed. The **mathematical prerequisites are college algebra and trigonometry**. Of course, the initial material can be covered much faster if the student has used other computer languages or software tools.

Course Structure

The material in these chapters was selected to provide the basis for a **one-term course in engineering computing**. These chapters contain the essential topics of mathematical computing, character data, control structures, functions, arrays, pointers, and structures. Students with a background in another computer language should be able to complete this material in less than a semester. A minimal course that provides only an introduction to C can be designed using the nonoptional sections of the text. (Optional sections are indicated in the table of contents.) There are three ways to use the text, along with the recommended chapter sections:

- **Introduction to C**. Many freshman courses introduce the student to several computer tools in addition to a language. For these courses, we recommend covering **the non-optional sections of Chapters 1 through 5**. This material introduces students to the fundamental capabilities of C, and they will be able to write substantial programs using mathematical computations, character data, control structures, functions, and arrays.

- **Problem solving with C**. In a semester course devoted specifically to teaching students to **master the C language**, we recommend covering **all non-optional sections of Chapters 1 through 7**. This material covers all the fundamental concepts of the C language, including mathematical computations, character data, control structures, functions, arrays, pointers, and structures.

- **Problem solving with C and numerical techniques**. A number of sections included in the text cover common numerical techniques, such as linear interpolation, linear modeling, finding roots of polynomials, and solutions to simultaneous equations. Including these along with the sections on the C language provides a strong combination for students who may need to use numerical techniques in their course work. This coverage would include **all sections of Chapters 1 through 7**.

Many students may be interested in reading about some of the additional **object-oriented features found in C++**. We recommend that students cover all non-optional sections of Chapters 1 through 7 before reading **Chapter 8**.

Problem-Solving Methodology

The emphasis on engineering and scientific problem solving is an integral part of the text. Chapter 1 introduces **a five-step process** for solving engineering problems using the computer. This five-step problem-solving process was developed by the author of this text early in her academic career, and it has been successfully used by the many thousands of students who were in her classes or used one of her textbooks. This successful process has also been adopted by a number of other authors. The five steps are:

1. **State the problem clearly.**
2. **Describe the input and output information.**
3. **Work a simple example by hand.**
4. **Develop an algorithm and convert it to a computer program.**
5. **Test the solution with a variety of data.**

To reinforce the development of problem-solving skills, each of these five steps is clearly identified each time that a complete engineering problem is solved. In addition, **top-down design and stepwise refinement are presented with the use of decomposition outlines, pseudocode, and flowcharts**.

Engineering and Scientific Applications

Throughout the text, emphasis is placed on incorporating real-world engineering and scientific examples and problems. Some examples to illustrate this **wide variety of engineering applications are**

- salinity of sea water
- velocity computation
- amino acid molecular weights
- wind tunnels
- ocean wave interactions
- ozone measurements
- sounding rocket trajectory
- suture packaging
- timber regrowth
- critical path analysis
- weather balloons
- iceberg tracking
- instrumentation reliability
- system stability
- component reliability
- flight simulator wind speeds
- hurricane categories
- molecular weights
- speech signal analysis
- terrain navigation
- electrical circuit analysis

- power plant data
- cryptography
- temperature distribution
- El Niño–Southern Oscillation
- seismic event detection
- tsunami analysis
- surface wind directions

In addition, each chapter begins with a discussion of some aspect of the **new theme**. Later in the chapter, we solve **a problem that relates to the introductory discussion on the technology behind crime scene investigation**. These problems address the following applications:

- forensic anthropology
- face recognition and surveillance video
- iris recognition
- speech analysis
- DNA analysis
- fingerprint recognition
- hand recognition

ANSI C

The statements presented and all programs developed use the **C standards** developed by the American National Standards Institute. By using ANSI C, students learn to write **portable code** that can be transferred from one computer system to another.

Software Engineering Concepts

Engineers and scientists are expected to develop and implement **user-friendly and reusable computer solutions**. Learning software engineering techniques is crucial to successfully developing these computer solutions. **Readability and documentation** are stressed in the development of programs. Additional topics that relate to software engineering issues are discussed throughout the text and include issues such as **software life cycle, portability, maintenance, modularity, recursion, abstraction, reusability, structured programming, validation, and verification**.

Four Types of Problems

Learning any new skill requires practice at several different levels of difficulty. Four types of exercises are used throughout the text to develop problem-solving skills. The first set of exercises is **Practice! problems**. These are short-answer questions that **relate to the section of the material just presented**. Most sections are immediately followed by a set of Practice! problems so that students can determine whether they are ready to continue to the next section. Complete solutions to all the Practice! problems are included at the end of the text.

The **Modify! problems** are designed to provide **hands-on experience with the programs developed in the Problem Solving Applied sections**. In these sections, we develop a complete C program using the five-step process. The Modify! problems ask students to run the program with different sets of data to test their understanding of how the program works and of the relationships among the engineering variables. These exercises also ask the students to make simple modifications to the program and then run the program to test their changes. Selected solutions to some of the Modify! problems are included at the end of the text.

Each chapter ends with two sets of problems. The **Short-Answer problems** include **true/false problems, multiple choice problems, matching problems, syntax problems, fill-in-the-blank problems, memory snapshot problems, program output problems, and program segment analysis problems**. Complete solutions to all the Short-Answer problems are included at the end of the text.

The final set of problems in each chapter (except for Chapter 1) are **Programming problems**. These are **new problems that relate to a variety of engineering applications**. The level of difficulty ranges from very straightforward to longer project assignments. Each problem requires that the students develop a complete C program or function. Selected solutions to the programming problems are included at the end of the text. Complete solutions to the programming problems are available for instructors. Instructor resources for the International Edition are accessible at www.pearsoninternationaleditions.com/etter.

Study and Programming Aids

Margin notes are used to help the reader not only identify the important concepts, but also to easily locate specific topics. In addition, margin notes are used to identify programming style guidelines and debugging information. **Style guidelines** show students how to write C programs that incorporate good software discipline; **debugging notes** help students recognize common errors so that they can avoid them. The programming style notes are indicated with a margin note, and the debugging notes are indicated with a bug icon. Each Chapter Summary contains a summary of the style notes and debugging notes, plus a list of the **Key Terms** from the chapter and a **C Statement Summary** of the new statements to make the book easier to use as a reference. The combined list of these key terms, along with their definitions, is included in a **Glossary** at the end of the text. In addition, the **inside of the front cover contains common functions and the precedence table; the inside of the back cover contains examples of most of the C statements**.

Optional Numerical Techniques

Numerical techniques that are commonly used in solving engineering problems are also discussed in the text, and they include **interpolation, linear modeling (regression), root finding, and the solution to simultaneous equations**. The concept of a matrix is also introduced and then illustrated using a number of examples. All of these topics are presented **assuming only a trigonometry and college algebra background**.

MATLAB and Visualization

The **visualization** of the information related to a problem and its solution is a **critical component in understanding and developing the intuition necessary to be a creative engineer**. Therefore, we have included a number of plots of data throughout the text to illustrate the relationships of the information needed to solve specific problems. All the plots were generated **using MATLAB, a powerful environment for numerical computations, data analysis, and visualization**. We have also included an appendix that shows how to generate a simple plot from data that have been stored in a text file; this text file could be generated with a word processor or it could be generated by a C program.

Appendices

To further enhance reference use, the appendices include a number of important topics. Appendix A contains a discussion of the components in the **ANSI C Standard Library**. Appendix B presents the **ASCII character codes**. Appendix C shows how to use **MATLAB to plot data** from ASCII files; this allows students to generate ASCII files with their C programs and to plot the values using MATLAB.

Nontechnical Skills

The engineer of the twenty-first century needs many skills and capabilities in addition to the technical ones learned in an engineering program. In Chapter 1, we present a brief discussion on some of these nontechnical skills that are so important to engineers. Specifically, we discuss developing **both oral and written communications skills**, understanding the **design/ process/manufacture path** that takes an idea and leads to a product, working in **interdisciplinary teams**, understanding the **world marketplace**, the importance of **synthesis as well as analysis**, and the importance of **ethics and other societal concerns** in engineering solutions. While this text is devoted primarily to teaching problem-solving skills and the C language, we have attempted to tie these other nontechnical topics into many of the problems and discussions in the text.

Additional Resources

All instructor and student resources can be accessed at www.pearsoninternationaleditions .com/etter. Here, students can access student data files for the book, and instructors can register for the password-protected Instructor's Resource Center. The IRC contains complete solutions to all of the Programming Projects found at the end of each chapter, and a complete set of PowerPoint lecture slides.

Acknowledgments

A number of people have made significant contributions to this text. Students are always the best judge of "what works" and "what doesn't work." I appreciate the feedback from students who had never used the computer when they started this text, to undergraduates who already knew another language, and to graduate students who wanted to use C to do their research analysis. The comments and suggestions from these students greatly improved the text.

A constructive, but critical, review is extremely important in improving a text. The many reviewers who provided this critical guidance included Murali Narayanan (Kansas State University), Kyle Squires (Arizona State University), Amelia Regan (University of California at Irvine), Hyeong-Ah Choi (George Washington University), George Friedman (University of Illinois, Champaign), D. Dandapani (University of Colorado, Colorado Springs), Karl Mathias (Auburn University), William Koffke (Villanova University), Paul Heinemann

(Pennsylvania State University), A. S. Hodel (Auburn University), Armando Barreto (Florida International University), Arnold Robbins (Georgia Technology College of Computing), Avelino Gonzalez (University of Central Florida), Thomas Walker (Virginia Polytechnic Institute and State University), Christopher Skelly (Insight Resource Inc.), Betty Barr (The University of Houston), John Cordero (University of Southern California), A. R. Marundarajan (Cal Poly, Pomona), Lawrence Genalo (Iowa State University), Karen Davis (University of Cincinnati), Petros Gheresus (General Motors Institute), Leon Levine (UCLA), Harry Tyrer (University of Missouri, Columbia), Caleb Drake (University of Illinois, Chicago), John Miller (University of Michigan, Dearborn), Elden Heiden (New Mexico State University), Joe Hootman (University of North Dakota), Nazeih Botros (Southern Illinois University), Mark C. Petzold (St. Cloud State University), Ali Saman Tosun (University of Texas at San Antonio), Turgay Korkmaz (University of Texas at San Antonio), Billie Goldstein (Temple University), Mark S. Hutchenreuther (California Polytechnic State University), Frank Friedman (Temple University), and Harold Mitchell Jr. (University of Houston).

The outstanding team at Pearson Education continues to be a delight to work with on my book projects. They include Marcia Horton, Tracy (Dunkelberger) Johnson, Emma Snider, Kayla Smith-Tarbox, and Eric Arima. I want to thank Jeanine Ingber (University of New Mexico) for her contributions as a co-author of the second edition; many of her contributions remain in this fourth edition.

DELORES M. ETTER
Department of Electrical Engineering
Southern Methodist University
Dallas, TX

The publishers would like to thank Mohit P. Tahiliani, of the National Institute of Technology Karnataka, Surathkal, for reviewing the content of the International Edition.

CONTENTS

---

*Optional section.

3 Control Structures and Data Files   101

*Crime Scene Investigation: Face Recognition
and Surveillance Video*

___
*Optional section.

*Optional section.

6 Programming with Pointers

Crime Scene Investigation: DNA Analysis

*Optional section.

---

*Optional section.

ENGINEERING APPLICATIONS_____

Aerospace Engineering

Wind Tunnel Data Analysis (Chapter 2 Problems, p. 98;
Chapter 5 Problems, p. 302)
Sounding Rockets (Chapter 3 Problems, p. 163)
Flight Simulator Wind Speed (Chapter 4 Problems, p. 228)

Biomedical Engineering

Suture Packaging (Chapter 3 Problems, p. 164)

Chemical Engineering

Temperature Conversions (Chapter 3 Problems, p. 163)
Molecular Weights (Section 5.3, p. 246)
Temperature Distribution (Chapter 5 Problems, p. 305)

Computer Engineering

Simulations (Chapter 4 Problems, p. 226)
Cryptography (Chapter 5 Problems, p. 304)
Pattern Recognition (Chapter 6 Problems, p. 352)

Crime Scene Investigation

Forensic Anthropology (Section 2.5, p. 68)
Face Recognition (Section 3.4, p. 117)
Iris Recognition (Section 4.3, p. 183)
Speech Recognition (Section 5.5, p. 256)
DNA Sequencing (Section 6.7, p. 338)
Fingerprint Analysis (Section 7.3, p. 362)
Hand Recogntion (Section 8.5, p. 402)

Electrical Engineering

Electrical Circuit Analysis (Section 5.12, p. 292)
Noise Simulations (Chapter 5 Problems, p. 302)
Power Plant Distribution (Chapter 5 Problems, p. 303)

Enviromental Engineering

Ozone Measurements (Section 3.9, p. 151)
Timber Regrowth (Chapter 3 Problems, p. 164)
Weather Balloons (Chapter 3 Problems, p. 165)
Seismic Event Detection (Section 6.5, p. 329)

Genetic Engineering

Manufacturing Engineering

Mechanical Engineering

Ocean Engineering

ENGINEERING PROBLEM SOLVING WITH C

CHAPTER ONE

Crime Scene Investigation

We are all familiar with the investigation of crime scenes, from movies, books, and TV shows. However, you may not be aware of the technology behind many aspects of crime scene investigation. Learning about this technology is not only very interesting, but it also provides a theme that we will use throughout the text as we are learning about the C language. Starting with Chapter 2, we present in each chapter an aspect of crime scene investigation and explain more about the technology behind it. We then present a problem related to that aspect of crime scene investigation and solve the problem using a C language program. Additional information related to crime scene investigation is included in the four-page color insert that defines biometrics and gives a number of examples of how biometrics and related technology are used to identify a person.

ENGINEERING PROBLEM SOLVING

CHAPTER OUTLINE

OBJECTIVES *In this chapter, we introduce you to*

- recent outstanding engineering achievements,
- the changing engineering environment and the nontechnical skills needed to successfully adapt to this environment,

- computer systems, in terms of both hardware and software, and
- a five-step problem-solving technique that we will use throughout the text.

1.1 Engineering in the 21st Century

Engineers solve real-world problems using scientific principles from disciplines that include computer science, mathematics, physics, biology, and chemistry. It is this variety of subjects, and the challenge of real problems, that makes engineering so interesting and so rewarding. In this section, we present some of the outstanding engineering achievements of recent years. We also consider some of the nontechnical skills and capabilities needed by the engineers of the twenty-first century.

Recent Engineering Achievements

10 engineering achievements

Since the development of the computer in the late 1950s, a number of significant engineering achievements have occurred. In 1989, the National Academy of Engineering selected the **10 engineering achievements** that it considered to be the most important accomplishments during the previous 25 years. These achievements illustrate the multidisciplinary nature of engineering and demonstrate how engineering has improved our lives and expanded the possibilities for the future while providing a wide variety of interesting and challenging careers. We now briefly discuss these 10 achievements.

Microprocessor

The development of the **microprocessor**, a tiny computer smaller than a postage stamp, is one of the top engineering achievements of the last 25 years. Microprocessors are used in electronic equipment, household appliances, toys, and games, as well as in automobiles, aircraft, and space shuttles, because they provide powerful yet inexpensive computing capabilities. Microprocessors also provide the computing power inside calculators and smart phones.

MICROPROCESSOR

Moon landing

Several of the top 10 achievements relate to the exploration of space. The **moon landing** was probably the most complex and ambitious engineering project ever attempted. Major breakthroughs were required in the design of the Apollo spacecraft, the lunar lander, and the three-stage Saturn V rocket. Even the design of the spacesuit was a major engineering project that resulted in a system that included a three-piece suit and backpack, which together weighed 190 pounds. The computer played a key role not only in the design of the various systems, but also in the communications required during an individual moon flight. A single flight required the coordination of over 450 people in the launch control center and over 7000 others on nine ships, in 54 aircraft, and at stations located around the earth.

MOON LANDING

Application
satellites

The space program also provided much of the impetus for the development of **application satellites** that are used to provide weather information, relay communication signals, map uncharted terrain, and provide environmental updates on the composition of the atmosphere. The Global Positioning System (GPS) is a constellation of 24 satellites that broadcasts position, velocity, and time information worldwide. GPS receivers measure the time it takes for signals to travel from the GPS satellite to the receiver. Using information received from four satellites, a microprocessor in the receiver can determine very precise measurements of the receiver's location; its accuracy varies from a few meters to centimeters, depending on the computation techniques used.

SATELLITE

Computer-aided
design and
manufacturing

Another of the top engineering achievements recognizes the contributions of **computer-aided design and manufacturing** (CAD/CAM). CAD/CAM has generated a new industrial revolution by increasing the speed and efficiency of many types of manufacturing processes. CAD allows the design to be done using the computer, which then produces the final schematics, parts lists, and computer simulation results. CAM uses design results to control machinery or industrial robots to manufacture, assemble, and move components.

COMPUTER-AIDED DESIGN

Jumbo jet

The **jumbo jet** originated from the U.S. Air Force C-5A cargo plane that began operational flights in 1969. Much of the success of the jumbo jets can be attributed to the high-bypass fanjet that allows them to fly farther with less fuel and with less noise than previous jet engines. The core of the engine operates like a pure turbojet, in which compressor blades pull air into the engine's combustion chamber. The hot expanding gas thrusts the engine forward, and at the same time spins a turbine that drives the compressor and the large fan on the front of the engine. The spinning fan provides the bulk of the engine's thrust.

JUMBO JET

Advanced composite materials

The aircraft industry was also the first industry to develop and use **advanced composite materials** that consist of materials that can be bonded together in such a way that one material reinforces the fibers of the other material. Advanced composite materials were developed to provide lighter, stronger, and more temperature-resistant materials for aircraft and spacecraft. New markets for composites now exist in sporting goods. For example, downhill snow skis use layers of woven Kevlar fibers to increase their strength and reduce weight, and golf club shafts of graphite and epoxy are stronger and lighter than the steel in conventional shafts. Composite materials are also used in the design of prosthetics for artificial limbs.

Computerized axial tomography

The areas of medicine, bioengineering, and computer science were teamed for the development of the CAT (**computerized axial tomography**) scanner machine. This instrument can generate three-dimensional images or two-dimensional slices of an object using X rays that are generated from different angles around the object. Each X ray measures a density from its angle, and complicated computer algorithms combine the information from all the X rays to reconstruct a clear image of the inside of the object. CAT scans are routinely used to identify tumors, blood clots, and brain abnormalities.

ADVANCED COMPOSITE MATERIALS

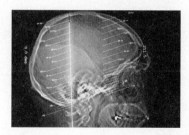

CAT SCAN

Genetic
engineering

   Genetic engineering, combining the work of geneticists and engineers, has resulted in many new products, ranging from insulin, to growth hormones, to infection-resistant vegetables. A genetically engineered product is produced by splicing a gene that produces a valuable substance from one organism into another organism that will multiply itself and the foreign gene along with it. The first commercial genetically engineered product was human insulin, which appeared under the trade name Humulin. Current work is investigating the use of genetically altered microbes to clean up toxic waste and to degrade pesticides.

GENETIC ENGINEERING

Lasers

Lasers are light waves that have the same frequency and travel in a narrow beam that can be directed and focused. CO_2 lasers are used to drill holes in materials that range from ceramics to composite materials. Lasers are also used in medical procedures to weld detached retinas, seal leaky blood vessels, vaporize brain tumors, and perform delicate inner-ear surgery. Three-dimensional pictures called holograms are also generated with lasers.

LASERS

Optical fiber

Fiber-optic communications use **optical fiber**, a transparent glass thread that is thinner than a human hair. This optical fiber can carry more information than either radio waves or electric waves in copper telephone wires, and it does not produce electromagnetic waves that

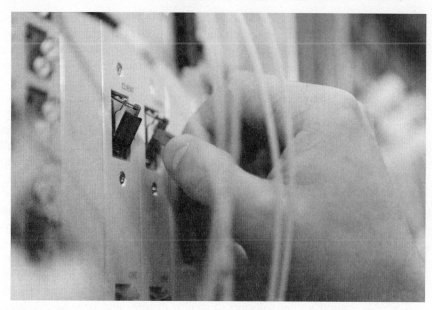

FIBER OPTICS

can cause interference on communication lines. Transoceanic fiber-optic cables provide communication channels between continents. Fiber optics are also used in medical instrumentation to allow surgeons to thread light into the human body for examination and laser surgery.

Changing Engineering Environment

The engineer of the twenty-first century works in an environment that requires many nontechnical skills and capabilities [2]. Although the computer will be the primary computational tool of most engineers, the computer will also be useful in developing additional nontechnical abilities.

Communication skills

Engineers need strong **communication skills** both for oral presentations and for the preparation of written materials. Computers provide the software to assist in writing outlines and developing materials, such as graphs, for presentations and technical reports. The problems at the end of this chapter include written and oral presentations to provide practice of these important skills.

Design/process/ manufacture path

The **design/process/manufacture path**, which consists of taking an idea from a concept to a product, is one that engineers must understand firsthand. Computers are used in every step of this process, from design analysis, machine control, robotic assembly, quality assurance, and market analysis. Several problems in the text relate to these topics. For example, in Chapter 4, programs are developed to simulate the reliability of systems that use multiple components.

Interdisciplinary teams

Engineering teams of the future will be **interdisciplinary teams**, just as the engineering teams of today are interdisciplinary teams. The discussions of the top 10 engineering achievements of the last 25 years clearly show the interdisciplinary nature of those achievements. Learning to interact in teams and to develop organizational structures for effective team communication is important for engineers. A good way to begin developing engineering team skills is to organize teams to study for exams. Assign specific topics to members of the team; then have them review these topics for the team, with examples and potential test questions.

World marketplace

Engineers need to understand the **world marketplace**. This involves understanding different cultures, political systems, and business environments. Courses in these topics and in foreign languages help provide some understanding, but exchange programs with international experiences provide invaluable knowledge in developing a broader world understanding.

Engineers are problem solvers, but problems are not always formulated carefully. An engineer must be able to extract a problem statement from a problem discussion and then determine the important issues. This involves not only developing order, but also learning to correlate chaos. It means not only **analyzing** the data, but also **synthesizing** a solution using many pieces of information. The integration of ideas can be as important as the decomposition of the problem into manageable pieces. A problem solution may involve not only abstract thinking about the problem, but also experimental learning from the problem environment.

Analyzing
Synthesizing

Societal context

Problem solutions must also be considered in their **societal context**. Environmental concerns should be addressed as alternative solutions to problems that are being considered. Engineers must also be conscious of ethical issues in providing test results, quality verifications, and design limitations. Ethical issues are never easy to resolve, and some of the exciting new technological achievements will bring more ethical issues with them. For example, the mapping of the genome will potentially provide ethical, legal, and social implications. Should the gene therapy that allows doctors to combat diabetes also be used to enhance athletic ability? Should prospective parents be given detailed information related to the physical and mental characteristics of an unborn child? What kind of privacy should an individual have over his or her genetic code? Very complicated issues arise with any technological advancement because the same capabilities that can do a great deal of good can often be applied in ways that are harmful.

The material presented in this text is only one step in building the knowledge, confidence, and understanding needed by engineers. We begin the process with an introduction to the range of computing systems available to engineers and an introduction to a problem-solving methodology that will be used throughout this text as we use C to solve engineering problems.

1.2   Computing Systems: Hardware and Software

Computer

Before we begin discussing the language C, a brief discussion on computing is useful, especially for those who have not had lots of experience with computers. A **computer** is a machine that is designed to perform operations that are specified with a set of instructions called a **program**. Computer **hardware** refers to the computer equipment, such as a notebook computer, a thumb drive, a keyboard, a flat-screen monitor, or a printer. Computer **software** refers to the programs that describe the steps we want the computer to perform. This can be software that we have written, or it can be programs that we download or purchase, such as computer games. Our computer hardware/software can be self-contained, as in a notebook computer. A computer can also access both hardware and software through a computer **network**, and through access to the **Internet**. In fact, **cloud computing** provides access to hardware, software, and large data sets through remote networks.

Program
Hardware
Software

Network
Internet
Cloud computing

Computer Hardware

Processor

All computers have a common internal organization as shown in Figure 1.1. The **processor** is the part of the computer that controls all the other parts. It accepts input values (from a device such as a keyboard or a data file) and stores them in memory. It also interprets the instructions

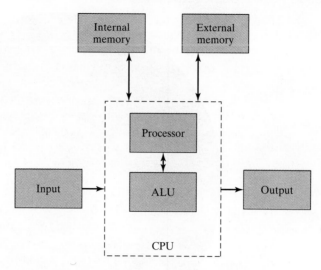

Figure 1.1 *Internal organization of a computer.*

Arithmetic logic unit (ALU)

Memory

Central processing unit

Microprocessor

in a computer program. If we want to add two values, the processor will retrieve the values from memory and send them to the **arithmetic logic unit (ALU)**. The ALU performs the addition, and the processor then stores the result in **memory**. The processing unit and the ALU use internal memory composed of read-only memory (ROM) and random access memory (RAM); data can also be stored in external storage devices such as external drives or thumb drives. The processor and the ALU together are called the **central processing unit (CPU)**. A **microprocessor** is a CPU that is contained in a single integrated-circuit chip, which contains millions of components in an area much smaller than a postage stamp.

Hard copy

Electronic copy

Soft copy

Smartphones

Many inexpensive printers today use ink-jet technology to print both color copies and black-and-white copies. We can also store information on a variety of digital memory devices, including CDs and DVDs. A printed copy of information is called a **hard copy**, and a digital copy of information is called an **electronic copy** or a **soft copy**. Many printers today can also perform other functions such as copying, faxing, and scanning.

Computers come in all sizes, shapes, and forms. In fact, most of our phones today contain CPUs and store programs that they can execute. **Smartphones** also contain a graphics processing unit, a significant amount of RAM, and are trending to multicore (or multiprocessor), low-power CPUs. Many homes today have personal computers that are used for a variety of applications, including e-mail, financial budgeting, and games; these computers are typically desktop computers with separate monitors and keyboards. Notebook computers contain all their hardware in a small footprint, and thus become very convenient. For some people, tablet computers (such as the iPad) and smartphones are even replacing the use of desktop and notebook computers.

Computer Software

Computer software contains the instructions or commands that we want the computer to perform. There are several important categories of software, including operating systems, software tools, desktop applications, and language compilers. Figure 1.2 illustrates the interaction among these categories of software and the computer hardware. We now discuss each of these software categories in more detail.

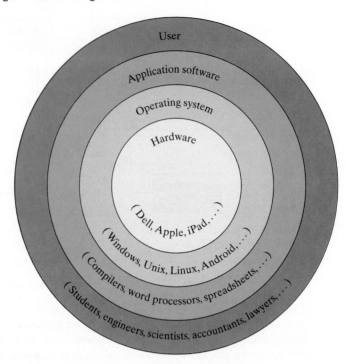

Figure 1.2   *Software interface to the computer.*

Operating system

Operating Systems.   Some software, such as an operating system, typically comes with the computer hardware when it is purchased. The **operating system** supplies an interface between you (the user) and the hardware by providing a convenient and efficient environment in which you can select and execute the software application on your system. The component of the operating system that manages the interface between the hardware and software applications is called a **kernel**. Examples of desktop operating systems include Windows, Mac OS, Unix, and Linux. Operating systems for smartphones include Android (a Linux variant) and iOS (a Unix variant).

Kernel

Utilities

Operating systems also contain a group of programs called **utilities** that allow you to perform functions such as printing files, copying files from one folder to another, and listing the files in a folder. Most operating systems today simplify the use of these utilities through icons and menus.

Word processors

Software Tools.   Software tools are programs that have been written to perform common operations. For example, **word processors** such as Microsoft Word are programs that allow you to enter and format text. They allow you to download information from the Internet into a file and allow you to enter mathematical equations. They can also check your grammar and spelling. Most word processors allow you to produce documents that have figures, contain images, and can print in two columns. These capabilities allow you to perform **desktop publishing**, from a notebook computer.

Desktop publishing

Spreadsheet programs

Spreadsheet programs such as Microsoft Excel are software tools that allow you to easily work with data that can be displayed in a grid of rows and columns. Spreadsheets were initially developed to be used for financial and accounting applications, but many science and engineering problems can also be easily solved with spreadsheets. Most spreadsheet packages include plotting capabilities, so they are especially useful in analyzing and displaying information in charts. **Database management tools** allow you to analyze and "mine" information from large data sets.

Database management tools

Mathematical computation tools

Another important category of software tools is **mathematical computation tools**. This category includes MATLAB, Mathematica, and Maple. Not only do these tools have very powerful mathematical commands, but they are also graphics tools that provide extensive capabilities for generating graphs. This combination of computational and visualization power makes them particularly useful tools for engineers.

If an engineering problem can be solved using a software tool, it is usually more efficient to use the software tool than to write a program in a computer language. The distinction between a software tool and a computer language is becoming less clear, as some of the more-powerful software tools include their own language in addition to having specialized operations. (In fact, many people would call MATLAB a programming language.)

Computer Languages.   Computer languages can be described in terms of generations. The first generation of computer languages (1GL) is machine languages. **Machine languages** are tied closely to the design of the computer hardware and are often written in **binary** strings consisting of 0s and 1s (also called **bits**). Therefore, machine language is also called binary language.

Machine languages

Binary

Bits

Assembly language

An **assembly language** is also unique to a specific computer design, but its instructions are written in symbolic statements instead of binary. Assembly languages usually do not have many statements; thus, writing programs in assembly language can be tedious. In addition, to use an assembly language you must also have information that relates to the specific hardware. Instrumentation that contains microprocessors often requires that the programs operate very fast; thus, the programs are called **real-time programs**. These real-time programs are usually written in assembly language to take advantage of specific computer hardware in order to perform the steps faster. Assembly languages are second-generation languages (2GL).

Real-time programs

High-level languages, or third-generation languages (3GL), use English-like commands. These languages include C, C++, C#, and Java. Writing programs in a high-level language is certainly easier than writing programs in machine language or in assembly language. However, a high-level language contains a large number of commands and an extensive set of **syntax** (or grammar) rules for using these commands.

High-level languages

C is a general-purpose language that evolved from two languages, BCPL and B, which were developed at Bell Laboratories in the late 1960s. In 1972, Dennis Ritchie developed and implemented the first C compiler at Bell Laboratories. The language became very popular for system development because it was hardware-independent. Because of its popularity in both industry and academia, it became clear that a standard definition for it was needed. A committee of the American National Standards Institute (ANSI) was created in 1983 to provide a machine-independent and unambiguous definition of C. In 1989, the **ANSI C** standard was approved; that language is described in this text.

ANSI C

Bjarne Stroustrup of AT&T Bell Laboratories developed C++ in the early 1980s. While C++ is a superset of C, it should really be considered a new language because of its expanded object-oriented features. Java is a pure object-oriented language that is especially useful for developing Internet-based applications and for devices that communicate over networks, due to its ease of use.

C is still the language of choice for many engineers and scientists because of its powerful commands and data structures and because it can easily be used for operating system operations. Once you learn C, it is much easier to move to C++ and Java. Therefore, our objective is to establish a solid foundation in C by covering the most important features of the language using a number of practical, real-world examples.

To finish our discussion of the different generations of language, fourth-generation languages (4GL) tend to be similar to human language (or natural language) and are usually domain-specific, such as in database development. The fifth generation of languages (5GL) is designed around constraints instead of algorithms and exists mainly in artificial intelligence (AI) research.

Executing a Computer Program.   A program written in a high-level language such as C must be translated into machine language before the instructions can be executed by the computer. A special program called a **compiler** is used to perform this translation. Thus, in order to write and execute C programs, we must have a C compiler. The C compilers are available as separate software packages for use with specific operating systems.

If any errors (often called **bugs**) are detected by the compiler during compilation, corresponding error messages are printed. We must correct our program statements and then perform the compilation step again. The errors identified during this stage are called **compiler errors** or compile-time errors. For example, if we want to divide the value stored in a variable called sum by 3, the correct expression in C is sum/3; if we incorrectly write the expression using the backslash, as in sum\3, we will get a compiler error. The process of compiling, correcting statements (or **debugging**), and recompiling is often repeated several times before the program compiles without compiler errors. When there are no compiler errors, the compiler generates a program in machine language that performs the steps specified by the original C program. The original C program is referred to as the **source program**, and the machine language version is called an object program. Thus, the source program and the object program specify the same steps; but the source program is written in a high-level language, and the **object program** is specified in machine language.

Once the program has compiled correctly, additional steps are necessary to prepare the object program for **execution**. This preparation involves linking other machine language statements to the object program and then loading the program into memory. After this linking/loading, the program steps are executed by the computer. New errors called execution errors, run-time errors, or **logic errors** may be identified in this stage; they are also called program bugs. Execution errors often cause the termination of a program. For example, the program statements may attempt to perform a division by zero, which generates an execution error. Some execution errors do not stop the program from executing, but they cause incorrect results to be computed. These types of errors can be caused by programmer errors in determining the correct steps in the solutions and by errors in the data processed by the program. When execution errors occur due to errors in the program statements, we must correct the errors in the source program and then begin again with the compilation step. Even when a program appears to execute properly, we must check the answers carefully to be sure that they are correct. The computer will perform the steps precisely as we specify, and if we specify the wrong steps, the computer will execute these wrong (but syntactically legal) steps and thus present us with an answer that is incorrect.

Compiler

Bugs

Compiler errors

Debugging

Source program

Object program

Execution

Logic errors

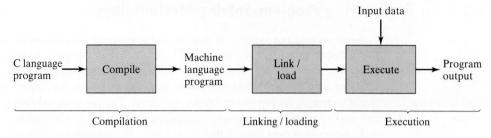

Figure 1.3   *Program compilation/linking/execution.*

The processes of compilation, linking/loading, and execution are outlined in Figure 1.3. The

Assembler

process of converting an assembly language program to binary is performed by an **assembler** program, and the corresponding processes are called assembly, linking/loading, and execution.

Software Life Cycle.   The cost of a computer solution to a problem can be estimated in terms of the cost of the hardware and the cost of the software. The majority of the cost in a computer solution today is in the cost of the software, and thus, a great deal of attention has been given to understanding the development of a software solution.

Software life cycle

The development of a software project generally follows definite steps or cycles that are collectively called the **software life cycle**. These steps typically include project definition, detailed specification, coding and modular testing, integrated testing, and maintenance. (These steps will be explained in more detail in later chapters.) Software maintenance is a significant part of the cost of a software system. This maintenance includes adding enhancements to the software, fixing errors identified as the software is used, and adapting the software to work with new hardware and software. The ease of providing maintenance is directly related to the original definition and specification of the solution, because these steps lay the foundation for the rest of the project. The problem-solving process that we present in the next section emphasizes the need to define and specify the solution carefully before beginning to code or test it.

Software prototypes

One of the techniques that has been successful in reducing the cost of software development both in time and cost is the development of **software prototypes**. Instead of waiting until the software system is developed and then letting the users work with it, a prototype of the system is developed early in the life cycle. This prototype does not have all the functions required of the final software, but it allows the user to use it early in the life cycle and to make desired modifications to the specifications. Making changes earlier in the life cycle is both cost- and time-effective. It is not uncommon for a software prototype to be developed using a software tool such as MATLAB and then for the final system to be developed in another language.

As an engineer, it is very likely that you will need to modify or add additional capabilities to existing software that has been developed using a software tool or a high-level language. These modifications will be much simpler if the existing software is well structured and readable and if the documentation that accompanies the software is up to date and clearly written. For these reasons, we stress developing good habits that make programs more readable and self-documenting.

1.3   An Engineering Problem-Solving Methodology

Problem solving is a key part of engineering courses, as well as courses in computer science, mathematics, physics, and chemistry. Therefore, it is important to have a consistent approach to solving problems. It is also helpful if the approach is general enough to work for all these different areas, so that we do not have to learn one technique for mathematics problems, a different technique for physics problems, and so on. The **problem-solving process** that we present works for engineering problems and can be tailored to solve problems in other areas as well; however, it does assume that we are using the computer to help solve the problem.

Problem-solving process

The process or methodology for problem solving that we will use throughout this text has five steps:

1. State the problem clearly.
2. Describe the input and output information.
3. Work the problem by hand (or with a calculator) for a simple set of data.
4. Develop a solution and convert it to a computer program.
5. Test the solution with a variety of data.

We now discuss each of these steps using an example of computing the distance between two points in a plane.

1. PROBLEM STATEMENT

The first step is to state the problem clearly. It is extremely important to give a clear, concise problem statement to avoid any misunderstandings. For this example, the problem statement is the following:

Compute the straight-line distance between two points in a plane.

2. INPUT/OUTPUT DESCRIPTION

The second step is to carefully describe the information that is given to solve the problem and then identify the values to be computed. These items represent the input and the output for the problem and collectively can be called input/output (I/O). For many problems, a diagram that shows the input and output is useful. At this point, the program is an "abstraction" because we are not defining the steps to determine the output; instead, we are only showing the information that is used to compute the output. The **I/O diagram** for this example follows:

I/O diagram

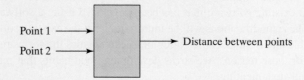

3. HAND EXAMPLE

The third step is to work the problem by hand or with a calculator, using a simple set of data. This is a very important step, and should not be skipped even for simple problems. This is the step in which you work out the details of the problem solution. If you cannot take a simple set of numbers and compute the output (either by hand or with a calculator), then you are not ready to move on to the next step; you should read the problem again and perhaps consult reference material. The solution by hand for this specific example is as follows:

Let the points p_1 and p_2 have the following coordinates:

$$p_1 = (1, 5); \quad p_2 = (4, 7).$$

We want to compute the distance between the two points, which is the hypotenuse of a right triangle, as shown in Figure 1.4. Using the Pythagorean theorem, we can compute the distance with the following equation:

$$\text{distance} = \sqrt{(\text{side}_1)^2 + (\text{side}_2)^2}$$
$$= \sqrt{(4 - 1)^2 + (7 - 5)^2}$$
$$= \sqrt{13}$$
$$= 3.61.$$

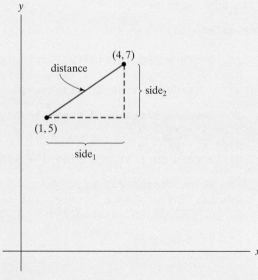

FIGURE 1.4   *Straight-line distance between two points.*

4. ALGORITHM DEVELOPMENT

Algorithm

Once you can work the problem for a simple set of data, you are ready to develop an **algorithm,** or a step-by-step outline, of the problem solution. For simple problems such as this one, the algorithm can be listed as operations that are performed one after another. This outline of steps decomposes the problem into simpler steps, as shown by the following outline of the steps required to compute and print the distance between two points:

Decomposition Outline

1. Give values to the two points.
2. Compute the lengths of the two sides of the right triangle generated by the two points.
3. Compute the distance between the two points, which is equal to the length of the hypotenuse of the triangle.
4. Print the distance between the two points.

Decomposition outline

This **decomposition outline** is then converted to C commands so that we can use the computer to perform the computations. From the following solution, you can see that the commands are very similar to the steps used in the hand example. The details of these commands are explained in Chapter 2.

```
/*-----------------------------------------------------------*/
/*  Program chapter1_1                                       */
/*                                                           */
/*  This program computes the                                */
/*  distance between two points.                             */

#include <stdio.h>
#include <math.h>

int main(void)
{
   /*  Declare and initialize variables.  */
   double x1=1, y1=5, x2=4, y2=7,
          side_1, side_2, distance;

   /*  Compute sides of a right triangle.  */
   side_1 = x2 - x1;
   side_2 = y2 - y1;
   distance = sqrt(side_1*side_1 + side_2*side_2);

   /*  Print distance.  */
   printf("The distance between the two points is "
          "%5.2f \n",distance);

   /*  Exit program.  */
   return 0;
}
/*-----------------------------------------------------------*/
```

5. TESTING

The final step in our problem-solving process is testing the solution. We should first test the solution with the data from the hand example because we have already computed the solution. When the C statements in this solution are executed, the computer displays the following output:

```
The distance between the two points is  3.61
```

This output matches the value that we calculated by hand. If the C solution did not match the hand-calculated solution, then we should review both solutions to find the error. Once the solution works for the hand-calculated example, we should also test it with additional sets of data to be sure that the solution works for other valid sets of data.

The set of steps demonstrated in this example are used in developing the programs in the Problem Solving Applied sections in the chapters that follow.

SUMMARY

A group of outstanding recent engineering achievements was presented to demonstrate the diversity of engineering applications. We also discussed some of the nontechnical skills required to be a successful engineer. Because most engineering problems will be solved by computer, we also presented a summary of the components of a computer system, from computer hardware to computer software. We also introduced a five-step problem-solving methodology that we will use to develop a computer solution to a problem. These five steps are as follows:

1. State the problem clearly.
2. Describe the input and output information.
3. Work the problem by hand (or with a calculator) for a simple set of data.
4. Develop an algorithm and convert it to a computer program.
5. Test the solution with a variety of data.

This process will be used throughout the text as we develop solutions to problems. Finally, we introduced you to the theme of this new edition—the technology behind crime scene investigation.

KEY TERMS

algorithm	database management tool
ANSI C	debug
arithmetic logic unit (ALU)	debugger
assembler	decomposition outline
assembly language	desktop publishing
binary	electronic copy
bit	execution
bug	hardware
central processing unit (CPU)	high-level language
cloud computing	I/O diagram
compiler	kernel
compiler error	linking/loading
computer	logic error

machine language   software
memory   software life cycle
microprocessor   software maintenance
network   software prototype
object program   software tool
operating system   source program
personal computer (PC)   spreadsheet
problem-solving process   syntax
processor   utility
program   word processor
real-time program

PROBLEMS

SHORT ANSWER PROBLEMS

True–False Problems

Indicate whether the following statements are true (T) or false (F):

1. The logical and arithmetic operations are performed in the control unit.   T   F
2. Linking/loading is the step that prepares the object program for execution.   T   F
3. Cloud computing provides access to hardware, software, and large data sets through remote networks.   T   F
4. A computer program is the implementation of an algorithm.   T   F
5. A utility program converts C to binary.   T   F
6. A microprocessor incorporates the functions of a CPU on a single integrated-circuit chip.   T   F
7. Data can be communicated between internal memory and external memory through an ALU.   T   F
8. Science and engineering problems cannot be solved with spreadsheets.   T   F
9. C is the language of choice for many engineers and scientists because of its object oriented capabilities.   T   F
10. A word processor allows us to enter and edit text.   T   F
11. Utilities are a group of programs that manage the interface between the hardware and software applications.   T   F
12. The errors detected by the compiler during compilation are called execution errors.   T   F
13. The compilation step identifies all the bugs in the program.   T   F
14. An algorithm gives the steps used to solve a problem.   T   F
15. A computer program is a set of instructions to solve a problem.   T   F
16. A program is completely tested if it works for one set of data.   T   F
17. A kernel is a piece of hardware.   T   F
18. C is a weakly typed, non-procedural language.   T   F
19. A software prototype may not have all the functions required in the final software, but it allows the user to use it early in the life cycle.   T   F
20. Machine languages are tied closely to the design of the computer hardware.   T   F

Multiple Choice Problems

Circle the letter for the best answer to complete each statement:

21. Instructions and data are stored in
    (a) the arithmetic logic unit (ALU).
    (b) the control unit (processor).
    (c) the central processing unit (CPU).
    (d) the memory.
    (e) the keyboard.

22. An operating system for smart phones is
    (a) Ubuntu
    (b) Windows Vista
    (c) Android
    (d) MacOS
    (e) DOS

23. Source code is
    (a) the result of compiler operations.
    (b) the process of getting information from the processor.
    (c) the set of instructions in a computer language that solves a specific problem.
    (d) the data stored in the computer memory.
    (e) the values entered through the keyboard.

24. Computer software
    (a) is an organized collection of related programs.
    (b) can be either system software or application software.
    (c) is created with programming languages and related utilities.
    (d) all of the above
    (e) none of the above

25. An algorithm refers to
    (a) a step-by-step solution to solve a specific problem.
    (b) a collection of instructions that the computer can understand.
    (c) a code that allows us to type in text materials.
    (d) stepwise refinement.
    (e) a set of math equations to derive the solution to a problem.

26. A hard copy is
    (a) the information stored on a hard disk.
    (b) the information printed out on paper.
    (c) the information shown on the screen.
    (d) a computer program.
    (e) all of the above.

27. Mathematical computation tools
    (a) have powerful mathematical commands.
    (b) have extensive capabilities for generating graphs.
    (c) include MATLAB, Mathematica, and Maple.
    (d) only (a) and (c)
    (e) (a), (b) and (c)

28. An example of software is
    (a) a printer.
    (b) a screen.
    (c) a computer code.
    (d) all of the above.

29. High-level languages are
    (a) good for real-time programming.
    (b) the second generation of computer languages.
    (c) called natural languages.
    (d) written in English-like words.

30. The difference between the source program and the object program is
    (a) the source program possibly contains some bugs, and the object program does not contain any bugs.
    (b) the source program is the original code, and the object program is a modified code.
    (c) the source program is specified in a high-level language, and the object program is specified in machine language.
    (d) the object program is also a source program.
    (e) the source program can be executed, and the object program cannot be executed.

31. The following is not a step in a software life cycle:
    (a) Requirement Analysis
    (b) Feasibility Study
    (c) Coding
    (d) Compilation
    (e) Testing

32. A hand example means
    (a) doing arithmetic on your hands.
    (b) working out the details of the problem solution using a simple set of data.
    (c) outlining a solution to a problem.
    (d) expanding the outline of a solution.
    (e) testing the algorithm step by step with a calculator.

33. A computer program is
    (a) a collection of components containing input and output devices.
    (b) a list of instructions needed to solve a problem.
    (c) a set of instructions to be performed by the computer and written in a language that the computer can understand.
    (d) a step-by-step procedure for solving a problem.
    (e) an outline that decomposes the problem into simpler steps.

Matching Problems

Select the correct term for each of the following definitions from this list:

algorithm	debugging
ANSI C	grammar
arithmetic logic unit (ALU)	hardware
central processing unit (CPU)	input devices
cloud computing	logic errors
compilation	machine language

memory
microprocessor
natural language
network
operating systems
output devices
program

software life cycle
software maintenance
spreadsheet
syntax
system software
word processor

34. A set of instructions that tells a computer what to do

35. The machinery that is part of the computer

36. The brain of the computer

37. Devices used to show the results of programs

38. Compilers and other programs that help run the computer

39. The steps to solve a problem

40. The process that converts a C program into machine language

41. A software tool designed to work with data stored in a grid or a table

42. The rules that define the punctuation and words that can be used in a program

43. The interface between the user and the hardware

44. The part of a computer that performs the mathematical computations

45. The process of removing errors from a program

46. Errors discovered during the execution of a program

47. The specific definition of the C language approved by the American National Standards Institute

48. A central processing unit contained in a single integrated-circuit chip

49. A technique for accessing large amounts of information remotely

50. The representation of a program in binary

ADDITIONAL PROBLEMS

The following problems combine an assignment in which you will learn more about one of the topics in this chapter with an opportunity to improve your written communication skills. (Perhaps your professor will even select some of the written reports for oral presentation in class.) Each report should include at least two references. Prepare your report using word processor software. If you do not already know how to use a word processor, ask your professor for guidance on locating manuals or seminars to teach you to use one of the word processors available on your university's computer systems.

51. Write a short report on one of these outstanding engineering achievements:

Moon landing	Application satellites
Microprocessors	CAD/CAM
CAT scans	Composite materials
Jumbo jets	Lasers
Fiber optics	Genetically engineered products

52. Write a short report discussing the environmental hazards related to engineering developments. Include the methods that can be incorporated to minimize them, like the concept of "green computing".

2

Crime Scene Investigation: Forensic Anthropology

Forensic anthropology is a field of study that combines the knowledge of physical and biological anthropology with the investigation of identify and the circumstances of death. Forensic anthropologists typically work with skeletal remains, but they may also work with decomposed, burned, or partial remains. Their work may be based on a single bone, or they may be working with many bones from mass fatalities caused by airplane crashes, explosions, or natural disasters. For example, the terrorist attack on the World Trade Centers killed 2,016 people; only 289 bodies were found intact. Another mass fatality was caused by the 2011 Joplin tornado that was more than a mile wide when it passed through Joplin, Missouri, killing 162 people. Forensic anthropologists often work with forensic odontologists (dental experts), radiologists, fingerprint experts, and law enforcement officers to identify remains and to provide clues relative to the circumstances of death. Later in this chapter, we discuss the relationship between the length of certain bones and the height of an individual. Then we develop a C program to estimate a person's height from the length of the femur (the leg bone between the hip and the knee) and from the humerus (the bone that connects the shoulder to the elbow).

SIMPLE C PROGRAMS

CHAPTER OUTLINE

OBJECTIVES

In this chapter, we develop problem solutions containing

- simple arithmetic computations,
- user-supplied information from the keyboard,
- information printed on the screen, and
- linear interpolation techniques.

2.1 Program Structure

In this section, we analyze the structure of a specific C program, and then we present the general structure of a C program. The program that follows was first introduced in Chapter 1; it computes and prints the distance between two points.

```
/*-------------------------------------------------------------*/
/*  Program chapter1_1                                         */
/*                                                             */
/*  This program computes the                                  */
/*  distance between two points.                               */

#include <stdio.h>
#include <math.h>
```

```
int main(void)
{
    /*  Declare and initialize variables.  */
    double x1=1, y1=5, x2=4, y2=7,
           side_1, side_2, distance;

    /*  Compute sides of a right triangle.  */
    side_1 = x2 - x1;
    side_2 = y2 - y1;
    distance = sqrt(side_1*side_1 + side_2*side_2);

    /*  Print distance.  */
    printf("The distance between the two points is "
           "%5.2f \n",distance);

    /*  Exit program.  */
    return 0;
}
/*-------------------------------------------------------------*/
```

We now briefly discuss the statements in this specific example; each of the statements is discussed in detail in later sections of this chapter.

Comments

The first five lines of this program contain **comments** that give the program a name (chapter1_1) and that document its purpose:

```
/*-------------------------------------------------------------*/
/*  Program chapter1_1                                         */
/*                                                             */
/*  This program computes the                                  */
/*  distance between two points.                               */
```

Comments begin with the characters /* and end with the characters */. A comment can be on a line by itself, or it can be on the same line as a command; a comment can also extend over several lines. Each of these comment lines is a separate comment because each line begins with /* and ends with */. Although comments are optional, good style requires that comments be used throughout a program to improve its readability and to document the computations. In the programs in this text, we always use initial comments to give a name to the program and to describe the general purpose of the program; additional explanation comments are also included throughout the program. ANSI C allows comments and statements to begin anywhere on a line; we begin the initial comments of a program in the first column.

Style

Preprocessor directives

Preprocessor directives provide instructions that are performed before the program is compiled. The most common directive inserts additional statements in the program; it contains the statement #include, followed by the name of the file containing the additional statements. This program contains the following two preprocessor directives:

```
#include <stdio.h>
#include <math.h>
```

These directives specify that statements in the files stdio.h and math.h should be included in place of these two statements before the program is compiled. The < and > characters around the file names indicate that the files are included with the **Standard C library**; this library is contained in the files that accompany an ANSI C compiler. The stdio.h file contains information related to the output statement used in this program, and the math.h file

Standard C library

contains information related to the function used in this program to compute the square root of a value. The `.h` extension on these filenames specifies that they are header files; more information on header files is included later in this chapter and in Chapter 4. Preprocessor directives are generally included after the initial comments describing the program's purpose.

Every C program contains a set of statements called a `main` function. The keyword `int` indicates that the function returns an integer value to the operating system. The keyword `void` indicates that the function is not receiving any information from the operating system. The body of the function is enclosed by braces, `{ }`. In order to easily identify the body of the function, we place these braces on lines by themselves. Thus, the two lines following the preprocessor directives specify the beginning of the main function:

```
int main(void)
{
```

Declarations
Initial values

The main function contains two types of commands: declarations and statements. The **declarations** define the memory locations that will be used by the statements and therefore must precede the statements. The declarations may or may not give **initial values** to be stored in the memory locations. A comment precedes the declaration statement in this program:

```
/*  Declare and initialize variables.  */
double x1=1, y1=5, x2=4, y2=7,
       side_1, side_2, distance;
```

These declarations specify that the program will use seven variables named `x1`, `y1`, `x2`, `y2`, `side_1`, `side_2`, and `distance`. The term `double` indicates that the variables will store double-precision floating-point values; these variables can store values such as 12.5 and −0.0005 with many digits of precision. In addition, this statement specifies that `x1` should be initialized (given an initial value) to the value 1, `y1` should be initialized to the value 5, `x2` should be initialized to the value 4, and `y2` should be initialized the value 7. The initial values of `side_1`, `side_2`, and `distance` are not specified and should not be assumed to be initialized to zero. Because the declaration was too long for one line, we split it over two lines; the indenting of the second line indicates that it is a continuation of the previous line. The indenting is a matter of style and readability. It is not required.

Style

Statements

The **statements** that specify the operations to be performed in the example program are the following:

```
/*  Compute sides of a right triangle.  */
side_1 = x2 - x1;
side_2 = y2 - y1;
distance = sqrt(side_1*side_1 + side_2*side_2);

/*  Print distance.  */
printf("The distance between the two points is "
       "%5.2f \n",distance);
```

These statements compute the lengths of the two sides of the right triangle formed by two points (see Figure 1.4), and then they compute the length of the hypotenuse of the right triangle. The details of the syntax of these statements are discussed later in the chapter. After the distance is computed, it is printed with the `printf` statement. This output statement is too long for a single line, so we separate the statement into two lines; the indenting of the second line again indicates that it is a continuation of the previous line. Additional comments were used to explain the computations and the output statement. Also, note that the declarations and statements are all required to end with a semicolon.

To end execution of the program and return control to the operating system, we use a return 0; statement:

```
/*  Exit program.   */
return 0;
```

This statement returns a value of 0 to the operating system. A value of zero indicates a successful end of execution.

The body of the main function then ends with the right brace on a line by itself and another comment line to delineate the end of the main function:

```
}
/*----------------------------------------------------------------*/
```

Style

Note that we have also included blank lines (also called white space) in the program to separate different components. These blank lines make a program more readable and easier to modify. The declarations and statements within the main function were indented to show the structure of the program. This spacing provides a consistent style, and makes our programs easier to read.

Now that we have closely examined the C program from Chapter 1, we can compare its structure to the **general form** of a C program:

general form

```
preprocessing directives
int main(void)
{
    declarations;
    statements;
}
```

This structure is evident in the programs developed in this chapter and in following chapters.

MODIFY!

1. Create a file containing the sample program discussed in this section. Use either an editor that is part of your C compiler or use a word processor.* Then compile and execute the program. You should get this output:

   ```
   The distance between the two points is   3.61
   ```

2. Change the values given to the two points to the coordinates $(-1, 6)$ and $(2, 4)$. Run the program with these new values. Did the distance change? Explain.

3. Change the values given to the two points to the coordinates $(1, 0)$ and $(5, 7)$. Check the program's answer with your calculator.

4. Change the values given to the two points to the coordinates $(2, 4)$ and $(2, 4)$. Does the program given the correct answer?

---

*If you use a word processor to generate the file, make sure you save it as a text file.

2.2   Constants and Variables

Constants

Variables
Identifier

Constants and variables represent values that we use in our programs. **Constants** are specific values, such as 2, 3.1416, −1.5, 'a', or "hello", that we include in the C statements, and **variables** are memory locations that are assigned a name or **identifier**. The identifier is used to reference the value stored in the memory location. A useful analogy for a memory location and its corresponding identifier is a mailbox that is associated with the name of an individual; the memory location (or mailbox) then contains a value. The following diagram shows the variables, their identifiers, and their initial values after the following declaration statement from program `Chapter1_1`:

```
double x1=1, y1=5, x2=4, y2=7,
       side_1, side_2, distance;
```

x1   [1]   y1   [5]   x2   [4]

y2   [7]   side_1 [?]   side_2 [?]

distance [?]

Garbage values

Memory snapshot

The values of variables that were not given initial values are unspecified, and thus indicated with a question mark; sometimes these values are called **garbage values** because they are values from the previous program. A diagram that shows a variable along with its identifier and its value is called a **memory snapshot**, because it shows the contents of a memory location at a specified point in the execution of the program. The preceding memory snapshot shows the variables and their contents as specified by the declaration statement. We frequently use memory snapshots to show the contents of a variable both before and after a statement is executed in order to show its effect.

The rules for selecting a valid identifier are as follows:

- an identifier must begin with an alphabetic character or the underscore character (_);
- alphabetic characters in an identifier can be lowercase or uppercase letters;
- an identifier can contain digits, but not as the first character; and
- an identifier can be of any length, but the first 31 characters of the identifier must be unique.

Case sensitive

Keywords

C is **case sensitive**, and thus uppercase letters are different from lowercase letters; thus, `Total`, `TOTAL` and `total` represent three different variables. Also, note that the variable `distance_in_miles_from_earth_to_mars` would not be distinguished from `distance_in_miles_from_earth_to_venus` because the first 31 characters are the same. C also includes **keywords** with special meaning to the C compiler that cannot be used for identifiers; a complete list of keywords is given in Table 2.1.

Examples of valid identifiers are `distance`, `x_1`, `X_sum`, `average_measurement`, and `initial_time`. Examples of invalid identifiers are `1x` (begins with a digit), `switch` (is a keyword), `$sum` (contains an invalid character, $), and `rate%` (contains an invalid character, %).

Table 2.1 Keywords

auto	double	ints	struct
break	else	long	switch
case	enum	register	typedef
char	extern	return	union
const	float	short	unsigned
continue	for	signed	void
default	goto	sizeof	volatile
do	if	static	while

Style　　An identifier name should be carefully selected so that it reflects the contents of the variable. If possible, the name should also indicate the units of measurement. For example, if a variable represents a temperature measurement in degrees Fahrenheit, use an identifier such as `temp_F` or `degrees_F`. If a variable represents an angle, name it `theta_rad` to indicate that the angle is measured in radians or `theta_deg` to indicate that the angle is measured in degrees.

The declarations at the beginning of the `main` function (and also at the beginning of other C functions that we write) must not only include all identifiers of the variables that we plan to use in our program, but also specify the types of values that will be stored in the variables. These data types are presented after a discussion on scientific notation.

PRACTICE!

Determine which of the following names are valid identifiers. If a name is not a valid identifier, give the reason that it is not acceptable, and suggest a valid replacement.

1. density	2. area	3. Time
4. xsum	5. x_sum	6. tax-rate
7. perimeter	8. sec^2	9. degrees_C
10. break	11. #123	12. x&y
13. count	14. void	15. f(x)
16. f2	17. Final_Value	18. w1.1
19. reference1	20. reference_1	21. m/s

Scientific Notation

Floating-point

Scientific notation

A **floating-point** value is one that can represent both integer and noninteger values, such as 2.5, –0.004, and 15.0. A floating-point value expressed in **scientific notation** is rewritten as a mantissa times a power of 10, where the mantissa has an absolute value greater than or equal to 1.0 and less than 10.0. For example, in scientific notation, 25.6 is written as 2.56×10^1,

Exponential notation

–0.004 is written as -4.0×10^{-3}, and 1.5 is written as 1.5×10^0. In **exponential notation**, the letter e is used to separate the mantissa from the exponent of the power of 10. Thus, in exponential notation, 25.6 is written as 2.56e1, –0.004 is written as –4.0e−3, and 1.5 is written as 1.5e0.

The number of digits allowed by the computer for the decimal portion of the mantissa determines the precision, and the number of digits allowed for the exponent determines the range. Thus, values with two digits of precision and an exponent range of −8 to 7 could include values such as 2.33×10^5 (233,000) and 5.92×10^{-8} (0.0000000592). This precision and exponent range would not be sufficient for many of the types of values that we use in engineering problem solutions. For example, the distance in miles from Mars to the Sun is 141,517,510 or 1.4151751×10^8; to represent this value, we would need at least seven digits of precision and an exponent range that included the integer 8.

PRACTICE!

In Problems 1 through 6, express the value in scientific notation.

1. 35.004
2. 0.00042
3. −50,000
4. 3.15723
5. −0.0999
6. 10,000,002.8

In Problems 7 through 12, express the value in floating-point notation.

7. 1.03e−5
8. −1.05e5
9. −3.552e6
10. 6.67e−4
11. 9.0e−2
12. −2.2e−2

Numeric Data Types

Numeric data types are used to specify the types of numbers that will be contained in variables. In C, numeric values are either integers or floating-point values. The following diagram shows the numeric data types that are discussed in the next few paragraphs:

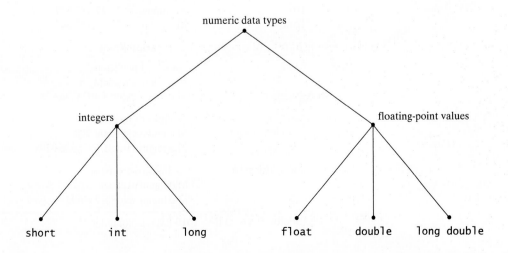

Type specifiers
System dependent

The **type specifiers** for signed integers are short, int, and long, for short integer, integer, and long integer, respectively. The specific ranges of values are **system dependent**, which means that the ranges can vary from one system to another. In the last section of this chapter, we present a program that you can use to determine the ranges of the numeric data types on your system. On many systems, the short integer and the integer data types range from –32,768 to 32,767, and the long integer type often represents values from –2,147,483,648 to 2,147,483,647. (The unusual limits such as 32,767 and 2,147,483,647 relate to conversions of binary values to decimal values.) C also allows an unsigned qualifier to be added to integer specifiers, where an unsigned integer represents only positive values. Signed and unsigned integers can represent the same number of values, but the ranges are different. For example, if an unsigned short has the range of values from 0 to 65,535, then a short integer has the range of values from –32,768 to 32,767; both variables can represent a total of 65,536 values.

The type specifiers for floating-point values are float (single precision), double (double precision), and long double (extended precision). The following statement from program chapter1_1 thus defines seven variables that all contain double-precision floating-point values:

```
double x1=1, y1=5, x2=4, y2=7,
       side_1, side_2, distance;
```

The difference between the float, double, and long double types relates to the precision (or accuracy) and the range of the values represented. The precision and range are system dependent. Table 2.2 contains precision and range information for integers and floating-point values used by the Microsoft Visual C++ compiler that can be used to run C programs, as well as C++ programs. (C++ is an extension of C.) A program given in Section 2.10 allows you to obtain this information for your computer system. On most systems, a double data type stores about twice as many decimal digits of precision as are stored with a float data type.

Table 2.2 Example Data-Type Limits*

Integers	
short	Maximum = 32,767
int	Maximum = 2,147,483,647
long	Maximum = 2,147,483,647
Floating Point	
float	6 digits of precision
	Maximum exponent 38
	Maximum value 3.402823e+38
double	15 digits of precision
	Maximum exponent 308
	Maximum value 1.797693e+308
long double	15 digits of precision
	Maximum exponent 308
	Maximum value 1.797693e+308

*Microsoft Visual C++ 2010 Express compiler.

In addition, a double value will have a wider range of exponent values than a float value. The long double value may have more precision and a still wider exponent range, but this is again system-dependent. A floating-point constant such as 2.3 is assumed to be a double constant. To specify a float constant or a long double constant, the letter (or suffix) F or L must be appended to the constant. Thus, 2.3F and 2.3L represent a float constant and a long double constant, respectively.

Character Data

Characters

Binary code
ASCII
EBCDIC

Character data is an important type of information to represent and to manipulate in engineering programming problems. In order to work with **characters**, we need to understand more about their representation in the computer's memory. Recall that all information stored in a computer is represented internally as sequences of binary digits (0 and 1). Each character corresponds to a **binary code** value. The most commonly used binary codes are **ASCII** (American Standard Code for Information Interchange) and **EBCDIC** (Extended Binary Coded Decimal Interchange Code). In the discussions that follow, we assume that ASCII code is used to represent characters. Table 2.3 contains a few characters, their binary form in ASCII, and the integer values that correspond to the binary values. The character 'a' is represented by the binary value 1100001, which is equivalent to the integer value of 97. A total of 128 characters can be represented in the ASCII code. A complete ASCII code table is given in Appendix B.

Character data can be represented by constants or by variables. A character constant is enclosed in single quotes, such as 'A', 'b', and '3'. A variable that is going to contain a character is defined as an integer or as a character data type. The type specifier for characters is char.

Once a character is stored in memory as a binary value, the binary value can be interpreted as a character or as an integer. Thus, when we define a variable for storing a character, we can define it as either a character variable to store the ASCII value or as an integer variable to store the integer equivalent value. However, it is important to note that the binary ASCII representation for a *character* digit is not equal to the binary representation for an *integer* digit. From Table 2.3, we see that the ASCII binary representation for the character digit 3 is 0110011, which is equivalent to the binary representation of the integer value 51. Thus, performing a computation with the character representation of a digit does not yield the same result as performing the computation with the integer representation of the digit. The following program illustrates the use of character data. The details of the program are discussed later in the chapter.

Table 2.3   Examples of ASCII Codes

Character	ASCII Code	Integer Equivalent
newline, \n	0001010	10
%	0100101	37
3	0110011	51
A	1000001	65
a	1100001	97
b	1100010	98
c	1100011	99

```
/*-------------------------------------------------------------*/
/*  Program chapter2_1                                         */
/*                                                             */
/*  This program prints two values                            */
/*  as characters and integers.                               */

#include <stdio.h>

int main(void)
{
    /*  Declare and initialize variables.  */
    char ch='a';
    int i=97;

    /*  Print both values as characters.  */
    printf("value of ch: %c; value of i: %c \n",ch,i);

    /*  Print both values as integers.  */
    printf("value of ch: %i; value of i: %i \n",ch,i);

    /*  Exit program.  */
    return 0;
}
/*-------------------------------------------------------------*/
```

The output from this program is

```
value of ch: a; value of i: a
value of ch: 97; value of i: 97
```

Symbolic Constants

Symbolic constant

A **symbolic constant** is defined with a preprocessor directive that assigns an identifier to the constant. The directive can appear anywhere in a C program; the compiler will replace each occurrence of the directive identifier with the constant value in all statements that follow the directive. Engineering constants such as π or g (the acceleration due to gravity) are good candidates for symbolic constants. For example, consider the following preprocessing directive to assign the value 3.141593 to the variable PI:

```
#define PI 3.141593
```

Statements that need to use the value of π would then use the symbolic constant identifier PI instead of 3.141593, as illustrated in the following statement which computes the area of a circle:

```
area = PI*radius*radius;
```

Symbolic constants are usually defined with uppercase identifiers (as in PI instead of pi) to indicate that they are symbolic constants, and, of course, the identifiers should be selected so that they are easy to remember. Finally, only one symbolic constant can be defined in a directive; if several symbolic constants are desired, separate directives are required. Note that preprocessor directives, which include the #define statement, do not end with a semicolon.

In the next section, we discuss C statements that allow us to assign values to variables. These assignment statements could be used to assign constant values to variables, but we will see later that there are often some special advantages to using symbolic constants.

PRACTICE!

Give the preprocessor directives to assign symbolic constants for these constants:

1. Speed of light, $c = 2.99792 \times 10^8$ m/s
2. Charge of an electron, $e = 1.602177 \times 10^{-19}$ C
3. Avogadro's number, $N_A = 6.022 \times 10^{23}$ mol^{-1}
4. Acceleration of gravity, $g = 9.8$ m/s^2
5. Acceleration of gravity, $g = 32$ ft/s^2
6. Mass of the Earth, $M_E = 5.98 \times 10^{24}$ kg
7. Radius of the Moon, $r = 1.74 \times 10^6$ m
8. Unit of length, Unit_Length = 'm'
9. Unit of time, Unit_Length = 's'

2.3   Assignment Statements

Assignment state-
ment

An **assignment statement** is used to assign a value to an identifier. The general form of the assignment statement is

```
identifier = expression;
```

Expression

where an **expression** can be a constant, another variable, or the result of an operation. Consider the following two sets of statements that declare and give values to the variables sum, x1, and ch:

```
double sum=10.5;        double sum;
int x1=3;               int x1;
char ch='a';            char ch;

                        ...

                        sum = 10.5;
                        x1 = 3;
                        ch = 'a';
```

After either set of statements is executed, the value of sum is 10.5, the value of x1 is 3, and the value of ch is 'a', as shown in the following memory snapshot:

sum | 10.5 | x1 | 3 | ch | 'a' |

In the previous example, the statements on the left define and initialize the variables at the same time; the assignment statements on the right could be used at any point in the program, and thus may be used to change (as opposed to initialize) the values in variables.

Multiple assignments are also allowed in C, as in the following statement, which assigns a value of zero to each of the variables x, y, and z:

```
x = y = z = 0;
```

Multiple assignments are discussed further at the end of this section.

We can also assign a value from one variable to another with an assignment statement:

```
rate = state_tax;
```

The equal sign should be read as "is assigned the value of"; thus, this statement is "rate is as-signed the value of state_tax." If state_tax contains the value 0.06, then rate also con-tains the value 0.06 after the statement is executed; the value in state_tax is not changed. Thus, the memory snapshots before and after this statement is executed are the following:

Before:	rate ?	state_tax 0.06
After:	rate 0.06	state_tax 0.06

If we assign a value to a variable that has a different data type, then a conversion must occur during the execution of the statement. Sometimes the conversion can result in informa-tion being lost. For example, consider the following declaration and assignment statement:

```
int a;
...
a = 12.8;
```

Because a is defined as an integer, it cannot store a value with a nonzero decimal portion. Therefore, in this case, the memory snapshot after executing the assignment statement is the following:

a   12

To determine whether a numeric conversion will work properly or not, we use the fol-lowing order (from high to low):

high: long double
          double
          float
          long integer
          integer
low: short integer

If a value is moved to a data type that is higher in order, no information will be lost; if a value is moved to a data type that is lower in order, information may be lost. Thus, moving an inte-ger to a double will work properly, but moving a float to an integer may result in the loss of some information or in an incorrect result. In general, use only assignments that do not cause potential conversion problems. (Unsigned integers were not included in the list because errors can occur in both directions.)

Arithmetic Operators

An assignment statement can be used to assign the result of an arithmetic operation to a vari-able, as shown in this statement that computes the area of a square:

```
area_square = side*side;
```

where * is used to indicate multiplication. The symbols + and − are used to indicate addition and subtraction, respectively, and the symbol / is used for division. Thus, each of the following statements is a valid computation for the area of a triangle:

```
area_triangle = 0.5*base*height;
area_triangle = (base*height)/2;
```

The use of parentheses in the second statement is not required, but is used for readability.

Consider this assignment statement:

```
x = x + 1;
```

In algebra, this statement is invalid, because a value cannot be equal to itself plus 1. However, this assignment statement should not be read as an equality; instead, it should be read as "x is assigned the value of x plus 1." With this interpretation, the statement indicates that the value stored in the variable x is incremented by 1. Thus, if the value of x is 5 before this statement is executed, then the value of x will be 6 after the statement is executed.

Modulus   C also includes a **modulus** operator (%) that is used to compute the remainder in a division between two integers. For example, 5%2 is equal to 1, 6%3 is equal to 0, and 2%7 is equal to 2. (The quotient of 2/7 is zero with a remainder of 2.) If a and b are integers, then the expression a/b computes the integer quotient, whereas the expression a%b computes the integer remainder. Thus, if a is equal to 9 and b is equal to 4, the value of a/b is 2, and the value of a%b is 1. An execution error occurs if the value of b is equal to zero in either a/b or a%b because the computer cannot perform division by zero. If either of the integer values in a and b is negative, the result of a%b is system-dependent.

The modulus operator is useful in determining if an integer is a multiple of another number. For example, if a%2 is equal to zero, then a is even; otherwise, a is odd. If a%5 is equal to zero, then a is a multiple of 5. We will use the modulus operator frequently in the development of engineering solutions.

Binary operators
Unary operators
     The five operators (+, −, *, /, %) discussed in the previous paragraphs are **binary operators**—operators that operate on two values. C also includes **unary operators**—operators that operate on a single value. For example, plus and minus signs can be unary operators when they are used in an expression such −x.

Truncated result
     The result of a binary operation with values of the same type is another value of the same type. For example, if a and b are double values, then the result of a/b is also a double value. Similarly, if a and b are integers, then the result of a/b is also an integer; however, an integer division can sometimes produce unexpected results, because any decimal portion of the integer division is dropped; thus, the result is a **truncated result**, not a rounded result. Thus, 5/3 is equal to 1, and 3/6 is equal to 0.

Mixed operation
     An operation between values with different types is a **mixed operation**. Before the operation is performed, the value with the lower type is converted or promoted to the higher type (as discussed in conversions within assignment statements), and thus the operation is performed with values of the same type. For example, if an operation is specified between an int and a float, the int will be converted to a float before the operation is performed; the result will be a float.

Suppose that we want to compute the average of a set of integers. If the sum and the count of the integers have been stored in the integer variables sum and count, it would seem that the following statements should correctly compute the average:

```
int sum, count;
```

```
float average;
...
average = sum/count;
```

However, the division between two integers gives an integer result that is then converted to a `float` value. Thus, if `sum` is 18, and `count` is 5, then the value of `average` is 3.0, not 3.6. To compute this sum correctly, we use a **cast operator**—a unary operator that allows us to specify a type change in the value before the next computation. In this example, the cast (`float`) is applied to `sum`:

Cast operator

```
average = (float)sum/count;
```

The value of `sum` is converted to a `float` value before the division is performed. The division is then a mixed operation between a `float` value and an integer, so the value of `count` is converted to a `float` value; the result of the division is then a `float` value that is stored in `average`. If the value of `sum` is 18 and the value of `count` is 5, the value of `average` is now correctly computed to be 3.6. Note that the cast operator affects only the value used in the computation; it does not change the value stored in the variable `sum`.

PRACTICE!

Give the value computed by each of the following sets of statements:

1. ```
   int a=27, b=6, c;
   ...
   c = b%a;
   ```

2. ```
   int a=27, b=6;
   float c;
   ...
   c = a/(float)b;
   ```

3. ```
   int a;
   float b=6, c=18.6;
   ...
   a = c/b;
   ```

4. ```
   int b=6;
   float a, c=18.6;
   ...
   a = (int)c/b;
   ```

Priority of Operators

Precedence

In an expression that contains more than one arithmetic operator, we need to be concerned about the order in which the operations are performed. Table 2.4 contains the **precedence** of the arithmetic operators, which matches the standard algebraic precedence. Operations within parentheses are always evaluated first; if the parentheses are nested, the operations within the innermost parentheses are evaluated first. Unary operators are evaluated before the binary operations *, /, and %; binary addition and subtraction are evaluated last. If there are several operators of the same precedence level in an expression, the variables or constants are grouped (or associated) with the operators in a specific order, as specified in Table 2.4. For example, consider the following expression:

```
a*b + b/c*d
```

Table 2.4   Precedence of Arithmetic Operators

Precedence	Operator	Associativity
1	Parentheses: ( )	Innermost first
2	Unary operators: + − (type)	Right to left
3	Binary operators: * / %	Left to right
4	Binary operators: + −	Left to right

Because multiplication and division have the same precedence level, and because the **associativity** (the order for grouping the operations) is from left to right, this expression will be evaluated as if it contained the following:

Associativity

```
(a*b) + ((b/c)*d)
```

The precedence order does not specify whether `a*b` is evaluated before `(b/c)*d`; the order of evaluation of these terms is system-dependent (but does not affect the final value).

The spacing within an arithmetic expression is a style issue. Some people prefer to put spaces around each operator. We prefer to put spaces only around binary addition and subtraction, because they are evaluated last. Choose the spacing style that you prefer, but then use it consistently.

Assume that we want to compute the area of a trapezoid and that we have declared four `double` variables: `base`, `height_1`, `height_2`, and `area`. Assume further that the variables `base`, `height_1`, and `height_2` already have values. A statement to correctly compute the area of the trapezoid is

```
area = 0.5*base*(height_1 + height_2);
```

Suppose that we omitted the parentheses in the expression:

```
area = 0.5*base*height_1 + height_2;
```

The statement would be executed as if it were this statement:

```
area = ((0.5*base)*height_1) + height_2;
```

Note that although an incorrect answer has been computed, there is no error message to alert us to the error. Therefore, it is important to be very careful when converting expressions into C. In general, use parentheses to indicate the order of operations in a complicated expression; this will avoid confusion and make sure that the expression is evaluated in the desired manner.

You may have noticed that there is no operator for exponentiation to compute values such as x^4. A special mathematical function will be discussed later in this chapter to perform exponentiations. Of course, exponentiations with integer exponents, such as a^2, can be computed with repeated multiplications, as in `a*a`.

Style

The evaluation of long expressions should be broken into several statements. For example, consider the following equation:

$$f = \frac{x^3 - 2x^2 + x - 6.3}{x^2 + 0.05005x - 3.14}.$$

If we try to evaluate the expression in one statement, it becomes too long to be easily read:

```
f = (x*x*x - 2*x*x + x - 6.3)/(x*x + 0.05005*x - 3.14);
```

We could break the statement into two lines:

```
f = (x*x*x - 2*x*x + x - 6.3)/
    (x*x + 0.05005*x - 3.14);
```

Another solution is to compute the numerator and denominator separately:

```
numerator = x*x*x - 2*x*x + x - 6.3;
denominator = x*x + 0.05005*x - 3.14;
f = numerator/denominator;
```

The variables x, numerator, denominator, and f must be floating-point variables in order to compute the correct value of f.

PRACTICE!

In Problems 1 through 3, give C statements to compute the indicated values. Assume that the identifiers in the expressions have been defined as double variables and have also been assigned appropriate values. Use the following constant:

Acceleration of gravity: $g = 9.80665$ m/s^2

1. Distance traveled:

   Distance $= x_0 + v_0 t + \frac{1}{2} a t^2$.

2. Tension in a cord:

   Tension $= \dfrac{2m_1 m_2}{m_1 + m_2} \times g$.

3. Fluid pressure at the end of a pipe:

   $P_2 = P_1 + \dfrac{\rho v_2^2 (A_2^2 - A_1^2)}{2A_1^2}$.

In Problems 4 through 6, give the mathematical equations computed by the C statements. Assume that the following symbolic constants have been defined, where the units of G are m^3/(kg · s^2):

```
#define PI 3.141593
#define G 6.67259e-11
```

4. Centripetal acceleration:

   ```
   centripetal = 4*PI*PI*r/(T*T);
   ```

5. Potential energy:

   ```
   potential_energy = -G*M_E*m/r;
   ```

6. Change in potential energy:

   ```
   change = G*M_E*m*(1/R_E - 1/(R_E + h));
   ```

Overflow and Underflow

The values stored in a computer have a wide range of allowed values. However, if the result of a computation exceeds the range of allowed values, an error occurs. For example, assume that the exponent range of a floating-point value is from −38 to 38. This range should accommodate most computations, but it is possible for the results of an expression to be outside of this range. For example, suppose that we execute the following commands:

```
x = 2.5e30;
y = 1.0e30;
z = x*y;
```

Overflow

The values of x and y are within the allowable range. However, the value of z should be 2.5e60, but this value exceeds the range. This error is called exponent **overflow**, because the exponent of the result of an arithmetic operation is too large to store in the memory assigned to the variable. The action generated by an exponent overflow is system dependent.

Underflow

Exponent **underflow** is a similar error that occurs when the exponent of the result of an arithmetic operation is too small to store in the memory assigned to the variable. Using the same allowable range as in the previous example, we obtain an exponent underflow with the following commands:

```
x = 2.5e-30;
y = 1.0e30;
z = x/y;
```

Again, the values of x and y are within the allowable range, but the value of z should be 2.5e−60. Because the exponent is less than the minimum value allowed, we have caused an exponent underflow. Again, the action generated by an exponent underflow is system dependent; on some systems, the result of an operation with exponent underflow is set to zero.

Increment and Decrement Operators

The C language contains unary operators for incrementing and decrementing variables; these operators cannot be used with constants or expressions. The increment operator ++ and the decrement operator − − can be applied either in a **prefix** position (before the identifier), as in ++count, or in a **postfix** position (after the identifier) as in count++. If an increment or decrement operator is used by itself, it is equivalent to an assignment statement that increments or decrements the variable. Thus, the statement

Prefix
Postfix

```
y--;
```

is equivalent to the statement

```
y = y - 1;
```

If the increment or decrement operator is used in an expression, then the expression must be evaluated carefully. If the increment or decrement operator is in a prefix position, the identifier is modified, and then the new value is used in evaluating the rest of the expression. If the increment or decrement operator is in a postfix position, the old value of the identifier is used to evaluate the rest of the expression, and then the identifier is modified. Thus, the execution of the statement

```
w = ++x - y;
```
(2.1)

is equivalent to the execution of this pair of statements:

```
x = x + 1;
w = x - y;
```

Similarly, the statement

```
w = x++ - y;
```
(2.2)

is equivalent to this pair of statements:

```
w = x - y;
x = x + 1;
```

When executing either Equation (2.1) or (2.2), if we assume that the value of x is equal to 5 and that the value of y is equal to 3, then the value of x increases to 6. However, after executing Equation (2.1), the value of w is 3; but after executing Equation (2.2), the value of w is 2.

The increment and decrement operators have the same precedence as the other unary operators. If several unary operators are in an expression, they are associated from right to left.

Abbreviated Assignment Operators

C allows simple assignment statements to be abbreviated. For example, each pair of statements contains equivalent statements:

```
x = x + 3;
x += 3;
```

```
sum = sum + x;
sum += x;
```

```
d = d/4.5;
d /= 4.5;
```

```
r = r%2;
r %= 2;
```

In fact, any statement of the form

```
identifier = identifier operator expression;
```

can be written in this form:

```
identifier operator = expression;
```

Abbreviated
assignment

Abbreviated assignment statements are usually used because they are shorter.

Earlier in this section, we used the following **multiple-assignment** statement:

Multiple-
assignment

```
x = y = z = 0;
```

The interpretation of this statement is clear, but the interpretation of the following statement is not as evident:

```
a = b += c + d;
```

To evaluate this properly, we use Table 2.5, which indicates that the assignment operators are evaluated last and their associativity is right to left. Thus, the statement is equivalent to the following:

```
a = (b += (c + d));
```

Table 2.5 **Precedence of Arithmetic and Assignment Operators**

Precedence	Operator	Associativity
1	Parentheses: ( )	Innermost first
2	Unary operators: + – ++ –– (type)	Right to left
3	Binary operators: * / %	Left to right
4	Binary operators: + –	Left to right
5	Assignment operators: = += –= *= /= %=	Right to left

If we replace the abbreviated forms with the longer forms of the operations, we have

```
a = (b = b + (c + d));
```

or

```
b = b + (c + d);
a = b;
```

Evaluating this statement was good practice with the precedence/associativity table, but statements used in a program should be more readable. Therefore, using abbreviated assignment statements in a multiple-assignment statement is not recommended. Also, note that the spacing convention that we use inserts spaces around abbreviated operators and multiple-assignment operators because these operators are evaluated after the arithmetic operators.

Style

PRACTICE!

Give a memory snapshot after each statement is executed, assuming that x is equal to 2 and that y is equal to 4 before the statement is executed. Also, assume that all the variables are integers.

1. z = x++*y;   2. z = ++x*y;
3. x += y;   4. y %= x;

2.4   Standard Input and Output

We have discussed statements for declaring variables and then using the variables to compute new values. We now present a statement that allows us to print these new values. In addition, we also discuss a statement that allows us to enter values from the keyboard when the program is executed. To use either of these statements in a program, we must include the following preprocessor directive:

```
#include <stdio.h>
```

This directive gives the compiler the information that it needs to check references to the input/output functions in the Standard C library.

printf **Function**

The printf function allows us to print values and explanatory text to the screen. For example, consider the following statement that prints the value of a double variable named angle along with the corresponding units:

```
printf("Angle = %f radians \n",angle);
```

Control string

Conversion specifier

This printf statement contains two arguments: a control string and an identifier to specify the value to be printed. A **control string** is enclosed in double quotation marks, and can contain text, conversion specifiers, or both. A **conversion specifier** describes the format to use in printing the value of a variable. In the previous example, the control string specifies that the characters Angle = are to be printed. The next group of characters (%f) represents a conversion specifier that indicates that a value is to be printed next, which will then be followed by the characters radians. The next combination of characters (\n) represents a new line indicator; it causes a skip to a new line on the screen after the information has been printed. The second argument in the printf statement is a variable angle; it is matched to the conversion specifier in the control string. Thus, the value in angle is printed according to the specification %f, which will be explained later. If the value of angle is 2.84, then the output generated by the previous statement is

```
Angle = 2.840000 radians
```

Now that we have analyzed a simple statement and its corresponding output, we are ready to take a closer look at the conversion specifiers.

To select a conversion specifier for a value to be printed, first select the correct type of specifier as indicated in Table 2.6. For example, to print a short or an int, use an %i (integer) or %d (decimal) specifier (either specifier gives the same results). To print a long, use an %li or %ld specifier. To print a float or a double, use an %f (floating-point form),

Table 2.6 **Conversion Specifiers for Output Statements**

Variable Type	Output Type	Specifier
Integer Values		
short, int	int	%i, %d
int	short	%hi, %hd
long	long	%li, %ld
int	unsigned int	%u
int	unsigned short	%hu
long	unsigned long	%lu
Floating-Point Values		
float, double	double	%f, %e, %E, %g, %G
long double	long double	%LF, %Le, %LE, %Lg, %LG
Character Values		
char	char	%c

%e (exponential form, as in 2.3e+02), or %E (exponential form, as in 2.3E+02). The %g (general) specifier prints the value using an %f or %e specifier, depending on the size of the value; the %G specifier is the same as the %g, except that it prints the value using an %f or %E specifier.

Field width
Precision

After selecting the correct specifier, additional information can be added. A minimum **field width** can be specified, along with an optional precision that controls the number of characters printed. The field width and the **precision** can be used together or separately. If the precision is omitted, a default of 6 is used for the %f specifier. The decimal portion of a value is rounded to the specified precision; thus, the value 14.51678 will be printed as 14.52 if a %.2f specification is used. The specification %5i indicates that a short or an int is to be printed with a minimum field width of 5. The field width will be increased if necessary to print the corresponding value. If the field width specifies more positions than are needed for the value, the value is **right justified**, which means that the extra positions are filled with blanks on the left of the value. To **left justify** a value, a minus sign is inserted before the field width, as in %-8i. If a plus sign is inserted before the field width, as in %+6f, a sign will always be printed with the value.

Right justified
Left justify

The following list shows several conversion specifiers and the resulting output fields for a given value; the character $_b$ is used to indicate the location of blanks within the field. In these examples, assume that the corresponding integer value is -145:

Specifier	Value Printed
%i	-145
%4d	-145
%3i	-145
%6i	$_{bb}$-145
%-6i	-145$_{bb}$

The next list shows several conversion specifiers and the resulting output fields for the double value 157.8926:

Specifier	Value Printed
%f	157.892600
%6.2f	157.89
%+8.2f	$_b$+157.89
%7.5f	157.89260
%e	1.578926e+02
%.3E	1.579E+02
%g	157.893

Note the rounding that occurred with the last two specifiers.

If a control argument contains three conversion specifiers, then three corresponding identifiers or expressions would need to follow the control string, as in this statement:

```
printf("Results: x = %5.2f, y = %5.2f, z = %5.2f \n",
       x,y,z+3);
```

An example output from this statement is

```
Results: x =  4.52, y =  0.15, z = -1.34
```

Note that the last conversion specifier matches to an arithmetic expression instead of a simple variable.

Escape character

The backslash (\) is called an **escape character** when it is used in a control string. The compiler combines it with the character that follows it and then attaches a special meaning to the combination of characters. For example, we have already seen that \n represents a skip to a new line. In addition, the sequence \\ is used to insert a single backslash in a control string, and the sequence \" will insert a double quote in a control string. Thus, the output of the statement

```
printf("\"The End.\"\n");
```

is a line containing

```
"The End."
```

The other escape sequences recognized by C are as follows:

Sequence	Character Represented
\a	alert (bell) character
\b	backspace
\f	formfeed
\n	newline
\r	carriage return
\t	horizontal tab
\v	vertical tab
\\	backslash
\?	question mark
\'	single quote
\"	double quote

If a printf statement is long, you should split it into two lines. In general, long lines should be split at a point that preserves readability. However, to split text that is contained in quotation marks, you should split the text into two separate pieces of text, each in its own set of quotation marks. The following statements show several different ways to correctly separate a statement:

```
printf("The distance between the points is %5.2f \n",
       distance);

printf("The distance between the points is"
       " %5.2f \n",distance);

printf("The distance between the "
       "points is %5.2f \n",distance);
```

Style

Conversion specifiers can be used to make the output of your program readable and usable. For engineering values, it is also very important to include the corresponding units in the output along with the numerical values.

Although the purpose of the printf function is to print information, it also returns a value that represents the number of characters printed.

PRACTICE!

Assume that the integer variable sum contains the value 65, the double variable average contains the value 12.368, and the char variable ch contains the value 'b'. Show the output line (or lines) generated by the following statements (use ᵦto indicate spaces):

1. printf("Sum = %5i; Average = %7.1f \n", sum, average);
2. printf("Sum = %4i \n Average = %8.4f \n", sum, average);
3. printf("Sum and Average \n\n %d %.1f \n", sum, average);
4. printf("Character is %c; Sum is %c \n", ch, sum);
5. printf("Character is %i; Sum is %i \n", ch, sum);
6. printf("%7.2f is the average; \n", average);
   printf("%8d is the sum \n", sum);
7. printf("%7.2f is the average; ", average);
   printf("%8d is the sum \n", sum);

scanf **Function**

The scanf function allows you to enter values from the keyboard when the program is executed. For example, suppose that a program computes the number of acres of new forest growth after a specified period elapses. If the time elapsed is a constant in the program, we would have to change the value of the constant, and then recompile and reexecute the program to obtain the output for a different period. Alternatively, if we use the scanf function to read the time period, we do not need to recompile the program; we only need to reexecute it and enter the desired period from the keyboard.

The first argument of the scanf function is a control string that specifies the types of the variables whose values are to be entered from the keyboard. The type specifiers are shown in Table 2.7; thus, for example, the specifiers for an integer variable are %i or %d; the specifiers

Table 2.7   Conversion Specifiers for Input Statements

Variable Type	Specifier
Integer Values	
int	%i, %d
short	%hi, %hd
long int	%li, %ld
unsigned int	%u
unsigned short	%hu
unsigned long	%lu
Floating-Point Values	
float	%f, %e, %E, %g, %G
double	%lf, %le, %lE, %lg, %lG
long double	%Lf, %Le, %LE, %Lg, %LG
Character Values	
char	%c

Address operator

for a `float` variable are %f, %e, and %g; and the specifiers for a `double` variable are %1f, %1e, and %1g. It is very important to use a correct specifier. For example, errors will occur if you use an %f specifier to read the value for a `double` variable. The remaining arguments in the `scanf` function are memory locations that correspond to the specifiers in the control string. These memory locations are indicated with the **address operator** &. This operator is a unary operator that determines the memory address of the identifier with which it is associated. Thus, if the value to be entered through the keyboard is an integer that is to be stored in the variable `year`, we could use this statement to read the value:

```
scanf("%i",&year);
```

The precedence level of the address operator is the same as the other unary operators; if there are several unary operators in the same statement, they are associated from right to left. A common error in `scanf` statements is to omit the address operator for the identifiers.

If we wish to read more than one value from the keyboard, we can use statements such as the following:

```
scanf("%1f %c",&distance,&unit_length);
```

Prompt

When this statement is executed, the program will read two values from the keyboard and convert them into one `double` value and one character value. The values must be separated by at least one blank; they can be on the same line or on different lines. In order to **prompt** the program user to enter the values, a `scanf` statement is usually preceded by a `printf` statement that describes the information that the user should enter from the keyboard:

```
printf("Enter the distance and the units (m for meters, f for "
       "feet): \n");
scanf("%1f %c",&distance,&unit_length);
```

The control string of the `printf` statement ended with a new line specifier, so the values entered by the user will be on the line (or lines) following the prompt text. Thus, after the previous statements are executed and the user has responded to the prompt, an example of the information on the screen is

```
Enter the distance and the units (m for meters, f for feet):
10 m
```

If the characters entered by the user cannot be successfully converted to the types of values indicated by conversion specifiers in the `scanf` statement, the result is system dependent. These conversion errors include entering values such as 14.2 for integer values, including commas in large values, and forgetting to separate values with blanks.

Although the main purpose of the `scanf` function is to read input from the keyboard, it also returns a value that is equal to the number of successful conversions. This value is used in programs in later chapters.

2.5   Problem Solving Applied: Estimating Height from Bone Lengths

In this section, we use the new statements presented in this chapter to solve a problem related to forensic anthropology. Recall from the chapter opening discussion that skeletal remains can be used to determine information about the identity of a person. We now consider estimating the height of an adult from bone lengths.

An adult skeleton typically contains 206 bones, ranging from the longest bone (the femur, the leg bone between the hip and the knee) to the small bones in the middle ear (the malleus, incus, and stapes). The hands alone contain 54 bones. Some of the bones commonly used to estimate height are the femur, the tibia (the larger of the two bones that connect the knee to the ankle), and the humerus (connecting the shoulder to the elbow). Bones are supported by ligaments, tendons, muscles, and cartilage, and together they provide protection to the body's organs such as the heart, lungs, and brain. Bones represent about a third of the body's weight.

The femur is the largest bone in a human body. Its length can be used to estimate the height of a person; height can also be estimated from the humerus. There are a number of different equations available for estimating an adult person's height from a bone length. These are generally derived from making many measurements from bones of people with known heights, and then finding the least-squares linear model to the data. (This technique is described in detail in the next chapter.) Many of such equations are derived specifically for males or for females. The ones we will use for this problem are the following, where the bone lengths are in inches:

Height estimation from femur length:
female height = femur length × 1.94 + 28.7
male height = femur length × 1.88 + 32

Height estimation from humerus length:
female height = humerus length × 2.8 + 28.2
male height = humerus length × 2.9 + 27.9

Write a C program that will ask the user to enter the lengths of a femur bone and a humerus bone from the same person. The program will compute the heights for both males and females from each bone and print those values.

I. PROBLEM STATEMENT

Estimate a person's height from the length of the femur and from that of the humerus.

2. INPUT/OUTPUT DESCRIPTION

The following diagram shows that the inputs to the program are the lengths of the two bones, and the outputs are the heights determined from each of the inputs. Because we don't know if the bones are from a male or female, we will compute both height estimates.

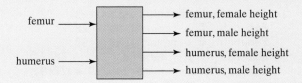

3. HAND EXAMPLE

Suppose that the length of the femur is 15 in, and the length of the humerus is 12 in. Then the height estimates are:

femur_height_female = femur_length × 1.94 + 28.7 = 57.8 in = 4.82 ft = 4 ft 9.8 in

femur_height_male = femur_length × 1.88 + 32 = 60.2 in = 5.02 ft = 5 ft .24 in

humerus_height_female = humerus_length × 2.8 + 28.2 = 61.8 in = 5.15 ft = 5 ft 1.8 in

humerus_height_male = humerus × 2.9 + 27.9 = 62.7 in = 5.23 ft = 5 ft 2.76 in

4. ALGORITHM DEVELOPMENT

The first step in the development of an algorithm is the decomposition of the problem solution into a set of sequentially executed steps.

Decomposition Outline

1. Read the lengths of the femur and the humerus.
2. Compute the height estimates.
3. Print the height estimates.

This program has a simple structure, so we can convert the decomposition directly into C.

```
/*-------------------------------------------------------------*/
/*  Program chapter2_2                                         */
/*                                                            */
/*  This program estimates a person's height from the length   */
/*  of the femur and from the length of the humerus.          */

#include <stdio.h>
#include <math.h>

int main(void)
{
   /*  Declare variables.  */
   double femur, femur_ht_f, femur_ht_m, humerus, humerus_ht_f,
   humerus_ht_m;

   /*  Get user input from the keyboard.  */
   printf("Enter Values in Inches. \n");
   printf("Enter femur length: \n");
   scanf("%lf",&femur);
   printf("Enter humerus length: \n");
   scanf("%lf",&humerus);

   /*  Compute height estimates.  */
   femur_ht_f = femur*1.94 + 28.7;
   femur_ht_m = femur*1.88 + 32;
   humerus_ht_f = humerus*2.8 + 28.2;
   humerus_ht_m = humerus*2.9 + 27.9;
```

```
        /* Print height estimates. */
        printf("\nHeight Estimates in Inches \n");
        printf("Femur Female Estimate:    %5.1f \n",femur_ht_f);
        printf("Femur Male Estimate:      %5.1f \n",femur_ht_m);
        printf("Humerus Female Estimate:  %5.1f \n",humerus_ht_f);
        printf("Humerus Male Estimate:    %5.1f \n",humerus_ht_m);
/* Exit program. */
return 0;
}
/*------------------------------------------------------------*/
```

5. TESTING

We first test the program using the data from the hand example. This generates the following interaction:

```
Enter Values in Inches.
Enter femur length:
15
Enter humerus length:
12

Height Estimates in Inches
Femur Female Estimate:        57.8
Femur Male Estimate:          60.2
Humerus Female Estimate:      61.8
Humerus Male Estimate:        62.7
```

The values computed match the hand example, so we can then test the program with additional lengths. If the values computed had not matched the result from the hand example, we would then need to determine if the error is in the hand example or in the C program.

MODIFY!

These problems relate to the program developed in this section for computing height estimates from bone lengths.

1. Modify the program so that it converts the output values from inches to feet. The input would still be entered in inches. Use a single value for the output, such as 6.5 feet.

2. Modify the program so that it asks the user to enter the values in feet. The program would then need to convert the values in feet to centimeters before doing the computations. The output would also print the output heights in feet, as in 6.5 feet.

3. Modify the program so that it reads the input values in inches and then estimates the output heights using feet and inches. (Note that you are printing two output values for each height estimate.)

4. Modify the program so that it reads the input values in feet and inches, and then estimates the output heights using feet and inches. (Note that you are reading two input values for each bone and you are printing two output values for each height estimate.)

5. Modify the program so that it reads the bone values in centimeters and outputs the height estimates in centimeters. (Recall that 1 in = 2.54 cm.)

2.6   Numerical Technique: Linear Interpolation

The collection of data from an experiment or from observing a physical phenomenon is an important step in developing a problem solution. These data points can generally be considered to be coordinates of points of a function $f(x)$. We would often like to use these data points to determine estimates of the function $f(x)$ for values of x that were not part of the original set of data. For example, suppose that we have data points $(a, f(a))$ and $(c, f(c))$. If we want to estimate the value of $f(b)$, where $a < b < c$, we could assume that a straight line

Linear interpolation   joined $f(a)$ and $f(c)$, and then use **linear interpolation** to obtain the value of $f(b)$. If we assume that the points $f(a)$ and $f(c)$ are joined by a cubic (third-degree) polynomial, we could use a cubic spline interpolation method to obtain the value of $f(b)$. Most interpolation problems can be solved using one of these two methods. Figure 2.1 contains a set of six data points that have been connected with straight-line segments and that have been connected with cubic degree polynomial segments. It should be clear that the values determined for the function between sample points depend on the type of interpolation that we select. In this section, we discuss linear interpolation.

MODIFY!

Write a short program that can be modified to include the following errors. How does your system respond to each of these errors?

1. Division by zero

2. Input conversion error:
      Enter 1,245 instead of 1245 for an %i specifier.

3. Input conversion error:
      Use %f for the input of an integer variable.

4. Input conversion error:
      Use %f for the input of a double variable.

5. Exponent overflow error:
      Use the example statements in the related discussion in this section.

6. Exponent underflow error:
      Use the example statements in the related discussion in this section.

A graph with two arbitrary data points $f(a)$ and $f(c)$ is shown in Figure 2.2. If we assume that the function between the two points can be estimated by a straight line, we can then compute the function value at any point $f(b)$ using an equation derived from similar triangles:

$$f(b) = f(a) + \frac{b-a}{c-a}[f(c) - f(a)].$$

Recall that we are also assuming that $a < b < c$.

To illustrate using this interpolation equation, assume that we have a set of temperature measurements taken from the cylinder head in a new engine that is being tested for possible use in a race car. These data are plotted with straight lines connecting the points in Figure 2.3, and they are also listed here:

Time, s	Temperature, °F
0.0	0.0
1.0	20.0
2.0	60.0
3.0	68.0
4.0	77.0
5.0	110.0

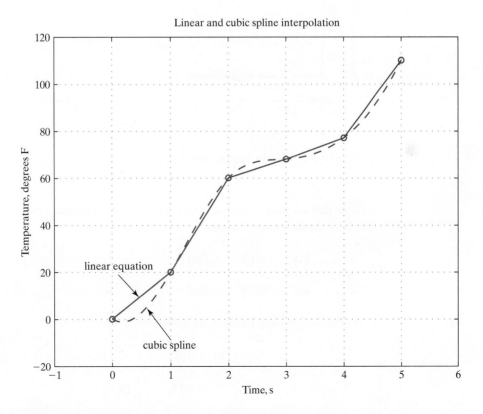

Figure 2.1 *Linear and cubic spline interpolation.*

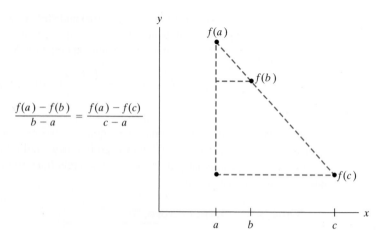

$$\frac{f(a) - f(b)}{b - a} = \frac{f(a) - f(c)}{c - a}$$

Figure 2.2 *Similar triangles.*

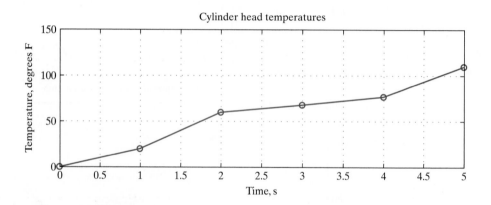

Figure 2.3 *Cylinder head temperatures.*

Assume that we want to interpolate a temperature to correspond to the value 2.6 seconds. We then have the following situation:

a	2.0	60.0	$f(a)$
b	2.6	?	$f(b)$
c	3.0	68.0	$f(c)$

Using the interpolation formula, we have

$$f(b) = f(a) + \frac{b - a}{c - a}[f(c) - f(a)]$$

$$= 60.0 + \frac{0.6}{1.0}(8.0)$$

$$= 64.8.$$

In this example, we used linear interpolation to find the temperature that corresponds to a specified time. We could also interchange the roles of temperature and time, so that we plot temperature on the *x*-axis and time on the *y*-axis. In this case, we can use the same process to compute the time that a specified temperature occurred, assuming that we have a pair of data points with temperatures below and above the specified temperature.

PRACTICE!

Assume that we have the following set of data points, which is also plotted in Figure 2.4:

Time, s	Temperature, °F
0.0	72.5
0.5	78.1
1.0	86.4
1.5	92.3
2.0	110.6
2.5	111.5
3.0	109.3

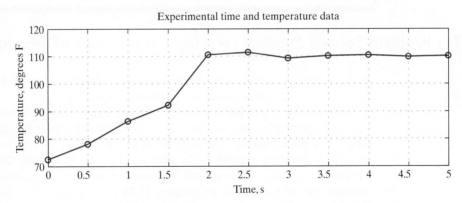

FIGURE 2.4 *Temperature values.*

Time, s	Temperature, °F
3.5	110.2
4.0	110.5
4.5	109.9
5.0	110.2

1. Use your calculator to compute temperatures at the following times using linear interpolation:

   0.3,   1.25,   2.36,   4.48.

2. Use your calculator to compute time values that correspond to the following temperatures using linear interpolation:

   81,   96,   100,   106.

3. Suppose Problem 2 asked you to compute the time value that corresponds to the temperature 110°F. What complicates this problem? How many time values correspond to the temperature 110°F? Find each of the corresponding time values using linear interpolation. (You may want to refer to Figure 2.5, which contains a plot of these data with the temperature data on the *x*-axis and the time values on the *y*-axis.)

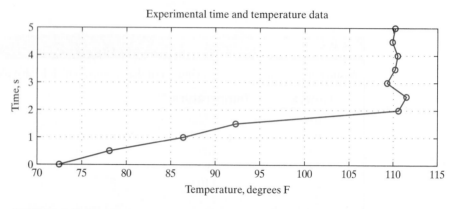

FIGURE 2.5 *Time values.*

2.7   Problem Solving Applied: Freezing Temperature of Seawater

Salinity

In this section, we use the new statements presented in this chapter along with linear interpolation to solve a problem related to the **salinity** of seawater.

The salinity of seawater is a measure of the amount of dissolved material in the seawater. Seawater is primarily water with about 3.5% dissolved materials (salts, metals, and gases) from volcanic eruptions and the weathering of rocks. The salinity of seawater is a measure of the amount of dissolved material in the seawater. Chlorine represents about 55% of the constituents in seawater, while sodium represents about 30.6%. The remaining primary constituents are sulfate (7.7%), magnesium (3.7%), calcium (1.2%), and potassium (1.1%). Salinity varies from one location to another in the ocean, but typically falls in the range of 33 to 38 parts per thousand (ppt), or a percentage of 3.3 to 3.8.

Salinity is often measured using an instrument that measures the electrical conductivity of the water; the more dissolved materials in the water, the better it conducts electricity. Measurements of salinity are especially important in colder regions because the temperature at which seawater freezes is dependent upon its salinity. The higher the salinity, the lower the temperature at which the seawater freezes. The following table [7] contains a set of salinity measurements and corresponding freezing temperatures:

Salinity (ppt)	Freezing Temperature (°F)
0 (fresh water)	32
10	31.1
20	30.1
24.7	29.6
30	29.1
35	28.6

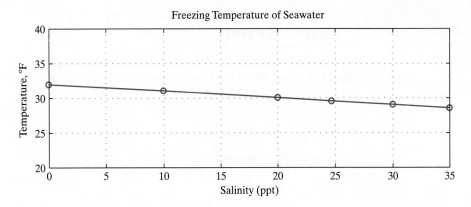

Figure 2.6 *Freezing temperature of seawater.*

The preceding values are also plotted in Figure 2.6.

Assume that we would like to use linear interpolation to determine the freezing temperature of water for which we have measured the salinity. Write a program that allows the user to enter the data for two points and a salinity measure between those points. The program should then compute the corresponding freezing temperature.

1. PROBLEM STATEMENT

Use linear interpolation to compute a new freezing temperature for water with a specified salinity.

2. INPUT/OUTPUT DESCRIPTION

The following diagram shows that the input to the program includes two consecutive points $(a, f(a))$ and $(c, f(c))$ and the new salinity measurement b, while the output is the new freezing temperature:

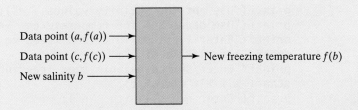

3. HAND EXAMPLE

Suppose that we want to determine the freezing temperature for water with a salinity measurement of 33 ppt. From the data, we see that this point falls between 30 and 35 ppt:

a	30	29.1	$f(a)$
b	33	?	$f(b)$
c	35	28.6	$f(c)$

Using the linear equation formula, we can compute $f(b)$:

$$f(b) = f(a) + (b - a)/(c - a) \cdot (f(c) - f(a))$$
$$= 29.1 + 3/5 \cdot (28.6 - 29.1)$$
$$= 28.8.$$

As expected, this value falls between $f(a)$ and $f(c)$.

4. ALGORITHM DEVELOPMENT

The first step in the development of an algorithm is the decomposition of the problem solution into a set of sequentially executed steps:

Decomposition Outline

1. Read the coordinates of the adjacent points and the new salinity value.
2. Compute the new freezing temperature.
3. Print the new freezing temperature.

This program has a simple structure, so we can convert the decomposition directly into C.

```
/*------------------------------------------------------------*/
/*   Program chapter2_3                                       */
/*                                                            */
/*   This program uses linear interpolation to               */
/*   compute the freezing temperature of seawater.           */

#include <stdio.h>
#include <math.h>

int main(void)
{
   /* Declare variables.  */
   double a, f_a, b, f_b, c, f_c;

   /* Get user input from the keyboard.  */
   printf("Use ppt for salinity values. \n");
   printf("Use degrees F for temperatures. \n");
   printf("Enter first salinity and freezing temperature: \n");
   scanf("%lf %lf",&a,&f_a);
   printf("Enter second salinity and freezing temperature: \n");
   scanf("%lf %lf",&c,&f_c);
   printf("Enter new salinity: \n");
   scanf("%lf",&b);
```

```
/*  Use linear interpolation to compute  */
/*  new freezing temperature.            */
f_b = f_a + (b-a)/(c-a)*(f_c - f_a);

/*  Print new freezing temperature.  */
printf("New freezing temperature in degrees F: %4.1f \n",f_b);

/*  Exit program.  */
return 0;
}
/*------------------------------------------------------------*/
```

5. TESTING

We first test the program using the data from the hand example. This generates the following interaction:

```
Use ppt for salinity values.
Use degrees F for temperatures.
Enter first salinity and freezing temperature:
30 29.1
Enter second salinity and freezing temperature:
35 28.6
Enter new salinity:
33
New freezing temperature in degrees F: 28.8
```

The value computed matches the hand example, so we can then test the program with other time values. If the new coefficient value had not matched the result from the hand example, we would then need to determine if the error is in the hand example or in the C program.

For the linear interpolation to work properly, the new salinity measurement must be between the first and second values that we entered. For this program, we assume that this relationship is maintained. In the next chapter, we learn how to use new C commands to make sure that the new salinity measurement is between the first and second measurements.

MODIFY!

These problems relate to the program developed in this section for computing new freezing points with linear interpolation.

1. Use the program to determine the freezing temperatures to go with the following salinity measurements in ppt:

    3   8.5   19   23.5   26.8   30.5

2. Modify the program so that it converts and prints the new temperature in degrees Centigrade. (Recall that $T_F = 9/5\ T_C + 32$, when T_F represents a temperature in degrees Fahrenheit and T_C represents a temperature in degrees Centigrade.)

3. Suppose that the data used with the program contained values with the degrees in Centigrade. Would the program need to be changed? Explain.

4. Modify the program so that it interpolates for a salinity, instead of a new freezing temperature. (You may want to refer to Figure 2.7, which contains a plot of this data with the freezing temperature on the *x*-axis and the salinity on the *y*-axis.)

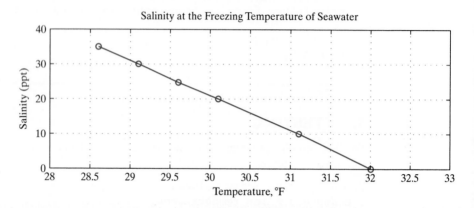

FIGURE 2.7 *Salinity at the freezing temperature of seawater.*

2.8   Mathematical Functions

Arithmetic expressions that solve engineering problems often require computations other than addition, subtraction, multiplication, and division. For example, many expressions require the use of exponentiation, logarithms, exponentials, and trigonometric functions. In this section, we discuss the mathematical functions that are available in the Standard C library. The following preprocessor directive should be used in programs referencing the mathematical functions:

```
#include <math.h>
```

This directive specifies that information be added to the program to aid the compiler when it converts references to the mathematical functions in the Standard C library.

Before we discuss the rules relating to functions, we present a specific example. The following statement computes the sine of an angle theta and stores the result in the variable b:

```
b = sin(theta);
```

The sin function assumes that the argument is in radians. If the variable theta contains a value in degrees, we can convert the degrees to radians with a separate statement. (Recall that $180° = \pi$ radians.) The following statements are illustrative:

```
#define PI 3.141593
...
theta_rad = theta*PI/180;
b = sin(theta_rad);
```

The conversion can also be specified within the function reference:

```
b = sin(theta*PI/180);
```

Performing the conversion with a separate statement is usually preferable, because it is easier to understand.

A function reference, such as `sin(theta)`, represents a single value. The parentheses following the function name contain the inputs to the function, which are called **parameters** or **arguments**. A function may contain no arguments, one argument, or many arguments, depending on its definition. If a function contains more than one argument, it is very important to list the arguments in the correct order. Some functions also require that the arguments be in specific units. For example, the trigonometric functions assume that arguments are in radians. Most of the mathematical functions assume that the arguments are `double` values; if a different type argument is used, it is converted to a `double` before the function is executed.

A function reference can also be part of the argument of another function reference. For example, the following statement computes the logarithm of the absolute value of x:

`b = log(fabs(x));`

When one function is used to compute the argument of another function, be sure to enclose the argument of each function in its own set of parentheses. This nesting of functions is also called **composition** of functions.

We now discuss several categories of functions that are commonly used in engineering computations. Other functions will be presented throughout the remaining chapters as we discuss relevant subjects. Tables of common functions are included on the inside front cover. Appendix A also contains more information on the functions included in the Standard C library.

Elementary Math Functions

The elementary **math functions** include functions to perform a number of common computations, such as computing the absolute value of a number and the square root of a number. In addition, they also include a group of functions used to perform rounding. These functions assume that the type of all arguments is `double`, and the functions all return a `double` value; if an argument is not a `double`, a conversion will occur using the rules described in Section 2.3. We now list these functions with a brief description:

`fabs(x)`	This function computes the absolute value of x.
`sqrt(x)`	This function computes the square root of x, where $x \geq 0$.
`pow(x,y)`	This function is used for exponentiation, and computes the value of x to the y power, or x^y. Errors occur if $x = 0$ and $y \leq 0$, or if $x < 0$ and y is not an integer.
`ceil(x)`	This function rounds x to the nearest integer toward ∞ (infinity). For example, `ceil(2.01)` is equal to 3.
`floor(x)`	This function rounds x to the nearest integer toward $-\infty$ (negative infinity). For example, `floor(2.01)` is equal to 2.
`exp(x)`	This function computes the value of e^x, where e is the base for natural logarithms, or approximately 2.718282.
`log(x)`	This function returns ln x, the natural logarithm of x to the base e. Errors occur if $x \leq 0$.
`log10(x)`	This function returns $\log_{10}x$, the common logarithm of x to the base 10. Errors occur if $x \leq 0$.

Marginal notes: Parameters / Arguments; Composition; Math functions

Remember that the logarithm of a negative value or zero does not exist, and thus an execution error occurs if you use a logarithm function with a negative value for its argument.

An additional mathematical function that you may find useful is the abs function. This function computes the absolute value of an integer, and returns an integer value. The header file containing information relative to this function is stdlib.h, and it should be included in programs referencing this function.

PRACTICE!

Evaluate the following expressions:

1. `floor(-2.6)`
3. `pow(2,-3)`
5. `fabs(-10*2.5)`
7. `log10(100) + log10(0.001)`

2. `ceil(-2.6)`
4. `sqrt(floor(10.7))`
6. `floor(ceil(10.8))`
8. `fabs(pow(-2,5))`

Trigonometric Functions

Trigonometric functions

The **trigonometric functions** assume that all arguments are of type double and that they return values of type double. In addition, as previously stated, the trigonometric functions also assume that angles are represented in radians. To convert radians to degrees, or degrees to radians, use the following statements:

```
#define PI 3.141593
...
angle_deg = angle_rad*(180/PI);
angle_rad = angle_deg*(PI/180);
```

The trigonometric functions are included in the Standard C library, and a preprocessor directive including the information in math.h should be used with these functions. Following is a brief summary of these functions:

`sin(x)`	This function computes the sine of x, where x is in radians.
`cos(x)`	This function computes the cosine of x, where x is in radians.
`tan(x)`	This function computes the tangent of x, where x is in radians.
`asin(x)`	This function computes the arcsine or inverse sine of x, where x must be in the range $[-1, 1]$. The function returns an angle in radians in the range $[-\pi/2, \pi/2]$.
`acos(x)`	This function computes the arccosine or inverse cosine of x, where x must be in the range $[-1, 1]$. The function returns an angle in radians in the range $[0, \pi]$.
`atan(x)`	This function computes the arctangent or inverse tangent of x. The function returns an angle in radians in the range $[-\pi/2, \pi/2]$.
`atan2(y,x)`	This function computes the arctangent or inverse tangent of the value y/x. The function returns an angle in radians in the range $[-\pi, \pi]$.

Note that the `atan` function always returns an angle in Quadrant I or IV, whereas the `atan2` function returns an angle that can be in any quadrant, depending on the signs of x and y. Thus, in many applications, the `atan2` function is preferred over the `atan` function.

The other trigonometric and inverse trigonometric functions can be computed with the use of the following equations:

$$\sec x = \frac{1}{\cos x} \qquad \mathrm{asec}\, x = \mathrm{acos}\left(\frac{1}{x}\right)$$

$$\csc x = \frac{1}{\sin x} \qquad \mathrm{acsc}\, x = \mathrm{asin}\left(\frac{1}{x}\right)$$

$$\cot x = \frac{1}{\tan x} \qquad \mathrm{acot}\, x = \mathrm{acos}\left(\frac{x}{\sqrt{1 + x^2}}\right)$$

Using degrees instead of radians is a common error in programs with trigonometric functions.

PRACTICE!

In Problems 1 through 3, give assignment statements for computing the indicated values, assuming that the variables have been declared and given appropriate values. Also assume that the following declarations have been made:

```
#define g 9.8
#define PI 3.141593
```

1. Velocity computation:

   $\text{Velocity} = \sqrt{v_0^2 + 2a(x - x_0)}.$

2. Length contraction:

   $\text{Length} = k\sqrt{1 - \left(\dfrac{v}{c}\right)^2}.$

3. Distance of the center of gravity from a reference plane in a hollow cylinder sector:

   $\text{Center} = \dfrac{38.1972(r^3 - s^3)\sin a}{(r^2 - s^2)\cdot a}.$

In Problems 4 through 6, give the equations that correspond to the assignment statement.

4. Electrical oscillation frequency:

   ```
   frequency = 1/sqrt(2*pi*c/L);
   ```

5. Range for a projectile:

   ```
   range = (v0*v0/g)*sin(2*theta);
   ```

6. Speed of a disk at the bottom of an incline:

   ```
   v = sqrt(2*g*h/(1 + I/(m*pow(r,2))));
   ```

Hyperbolic Functions*

Hyperbolic functions are functions of the natural exponential function e^x; the inverse hyperbolic functions are functions of the natural logarithm function $\ln x$. These functions are useful in specialized applications, such as the design of some types of digital filters. C includes several hyperbolic functions, as shown in the following descriptions:

sinh(x)   This functions computes the hyperbolic sine of x, which is equal to $\dfrac{e^x - e^{-x}}{2}$.

cosh(x)   This function computes the hyperbolic cosine of x, which is equal to $\dfrac{e^x + e^{-x}}{2}$.

tanh(x)   This function computes the hyperbolic tangent of x, which is equal to $\dfrac{\sinh x}{\cosh x}$.

Additional hyperbolic functions and the inverse hyperbolic functions can be computed with these relationships:

$$\coth x = \frac{\cosh x}{\sinh x} \quad (\text{for } x \neq 0).$$

$$\operatorname{sech} x = \frac{1}{\cosh x}.$$

$$\operatorname{csch} x = \frac{1}{\sinh x}.$$

$$\operatorname{asinh} x = \ln(x + \sqrt{x^2 + 1}).$$

$$\operatorname{acosh} x = \ln(x + \sqrt{x^2 - 1}) \quad (\text{for } x \geq 1).$$

$$\operatorname{atanh} x = \frac{1}{2}\ln\left(\frac{1 + x}{1 - x}\right) \quad (\text{for } |x| < 1).$$

$$\operatorname{acoth} x = \frac{1}{2}\ln\left(\frac{x + 1}{x - 1}\right) \quad (\text{for } |x| > 1).$$

$$\operatorname{asech} x = \ln\left(\frac{1 + \sqrt{1 - x^2}}{x}\right) \quad (\text{for } 0 < x \leq 1).$$

$$\operatorname{acsch} x = \ln\left(\frac{1}{x} + \frac{\sqrt{1 + x^2}}{|x|}\right) \quad (\text{for } x \neq 0).$$

Many of the hyperbolic functions and inverse trigonometric functions have restrictions on the range of acceptable values for arguments. If the arguments are entered from the keyboard, remind the user of the range restrictions. In the next chapter, we introduce C statements that allow you to determine if a value is in the proper range within the program.

---

*Optional section.

PRACTICE!

Give assignment statements for calculating the following values, given the value of x. (Assume that the value of x is in the proper range of values for the calculations.)

1. coth x 2. sec x

3. csc x 4. acoth x

5. acosh x 6. acsc x

2.9   Character Functions

Numerous functions are available to use with character data. There are functions designed for character input and output, functions to convert characters between uppercase and lowercase, and functions to perform character comparisons. In this section, we will discuss these character functions.

Character I/O

Although printf and scanf functions can be used to print and read characters using the %c specifier, C also contains special functions for reading and printing characters. The getchar function reads the next character from the keyboard and returns the integer value of the character; the putchar function prints a character to the computer screen.

The putchar function takes one integer argument and returns an integer value. The execution of the putchar function causes the character that corresponds to the integer argument to be written to the computer screen. If several putchar function references are made in a row, the characters are printed one after another on the same line, until a newline character is printed. Thus, the following statements cause the characters ab to be printed on one line, followed by the character c on the next line:

```
putchar('a');
putchar('b');
putchar('\n');
putchar('c');
```

The same information could be printed using the integer values that correspond to the characters (see Table 2.3):

```
putchar(97);
putchar(98);
putchar(10);
putchar(99);
```

The getchar function reads the next character from the keyboard and returns the integer value of the character. If several getchar function references are made in a row, the processing of characters continues until an **EOF** is received. The EOF is a special value defined as a symbolic constant in stdio.h. Recall that a variable defined with the char data type can represent only the 128 ASCII characters. The value of EOF is outside of this range. Because an integer variable can represent more than 128 values, we use an int type instead of a char

EOF

type to receive the value returned by the `getchar` function so that we can store the EOF and detect the end of data. The following statements cause one character to be read from the keyboard and printed on a new line:

```
int c;
...
putchar('\n');
c = getchar();
putchar(c);
putchar('\n');
```

The actual value of the EOF is system dependent. The EOF character is often represented by the control-z character. A control-z is entered from the keyboard by pressing the control (ctrl) key, and then pressing the z key while the control key is still pressed. This combination of characters is often written as ^z (control-z), but is not the same as pressing the ^ key with the z key. The use of EOF will be discussed in more detail in Chapter 3.

Character Comparisons

The Standard C library contains additional functions for use with characters. These **character functions** fall into two categories; one set of functions is used to convert characters between uppercase and lowercase, and the other set is used to perform character comparisons. The following preprocessor directive should be used in programs referencing these character functions:

```
#include <ctype.h>
```

The following statement converts the lowercase letter stored in the variable `ch` to an uppercase character and stores the result in the character variable `upper`:

```
upper = toupper(ch);
```

If `ch` is a lowercase letter, the function `toupper` returns the corresponding uppercase letter; otherwise, the function returns `ch`. Note that no change is made to the variable `ch`.

Each character function requires an integer argument, and each function returns an integer value. The character comparison functions return a nonzero value if the comparison is true; otherwise, they return a zero. We now list these functions with a brief explanation:

`tolower(ch)`	If `ch` is an uppercase letter, this function returns the corresponding lowercase letter; otherwise, it returns `ch`.
`toupper(ch)`	If `ch` is a lowercase letter, this function returns the corresponding uppercase letter; otherwise, it returns `ch`.
`isdigit(ch)`	This function returns a nonzero value if `ch` is a decimal digit; otherwise, it returns a zero.
`islower(ch)`	This function returns a nonzero value if `ch` is a lowercase letter; otherwise, it returns a zero.
`isupper(ch)`	This function returns a nonzero value if `ch` is an uppercase letter; otherwise, it returns a zero.
`isalpha(ch)`	This function returns a nonzero value if `ch` is an uppercase letter or a lowercase letter; otherwise, it returns a zero.
`isalnum(ch)`	This function returns a nonzero value if `ch` is an alphabetic character or a numeric digit; otherwise, it returns a zero.

Control characters

`iscntrl(ch)`	This function returns a nonzero value if `ch` is a control character; otherwise, it returns a zero. (The **control characters** have integer codes of 0 through 21, and 127.)
`isgraph(ch)`	This function returns a nonzero value if `ch` is a character that can be printed, as opposed to a character that cannot be printed, such as a control character or a tab; otherwise, it returns a zero. (The printing characters have integer codes from 32 through 126.)
`isprint(ch)`	This function returns a nonzero value if `ch` is a printing character (does include a space); otherwise, it returns a zero.
`ispunct(ch)`	This function returns a nonzero value if `ch` is a printing character (with the exception of a space or a letter or a digit); otherwise, it returns a zero.
`isspace(ch)`	This function returns a nonzero value if `ch` is a space, formfeed, newline, carriage return, horizontal tab, or vertical tab (these characters are also referred to as white space); otherwise, the function returns a zero.
`isxdigit(ch)`	This functions returns a nonzero value if `ch` is a hexadecimal digit, which is a decimal digit or an alphabetic character A through F (or *a* through *f*); otherwise, it returns a zero.

PRACTICE!

Show the output line (or lines) generated by the following statements:

1. `putchar('x');`

2. `putchar(65);`

3. `putchar(tolower(65));`

4. `putchar('\n');`
   `putchar(97);`
   `putchar('c');`
   `putchar('\n');`
   `putchar(toupper('c'));`

2.10 Problem Solving Applied: Velocity Computation

Open-rotor jet engines are a promising propulsion technology with many potential benefits. For example, test programs demonstrated a significant reduction in fuel consumption. However, there are also technical challenges. For example, there are strict guidelines on the noise produced by a commercial engine, because airports are often near residential areas, so additional noise reduction may be required. Testing on scale-model and full-size engines is done in wind tunnels, before testing moves to actual flight tests. (Problems related to wind tunnel analysis are included at the end of this chapter.)

During a test flight of an open-rotor aircraft, the test pilot has set the engine power level at 40,000 newtons, which causes the 20,000-kg aircraft to attain a cruise speed of 180 m/s (meters/second). The engine throttles are then set to a power level of 60,000 newtons, and the aircraft begins to accelerate. As the speed of the plane increases, the aerodynamic drag increases in proportion to the square of the airspeed. Eventually, the aircraft reaches a new cruise speed where the thrust from the engines is just offset by the drag. The equations used to estimate the velocity and acceleration of the aircraft from the time that the throttle is reset until the plane reaches its new cruise speed (at approximately 120 s) are as follows:

$$\text{Velocity} = 0.00001 \text{ time}^3 - 0.00488 \text{ time}^2$$
$$+ 0.75795 \text{ time} + 181.3566,$$

$$\text{Acceleration} = 3 - 0.000062 \text{ velocity}^2.$$

Plots of these functions are shown in Figure 2.8. Note that the acceleration approaches zero as the velocity approaches its new cruise speed.

Write a program that asks the user to enter a time value that represents the time elapsed (in seconds) since the power level was increased. Compute and print the corresponding acceleration and velocity of the aircraft at the new time value.

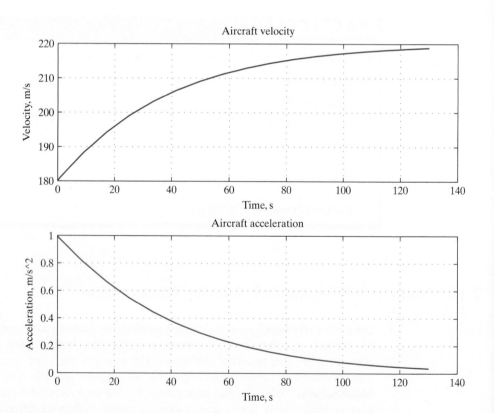

Figure 2.8 *Aircraft velocity and acceleration.*

I. PROBLEM STATEMENT

Compute the new velocity and acceleration of the aircraft after a change in power level.

2. INPUT/OUTPUT DESCRIPTION

The following diagram shows that the input to the program is a time value, and that the output of the program is the pair of new velocity and acceleration values:

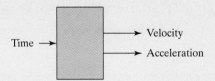

3. HAND EXAMPLE

Suppose that the new time value is 50 seconds. Using the equations given for the velocity and acceleration, we can compute these values:

$$\text{Velocity} = 208.3 \text{ m/s},$$
$$\text{Acceleration} = 0.31 \text{ m/s}^2.$$

4. ALGORITHM DEVELOPMENT

The first step in the development of an algorithm is the decomposition of the problem solution into a set of sequentially executed steps:

Decomposition Outline

1. Read new time value.
2. Compute corresponding velocity and acceleration values.
3. Print new velocity and acceleration.

Because this program is a very simple program, we can convert the decomposition directly to C:

```
/*-------------------------------------------------------------*/
/*  Program chapter2_4                                         */
/*                                                            */
/*  This program estimates new velocity and                   */
/*  acceleration values for a specified time.                 */

#include <stdio.h>
#include <math.h>

int main(void)
{
   /*  Declare variables.  */
   double time, velocity, acceleration;
```

```
/*  Get time value from the keyboard.  */
printf("Enter new time value in seconds: \n");
scanf("%lf",&time);

/*  Compute velocity and acceleration.  */
velocity = 0.00001*pow(time,3) - 0.00488*pow(time,2)
          + 0.75795*time + 181.3566;
acceleration = 3 - 0.000062*velocity*velocity;

/*  Print velocity and acceleration.  */
printf("Velocity = %8.3f m/s \n",velocity);
printf("Acceleration = %8.3f m/s^2 \n",acceleration);

/*   Exit program.  */
return 0;
}
/*-----------------------------------------------------------------*/
```

5. TESTING

We first test the program using the data from the hand example. This generates the following interaction:

```
Enter new time value in seconds:
50
Velocity =  208.304 m/s
Acceleration =    0.310 m/s^2
```

Because the values computed match the hand example, we can then test the program with other time values. If the values had not matched the hand example, we would need to determine if the error is in the hand example or in the program.

MODIFY!

These problems relate to the program developed in this section for computing velocity and acceleration values.

1. Enter different values of time until you find one that gives a velocity between 210 and 211 m/s.

2. Enter different values of time until you find one that gives an acceleration between 0.5 m/s^2 and 0.6 m/s^2.

3. Modify the program so that the input values are entered in minutes instead of seconds. Remember that the equations will still assume that the time values are in seconds.

4. Modify the program so that the output values are printed in feet per second, and feet per second^2. (Recall that 1 meter $= 39.37$ inches.)

2.11 System Limitations

In Section 2.2, we presented a table that contained the maximum values for the various types of integers and floating-point values for the Microsoft Visual C++ 2010 Express compiler. To print a similar table for your system, use the following program. Note that the program includes three header files. The stdio.h header file is necessary because the program references output functions; the limits.h header file is necessary because it contains information relative to the ranges of integer types; and the float.h header file is necessary because it contains information relative to the ranges of floating-point types. Appendix A contains more information on the constants and limits that are system dependent.

```c
/*-------------------------------------------------------------*/
/*   Program chapter2_5                                        */
/*                                                             */
/*   This program prints the system limitations.              */

#include <stdio.h>
#include <limits.h>
#include <float.h>

int main(void)
{
   /*  Print integer type maximums.  */
   printf("short maximum: %i \n",SHRT_MAX);
   printf("int maximum: %i \n",INT_MAX);
   printf("long maximum: %li \n\n",LONG_MAX);

   /*  Print float precision, range, maximum.  */
   printf("float precision digits: %i \n",FLT_DIG);
   printf("float maximum exponent: %i \n",
          FLT_MAX_10_EXP);
   printf("float maximum: %e \n\n",FLT_MAX);

   /*  Print double precision, range, maximum.  */
   printf("double precision digits: %i \n",DBL_DIG);
   printf("double maximum exponent: %i \n",
          DBL_MAX_10_EXP);
   printf("double maximum: %e \n\n",DBL_MAX);

   /*  Print long precision, range, maximum.  */
   printf("long double precision digits: %i \n",LDBL_DIG);
   printf("long double maximum exponent: %i \n",
          LDBL_MAX_10_EXP);
   printf("long double maximum: %Le \n\n",LDBL_MAX);

   /*  Exit program.  */
   return 0;
}
/*-------------------------------------------------------------*/
```

MODIFY!

Once you have run this program, change the conversion specifiers so that the following values are printed with full precision, instead of the default six digits of precision:

1. float maximum,
2. double maximum,
3. long double maximum.

SUMMARY

In this chapter, we presented the C statements necessary to write simple programs that compute and print new values. We also presented the statement that allows us to enter information through the keyboard when the program is executing. The computations that were presented included the standard arithmetic operations and a large number of functions that can be used to perform the types of computations needed for engineering solutions. We also included a discussion and example program using linear interpolation.

KEY TERMS

abbreviated assignment
address operator
argument
ASCII code
assignment statement
associativity
binary code
binary operator
case sensitive
cast operator
character
character function
comment
composition
constant
control character
control string
conversion specifier
declaration
EBCDIC code
EOF character
escape character
exponential notation
expression
field width
floating-point value
garbage value
hyperbolic function
identifier

initial value
keyword
left justify
linear interpolation
math function
memory snapshot
mixed operation
modulus
multiple assignment
overflow
parameter
postfix
precedence
precision
prefix
preprocessor directive
prompt
right justify
scientific notation
Standard C library
statement
symbolic constant
system dependent
trigonometric function
truncate
type specifier
unary operator
underflow
variable

C STATEMENT SUMMARY

Preprocessor directives to include information from the files in the Standard C library:

```
#include <stdio.h>
#include <math.h>
#include <ctype.h>
```

Preprocessor directive to define a symbolic constant:

```
#define PI 3.141593
```

Declarations for integers:

```
short sum=0;
int year_1, year_2;
long k;
```

Declarations for floating-point values:

```
float height_1, height_2;
double length=10, side1, side2;
long double distance, velocity;
```

Assignment statement:
```
area = 0.5*base*(height_1 + height_2);
```

Keyboard input statement:

```
scanf("%i",&year);
```

Screen output statement:

```
printf("The area is %f square feet. \n",area);
```

Program exit statement:

```
return 0;
```

Directive to read a character from keyboard:

```
c = getchar();
```

Directive to print a character to the screen:

```
putchar(c);
```

Style NOTES

1. Use comments throughout a program to improve the readability and to document the steps in it.
2. Use blank lines and indenting to identify the structure of a program.
3. Use the units in a variable name when possible.
4. Symbolic constants should be used for engineering constants such as π, and they should be uppercase so that they are easily identified.
5. Use consistent spacing around arithmetic and assignment operators.
6. Use parentheses in complicated expressions to improve readability.

7. The evaluation of long expressions should be broken into several statements.
8. Be sure to include units along with numerical values in the output of a program.
9. Use a prompt to the user to describe the information and units for values to be entered from the keyboard.

DEBUGGING NOTES

1. Remember that declarations and C statements must end with a semicolon.
2. Preprocessor directives do not end with a semicolon.
3. If possible, avoid assignments that could potentially cause information to be lost.
4. Use parentheses in a long expression to be sure that it is evaluated as desired.
5. Use double precision or extended precision to avoid problems with exponent overflow or underflow.
6. Be sure that the specifier matches the variable type in a `scanf` statement.
7. Errors can occur if user input values cannot be converted correctly to the specifier variable type in a `scanf` statement.
8. Do not forget the address operator with identifiers in the `scanf` statement.
9. Remember that symbolic constant definitions do not end with a semicolon.
10. In nested function references, each set of arguments must be in its own set of parentheses.
11. Remember that the logarithm functions cannot be used with negative values for arguments.
12. Be sure to use angles in radians with the trigonometric functions.
13. Remember that many of the inverse trigonometric functions and hyperbolic functions have restrictions on the ranges of allowable input values.
14. Remember that the integer representation for a character digit is not the same as the integer representation of the numerical digit.
15. Store the value returned by the `getchar` function in an integer variable so that the EOF character can be stored.

PROBLEMS

SHORT-ANSWER PROBLEMS

True–False Problems
Indicate whether the following statements are true (T) or false (F):

1. Preprocessor directives provide instructions that are performed before the program is compiled.   T   F
2. The statement x=(int)y;, where x and y are integer and float variables respectively, may result in loss of information.   T   F
3. Declarations can be placed anywhere in the program.   T   F
4. Statements and declarations must end with a semicolon.   T   F
5. The result of an integer division is a rounded result.   T   F

Syntax Problems
Indicate whether the declaration statements that follow are correct or not. If the statement is incorrect, modify it so that it is a correct statement:

6. `float register=9.04f;`

7. `unsigned long int x=900000;`
8. `LONG DOUBLE d = 12.78909876;`
9. `float a1=a2;`
10. `int n, m_m;`

Multiple Choice

Circle the letter of the best answer to complete the statement or answer the question:

11. Assuming that the initial value of y is 11, what will be the values of x and y after the statement x = y++ + ++y; is executed?
    (a) x=24, y=12
    (b) x=23, y=13
    (c) x=24, y=13
    (d) x=25, y=13
    (e) syntax error

12. In a declaration, the type specifier and the variable name are separated by
    (a) a period.   (b) a space.
    (c) an equal sign.   (d) a semicolon.
    (e) none of the above.

13. Which of the following declarations would properly define x, y, and z as double variables?
    (a) `double x, y, z;`
    (b) `long double x, y, z;`
    (c) `double x, y, z;`
    (d) `double x=y=z;`
    (e) `double X, Y, Z;`

14. In C, the binary operator % is applied to compute
    (a) integer division.
    (b) floating-point division.
    (c) the remainder of integer division.
    (d) the remainder of floating-point division.
    (e) none of the above.

15. The function `isxdigit(ch)`, accepts a variable ch and returns
    (a) a nonzero value if ch is a decimal digit.
    (b) a nonzero value if ch is a control character.
    (c) a nonzero value if ch is a character that can be printed, as opposed to a character that cannot be printed, such as a control character or a tab.
    (d) a nonzero value if ch is a hexadecimal digit.
    (e) a nonzero value if ch is a octal digit.

Memory Snapshot Problems
Give the corresponding snapshots of memory after each of the following sets of statements are executed:
16. `int x,y=9,z;`
    `x=pow((z=sqrt(y--)),3);`

17.
```
double a=3.8, x;
    int n=2, y;
    ...
    x = (y = a/n)*2;
```

Program Output Problems

Give the output generated by the following sets of statements:

18.
```
float f=8.09193f;
    ...
    printf("\nf= % + 4.3f \n",f);
```

19.
```
double value_4=66.45832;
    ...
    printf ("value_4 = %10.2e",value_4);
```

20.
```
int x=65;
    ...
    printf("%d    %c",x,x=x+32);
```

PROGRAMMING PROBLEMS

Conversions. This set of problems involves converting a value in one unit to a value in another unit. Each program should prompt the user for a value in the specified units and then print the converted value, along with the new units.

21. Write a program to convert miles to kilometers. (Recall that 1 mi = 1.6093440 km.)

22. Write a program to convert feet-inch to millimeters. (Recall that 1 feet = 12 inches and 1 inch = 25.4mm.)

23. Write a program to convert gallons to liters. (Recall that 1 gallon = 4.546 liters.)

24. Write a program to convert pound to tons. (Recall that 1 ton = 2240 pounds.)

25. Write a program that converts degrees Fahrenheit (T_F) to degrees Rankin (T_R). (Recall that $T_F = T_R - 459.67°R$.)

26. Write a program that converts degrees Celsius (T_C) to degrees Rankin (T_R). (Recall that $T_F = T_R - 459.67°R$ and that $T_F = (9/5) T_C + 32°F$.)

27. Write a program that converts degrees Kelvin (T_K) to degrees Fahrenheit (T_F). (Recall that $T_R = (9/5) T_K$ and that $T_F = T_R - 459.67°R$.)

Areas and Volumes. These problems involve computing an area or a volume using input from the user. Each program should include a prompt to the user to enter the variables needed.

28. Write a program to compute the area of a rectangle with sides a and b. (Recall that $A = a \times b$.)

29. Write a program to compute the area of a scalene triangle with sides a, b and c. (Recall that $A = \sqrt{s\,(s-a)\,(s-b)\,(s-c)}$, where $s = \dfrac{a+b+c}{2}$.)

30. Write a program to compute the radius of a circle having the same area as that of a rhombus with diagonals d_1 and d_2. (Recall that area of a circle is $A = \pi r^2$ and area of a rhombus is $A = \dfrac{1}{2} d_1 d_2$.)

31. Write a program to compute the volume of a cone of radius r and height h. (Recall that $A = \dfrac{1}{3}\pi r^2 h \cdot$)

32. Write a program to compute surface area and volume of a sphere of diameter d. (Recall that surface area $A = 4\pi r^2$ and volume $v = 4/3\pi r^3$ of a sphere with radius r.)

33. Write a program to compute the area of an ellipse with semiaxes a and b. (Recall that $A = \pi a \times b$.)

34. Write a program to compute the area of the surface of a sphere of radius r. (Recall that $A = 4\pi r^2$.)

35. Write a program to compute the volume of a sphere of radius r. (Recall that $V = (4/3)\pi r^3$.)

36. Write a program to compute the volume of a cylinder of radius r and height h. (Recall that $V = \pi r^2 h$.)

Amino Acid Molecular Weights. The amino acids in proteins are composed of atoms of oxygen, carbon, nitrogen, sulfur, and hydrogen, as shown in Table 2.8. The molecular weights of the individual elements follow:

Element	Atomic Weight
Oxygen	15.9994
Carbon	12.011
Nitrogen	14.00674
Sulfur	32.066
Hydrogen	1.00794

Table 2.8 Amino Acid Molecules

Amino Acid	O	C	N	S	H
Alanine	2	3	1	0	7
Arginine	2	6	4	0	15
Asparagine	3	4	2	0	8
Aspartic	4	4	1	0	6
Cysteine	2	3	1	1	7
Glutamic	4	5	1	0	8
Glutamine	3	5	2	0	10
Glycine	2	2	1	0	5
Histidine	2	6	3	0	10
Isoleucine	2	6	1	0	13
Leucine	2	6	1	0	13
Lysine	2	6	2	0	15
Methionine	2	5	1	1	11
Phenylalanine	2	9	1	0	11
Proline	2	5	1	0	10
Serine	3	3	1	0	7
Threonine	3	4	1	0	9
Tryptophan	2	11	2	0	11
Tyrosine	3	9	1	0	11
Valine	2	5	1	0	11

37. Write a program to compute and print the molecular weight of glycine.

38. Write a program to compute and print the individual molecular weights and average molecular weight of lysine, methionine, and threonine.

39. Write a program that asks the user to enter the number of atoms of each of the five elements for an amino acid. Then compute and print the molecular weight for this amino acid.

40. Write a program that asks the user to enter the number of atoms of each of the five elements for an amino acid. Then compute and print the average weight of the atoms in the amino acid.

Logarithms to the Base b. To compute the logarithm of x to base b, we can use the following relationship:

$$\log_b x = \frac{\log_e x}{\log_e b}.$$

41. Write a program that reads a positive number and then computes and prints the logarithm of the value to base 2. For example, the logarithm of 8 to base 2 is 3 because $2^3 = 8$.

42. Write a program that reads two positive number x and b, and then computes and prints the logarithm of x to base b. For example, the logarithm of 100 to base 8 is 2.214619 because $8^{2.214619}$ is 100.

Wind Tunnels. A wind tunnel is a test chamber built to generate different wind speeds, or Mach numbers (which is the wind speed divided by the speed of sound). Accurate scale models of aircraft can be mounted on force-measuring supports in the test chamber, and then measurements of the forces on the model can be made at many different wind speeds and angles. At the end of an extended wind tunnel test, many sets of data have been collected and can be used to determine the coefficient of lift, drag, and other aerodynamic performance characteristics of the new aircraft at its various operational speeds and positions. Data collected from a wind tunnel test are plotted in Figure 2.9 and are listed in the following table:

Flight-Path Angle (degrees)	Coefficient of Lift
−4	−0.182
−2	−0.056
0	0.097
2	0.238
4	0.421
6	0.479
8	0.654
10	0.792
12	0.924
14	1.035
15	1.076
16	1.103
17	1.120
18	1.121
19	1.121
20	1.099
21	1.059

43. Assume that we would like to use linear interpolation to determine the coefficient of lift for additional flight-path angles that are between –4 degrees and 21 degrees. Write a program that allows the user to enter the data for two points and a flight-path angle between those points. The program should then compute the corresponding coefficient of lift.

44. Modify the program developed in Problem 43 so that it prints the new angle in radians. The input range should also be in radians. (Recall that 180 degrees $= \pi$ radians.)

45. Modify the program developed in Problem 43 so that it interpolates for a new angle, instead of a new coefficient. Therefore, the user would enter the data for two points and a coefficient of lift between those two points. The program should then compute the corresponding angle in degrees.

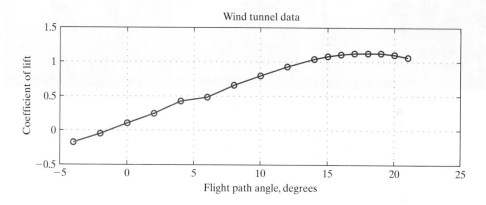

Figure 2.9 *Wind tunnel data.*

CHAPTER THREE

Crime Scene Investigation: Face Recognition and Surveillance Video

Face recognition is a commonly used technique to identify an individual from an image or from a single frame (also an image) taken from a surveillance video. By itself, face recognition is not as accurate as fingerprint recognition. As a result, face recognition is often used with another biometric or another security measure for access to secure areas. For example, a security system may use a face image to select the three most likely matches to faces in a master database and present those faces to a security guard. The security guard then makes the decision as to whether the person requesting entry is authorized for access. A number of commercial applications now use face recognition. Google's Picasa digital image organizer uses face recognition to search a group of images, looking for faces that match a specified image. Other applications that use face recognition include Apple's iPhoto (a photo organizer), Sony's Picture Motion Bowser (tags images with identical faces), and Facebook. Face recognition has also been used to solve crimes and to identify terrorists. It was used to help identify some of the 9/11 terrorists using airport surveillance video. Face recognition was also used in the identification of the subway bomber in London in 2005. Facial recognition was used with surveillance cameras at Super Bowl XXXV in January 2001 in Tampa Bay, Florida, to identify people with criminal records as a test of the system; 19 people with minor criminal backgrounds were potentially identified. Later in this chapter, we discuss the details of one facial recognition algorithm, and then we develop a C program to compare information from two faces, to determine if they might be the same person.

CONTROL STRUCTURES AND DATA FILES

CHAPTER OUTLINE

OBJECTIVES *In this chapter, we develop problem solutions containing*

- selection structures that allow us to provide alternative paths in a program,
- repetition structures that allow us to repeat a set of steps as long as a condition is true,
- information read from data files and information written to data files, and
- linear modeling techniques.

3.1 Algorithm Development

In Chapter 2, the C programs that we developed were very simple. The steps were sequential and typically involved reading information from the keyboard, computing new information, and then printing the new information. In solving engineering problems, most of the solutions require more complicated steps, and thus we need to expand the algorithm development part of our problem-solving process.

Top-Down Design

Top-down design

Top-down design presents a "big picture" description of the problem solution in sequential steps. This overall description of the problem is then refined until the steps are detailed enough to translate to language statements.

---

*Optional sections.

Decomposition
outline

Decomposition Outline. We used **decomposition outlines** in Chapters 1 and 2 to provide the first definition of a problem solution. This outline is written in sequential steps and can be shown in a diagram or a step-by-step outline. For very simple problems, such as the one that was developed in Chapter 2, we can go from the decomposition outline directly to the C statements:

Decomposition Outline

1. Read the new time value.
2. Compute the corresponding velocity and acceleration values.
3. Print the new velocity and acceleration.

Divide-and-conquer

Stepwise refinement

However, for most problem solutions, we need to refine the decomposition outline into a description with more detail. This process is often referred to as a **divide-and-conquer** strategy, because we keep breaking the problem solution into smaller and smaller portions. To describe this **stepwise refinement**, we use pseudocode or flowcharts.

Pseudocode
Flowchart

Refinement with Pseudocode and Flowcharts. The refinement of an outline into more detailed steps can be done with pseudocode or a flowchart. **Pseudocode** uses English-like statements to describe the steps in an algorithm, and a **flowchart** uses a diagram to describe the steps in an algorithm. The fundamental steps in most algorithms are shown in Figure 3.1, along with the corresponding notation in pseudocode and flowcharts.

Pseudocode and flowcharts are tools to help us determine the order of steps to solve a problem. Both tools are commonly used, although they are not generally both used with the same problem. In order to give examples of both tools, some problem solutions will use pseudocode and others will use flowcharts; the choice between pseudocode and flowcharts is usually a personal preference. Sometimes we need to go through several levels of pseudocode or flowcharts to develop complex problem solutions; this is the stepwise refinement that we previously discussed. Decomposition outlines, pseudocode, and flowcharts are working models of the solution, and thus are not unique. Each person working on a solution will have a different decomposition outline and pseudocode or a flowchart description, just the C programs developed by different people will vary, although they solve the same problem.

Structured Programming

Structured program

A **structured program** is written using simple control structures to organize the solution to a problem. A simple structure is usually defined to be a sequence, a selection, or a repetition. A **sequence** structure contains steps that are performed one after another; a **selection** structure contains one set of steps that is performed if a condition is true, and another set of steps that is performed if the condition is false; and a **repetition** structure contains a set of steps that is repeated as long as a condition is true. We now discuss each of these simple structures, and use pseudocode and flowcharts to give specific examples.

Sequence
Selection
Repetition

Sequence. A sequence contains steps that are performed one after another. All the programs developed in Chapter 2 have a sequence structure. For example, the pseudocode for the program that performed the linear interpolation is as follows:

Refinement in Pseudocode
main: *read a, f_a*
          read c, f_c
          read b
          set f_b to f_a $+ \dfrac{b-a}{c-a} \cdot (f\_c - f\_a)$
          print f_b

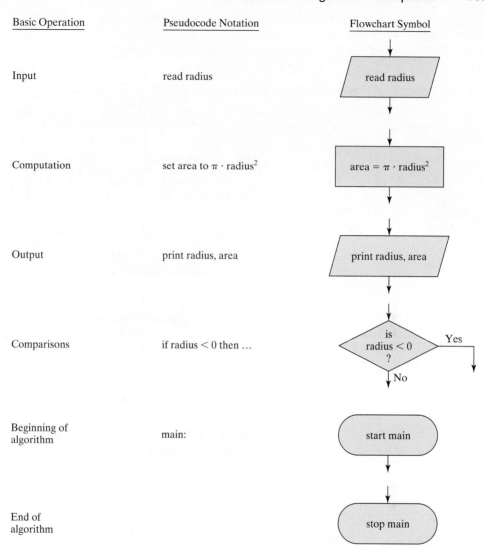

Basic Operation	Pseudocode Notation	Flowchart Symbol
Input	read radius	
Computation	set area to $\pi \cdot radius^2$	
Output	print radius, area	
Comparisons	if radius < 0 then ...	
Beginning of algorithm	main:	
End of algorithm		

Figure 3.1 *Pseudocode notation and flowchart symbols.*

The flowchart for the program that computed the velocity and acceleration of the aircraft with an open-rotor engine is shown in Figure 3.2.

Condition

Selection.   A selection structure contains a **condition** that can be evaluated as either true or false. If the condition is true, then one set of statements is executed; if the condition is false, then another set of statements is executed. For example, suppose that we have computed values for the numerator and denominator of a fraction. Before we compute the division, we want to be sure that the denominator is not close to zero. Therefore, the condition that we want to test is "denominator close to zero." If the condition is true, then we want to print a message indicating that we cannot compute the value. If the condition is false, which means that the denominator is not close to zero, then we compute and print the value of the fraction. In defining this condition, we need to define "close to zero." For this example, we will

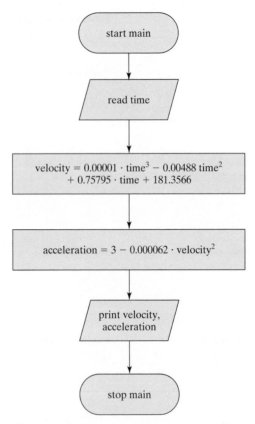

Figure 3.2 *Flowchart for open-rotor problem solution from Section 2.10.*

assume that "close to zero" means that the absolute value is less than 0.0001. A pseudocode description is as follows:

> *if |denominator| < 0.0001*
> > *print "Denominator close to zero"*
>
> *else*
> > *set fraction to numerator/denominator*
> > *print fraction*

A flowchart description of this structure is shown in Figure 3.3. Note that the structure also contains a sequence structure (e.g., compute a fraction and then print the fraction) that is executed when the condition is false. We will give more variations of the selection structure later in this chapter.

Loop

Repetition.   The repetition structure allows us to repeat (or **loop** through) a set of steps as long as a condition is true. For example, we might want to compute a set of velocity values that correspond to time values of 0, 1, 2, …, 10 seconds. We do not want to develop a sequential structure that has a statement to compute the velocity for a time of 0, then another statement to compute the velocity for a time of 1, and then another statement to compute the velocity for a time of 2, and so on. Although this structure would require only 11 statements

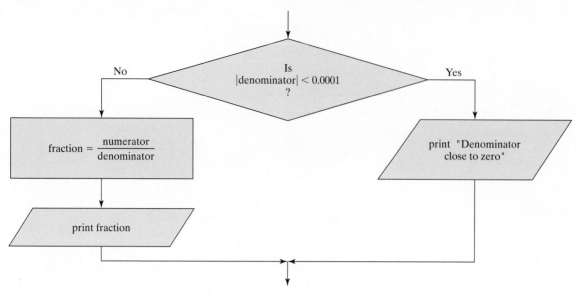

Figure 3.3 *Flowchart for selection structure.*

in this case, it could require hundreds of statements if we wanted to compute the velocity values over a long period. If we use the repetition structure, we can develop a solution in which we initialize the time to 0. Then, as long as the time value is less than or equal to 10, we compute and print a velocity value and increment the time value by 1. When the time value is greater than 10, we exit the structure. Figure 3.4 contains the flowchart for this repetition structure, and the pseudocode is as follows:

> *set time to 0*
> *while time ≤ 10*
> > *compute velocity*
> > *print velocity*
> > *increment time by 1*

In the remaining sections of this chapter, we present the C statements for performing selections and repetitions, and then we develop example programs that use these structures.

Evaluation of Alternative Solutions

There are usually many ways to solve the same problem. In most cases, there is not a single best solution, although some solutions are better than others. Selecting a good solution becomes easier with experience. In this text, we will give examples of the elements that contribute to a good solution. For example, a good solution is readable; therefore, it is not necessarily the shortest solution, because short solutions are often not very readable. We will strive to avoid subtle or clever steps that shorten a program but are difficult to understand.

As you begin to develop a solution to a problem, it is a good idea to try to think of several ways to solve it. Sketch the decomposition outline and pseudocode or flowchart for several solutions. Then choose the solution that you think will be the easiest to translate into C statements. Some algorithms fit different languages better than others, so you also want to pick a solution that is a good fit to the C language. Occasionally, other aspects of a solution must also be considered, such as execution speed and memory requirements.

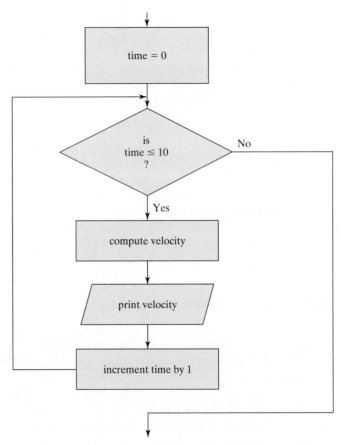

Figure 3.4 *Flowchart for repetition structure.*

Error Conditions

As we develop an algorithm, we usually assume that the input data are correct. However, in real applications, there are often errors in the input data. Therefore, it may be important to test the input data for errors that could occur and would cause the program to work incorrectly. As we compute new values, there may also be conditions that could arise and could cause problems. For example, suppose that we are performing a computation in which the denominator value for a division operation turns out to be zero. Or suppose that the result of an altitude computation is negative. These are examples of **error conditions** (which are separate from errors in the algorithm) that could occur when a program is run.

Error conditions

Some error conditions can be checked within the program itself using statements that we present in this chapter. But if we check for every possible error condition, our programs become long and a large percentage of the statements are checking for error conditions. Therefore, how do we decide which error conditions to check in our programs? Sometimes, the problem statement will include information on error conditions that could occur, as well as the response to take if they are detected. Usually, though, error conditions are not mentioned. In these cases, we suggest that you develop an algorithm based on the problem statement, and then generate a list of potential error conditions that could arise. If possible, discuss these with the person or group that will be using the program. Otherwise, include error checks that seem to catch the most common types of errors, and then include written documentation with the program. This documentation should describe the error conditions that your program will catch and the ones that your program will not catch.

Once you have decided which error conditions you will incorporate in your algorithm, you still need to decide what to do if one of the error conditions occurs. There are usually two possibilities—you can exit the program or you can attempt to correct the error and continue with the program. In either case, you should probably print an error message that describes the error condition that occurred and the action that you are taking. Make sure that the error message gives as much information as possible. Instead of printing "Error occurred in input data," print messages such as "Temperature out of bounds," "Time value is negative," or "Pressure exceeds safety limits."

Data file

In this chapter, we discuss how to read and write **data files**, which are files (similar to program files) that contain information used in other programs. Sometimes programs called data filters are written that check the information in the data files for error conditions. Then, programs that use the data files do not need to check for the same error conditions.

Generation of Test Data

Test data

The generation of **test data** is a very important part of developing problem solutions. Test data should include data to test each of the error conditions that is checked in our programs. Test data should also test each path through our program. As programs become longer, generating test data to completely test the program becomes very difficult. Entire courses and

Validation and verification

books are based on this **validation and verification** topic.

We now give some suggestions on generating test data sets. First, use the data from the hand example. If this does not work properly, we are not off to a good start! Once the program works correctly for these data, begin using test data that cover different ranges of values. Be sure to use test data that test the boundary conditions, or limits, if the data are supposed to be in certain ranges. Once the program seems to work for valid data, then begin including the error conditions to see if the program handles them properly. In general, use many small sets of data instead of one large set of data to test the program.

If you find an error in testing the program, go back to the algorithm development step. Correct the error in the decomposition outline and the pseudocode or flowchart, and then correct the C program. When you make a major change in the program, you should completely retest the program. Sometimes changes affect parts of the program that we had not anticipated. This retesting is easier if we keep a log of the test sets used so that we can repeat them.

Program walkthrough

Finally, we want to mention a technique called a **program walkthrough** that is commonly used in industry in the development of large programs. In a program walkthrough, the people who have developed an algorithm for a complicated problem present their solution to a small group of people who are knowledgeable about the problem, but did not take part in the algorithm development. The interaction between the people who developed the algorithm and the people who are analyzing it usually results in identifying potential problems with the algorithm and the generation of potential test data for the software after it is coded. The result is that the final program is completed sooner with more confidence in its accuracy. You might try simple program walkthroughs with other students in your class as you solve more complicated problems.

3.2   Conditional Expressions

Because both selection and repetition structures use conditions, we must discuss conditions before presenting the statements that implement selection and repetition structures. A condition is an expression that can be evaluated to be true or false, and it is composed of expressions combined with relational operators; a condition can also include logical operators. In this section, we present relational operators and logical operators and discuss the evaluation order when they are combined in a single condition.

Relational Operators

Relational operators   The **relational operators** that can be used to compare two expressions in C are shown in the following list:

Relational Operator	Interpretation
<	is less than
<=	is less than or equal to
>	is greater than
>=	is greater than or equal to
==	is equal to
!=	is not equal to

Blanks can be used on either side of a relational operator, but blanks cannot be used to separate a two-character operator, such as ==.

Example of conditions are the following:

```
a < b
x+y >= 10.5
fabs(denominator) < 0.0001
```

Given the values of the identifiers in these conditions, we can evaluate each one to be true or false. For example, if a is equal to 5 and b is equal to 8.4, then a<b is a true condition. If x is equal to 2.3 and y is equal to 4.1, then x+y >= 10.5 is a false condition. If denominator is equal to −0.0025, then fabs(denominator) < 0.0001 is a false condition. Note that

Style   we use spaces around the relational operator in a logial expression, but not around the arithmetic operators in the conditions.

In C, a true condition is assigned a value of 1 and a false condition is assigned a value of zero. Therefore, the following statement is valid:

```
d = b>c;
```

If b>c, then the value of d is 1; otherwise, the value of d is zero. A single value can be used in place of a condition. For example, consider the following statement:

```
if (a)
    count++;
```

If the condition value is zero, then the condition is assumed to be false; if the value is nonzero, then the condition is assumed to be true. Therefore, in the previous statement, the value of count will be incremented if a is nonzero.

Logical Operators

Logical operators can also be used within conditions. However, logical operators compare conditions, not expressions. C supports three **logical operators**: and, or, and not. These logical operators are represented by the following symbols:

Logical Operator	Symbol
and	&&
or	||
not	!

For example, consider the following condition:

a<b && b<c

The relational operators have higher precedence than logical operators; therefore, this condition is read "a is less than b, and b is less than c." In order to make a logical statement more readable, we insert spaces around the logical operator, but not around the relational operators. Given values for a, b, and c, we can evaluate this condition as true or false. For example, if a is equal to 1, b is equal to 5, and c is equal to 8, then the condition is true. If a is equal to −2, b is equal to 9, and c is equal to 2, then the condition is false.

If A and B are conditions, then the logical operators can be used to generate new conditions A && B, A || B, !A, and !B. The condition A && B is true only if both A and B are true. The condition A || B is true if either or both of A and B are true. The ! operator changes the value of the condition with which it is used. Thus, the condition !A is true only if A is false, and the condition !B is true only if B is false. These definitions are summarized in Table 3.1.

When expressions with logical operators are executed, C will only evaluate as much of the expression as necessary to evaluate it. For example, if A is false, then the expression A && B is also false, and there is no need to evaluate B. Similarly, if A is true, then the expression A || B is true, and there is no need to evaluate B.

Precedence and Associativity

A condition can contain several logical operators, as in the following:

!(b==c || b==5.5)

The hierarchy, from highest to lowest, is !, &&, and ||, but parentheses can be used to change the hierarchy. In the previous example, the expressions b==c and b==5.5 are evaluated first. Suppose b is equal to 3 and c is equal to 5. Then neither expression is true, so the expression b==c || b==5.5 is false. We then apply the ! operator to the false condition, which gives a true condition. Blanks cannot be used to separate the characters in either the || or &&. A common error is to use = instead of == in a logical expression.

Table 3.1 Logical Operators

A	B	A && B	A || B	!A	!B
False	False	False	False	True	True
False	True	False	True	True	False
True	False	False	True	False	True
True	True	True	True	False	False

Table 3.2 Operator Precedence for Arithmetic, Relational, and Logical Operators

Precedence	Operation	Associativity
1	( )	Innermost first
2	++ -- + - ! (type)	Right to left (unary)
3	* / %	Left to right
4	+ -	Left to right
5	< <= > >=	Left to right
6	== !=	Left to right
7	&&	Left to right
8	||	Left to right
9	= += -= *= /= %=	Right to left

A condition can contain both arithmetic operators and relational operators, as well as logical operators. Table 3.2 contains the precedence and the associativity order for the elements in a condition.

PRACTICE!

Determine if the following conditions in Problems 1 through 8 are true or false. Assume that the following variables have been declared and given these values:

$$a = 5.5 \qquad b = 1.5 \qquad k = -3$$

1. a < 10.0+k
2. !(a == 3*b)
3. a+b >= 6.5
4. -k <= k+6
5. k != a-b
6. a<10 && a>5
7. b-k > a
8. fabs(k)>3 || k<b-a

3.3   Selection Statements

The if statement allows us to test conditions and then perform statements based on whether the conditions are true or false. C contains two forms of if statements—the simple if statement and the if/else statement. C also contains a switch statement that allows us to test multiple conditions and then execute groups of statements based on whether the conditions are true or false.

Simple if Statement

The simplest form of an if statement has the following general form:

```
if (condition)
    statement 1;
```

If the condition is true, we execute statement 1; if the condition is false, we skip statement 1. The statement within the if statement is indented so that it is easier to visualize the structure of the program from the statements.

Compound
statement

If we wish to execute several statements (or a sequence structure) when the condition is true, we use a **compound statement**, or block, which is composed of a set of statements enclosed in braces. The location of the braces is a matter of style; two common styles are shown:

Style 1	*Style 2*
`if` (condition)	`if` (condition) {
{	statement 1;
statement 1;	statement 2;
statement 2;	...
...	statement n;
statement n;	}
}	

In the text solutions, we use the first style convention; thus, both braces are on lines by themselves. Although this makes the program a little longer, it also makes it easier to notice if a brace has been mistakenly omitted. Figure 3.5 contains flowcharts of the control flow with simple `if` statements containing either one statement to execute, or several statements to execute, if the condition is true.

A specific example of an `if` statement follows:

```
if (a < 50)
{
    ++count;
    sum += a;
}
```

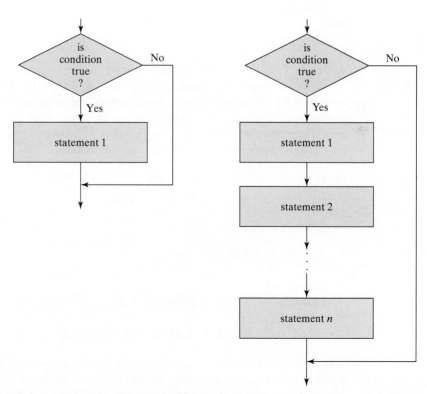

Figure 3.5 *Flowcharts for selection statements.*

If a is less than 50, then `count` is incremented by 1 and a is added to `sum`; otherwise, these two statements are skipped.

If statements can also be nested; the following example includes an `if` statement within an `if` statement:

```
if (a < 50)
{
    ++count;
    sum += a;
    if (b > a)
        b = 0;
}
```

Style

If a is less than 50, we increment `count` by 1 and add a to `sum`. In addition, if b is greater than a, then we also set b to zero. If a is not less than 50, then we skip all these statements. Be sure to indent the statements in each `if` statement when they are nested.

if/else Statement

An `if/else` statement allows us to execute one set of statements if a condition is true and a different set if the condition is false. The simplest form of an `if/else` statement is the following:

```
if (condition)
    statement 1;
else
    statement 2;
```

Empty statement

Statements 1 and 2 can also be replaced by compound statements. Statement 1 or statement 2 can also be an **empty statement**, which is just a semicolon. If statement 2 is an empty statement, then the `if/else` statement should probably be posed as a simple `if` statement. There are situations in which it is convenient to use an empty statement for statement 1; however, these statements can also be rewritten as simple `if` statements with the conditions reversed. For example, the following two statements are equivalent:

```
if (a < b)              if (a >= b)
    ;                       count++;
else
    count++;
```

Consider this `if/else` statement:

```
if (d <= 30)
    velocity = 0.425 + 0.00175*d*d;
else
    velocity = 0.625 + 0.12*d - 0.0025*d*d;
```

In this example, `velocity` is computed with the first assignment statement if the distance d is less than or equal to 30; otherwise, `velocity` is computed with the second assignment statement. A flowchart for this `if/else` statement is shown in Figure 3.6.

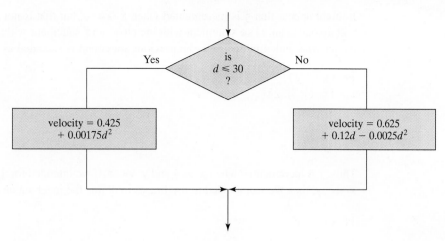

Figure 3.6 *Flowchart for* if/else *statement.*

Another example of the if/else statement is the following:

```
if (fabs(denominator) < 0.0001)
   printf("Denominator close to zero");
else
{
   x = numerator/denominator;
   printf("x = %f \n",x);
}
```

In this example, we examine the absolute value of the variable denominator. If this value is close to zero, we print a message indicating that we cannot perform the division. If the value of denominator is not close to zero, we compute and print the value of x. The flowchart for this statement was shown in Figure 3.3.

Consider the following set of nested if/else statements:

```
if (x > y)
   if (y < z)
      k++;
   else
      m++;
else
   j++;
```

The value of k is incremented when x > y and y < z. The value of m is incremented when x > y and y >= z. The value of j is incremented when x <= y. With careful indenting, this statement is easy to follow. Suppose that we now eliminate the else portion of the inner if statement. If we keep the same indention, the statements become the following:

```
if (x > y)
   if (y < z)
      k++;
else
   j++;
```

It might appear that j is incremented when x <= y, but that is not correct. The C compiler will associate an else statement with the closest if statement within a block. Therefore, no matter what indenting is used, the previous statement is executed as if it were the following:

```
if (x > y)
    if (y < z)
        k++;
    else
        j++;
```

Thus, j is incremented when x > y and y >= z. If we intended for j to be incremented when x <= y, then we would need to use braces to define the inner statement as a block:

```
if (x > y)
{
    if (y < z)
        k++;
}
else
    j++;
```

To avoid confusion and possible errors when using nested if/else statements, you should routinely use braces to clearly define the blocks of statements that go together.

Conditional operator

C allows a **conditional operator** to be used in place of a simple if/else statement. This conditional operator is a ternary operator, because it has three arguments—a condition, a statement to perform if the condition is true, and a statement to perform if the condition is false. The operation is indicated with a question mark following the condition, and with a colon between the two statements. To illustrate, the following two statements are equivalent:

```
if (a<b)                    a<b ? count++ : c = a + b;
    count++;
else
    c = a + b;
```

The conditional operator (specified as ?:) is evaluated before assignment operators, and if there is more than one conditional operator in an expression, they are associated from right to left.

In this section, we have presented a number of ways to compare values in selection statements. A caution is necessary when comparing floating-point values. As we saw in Chapter 2, floating-point values can sometimes be slightly different than we expect them to be because of the conversions between binary and decimal values. For example, earlier in this section, we did not compare denominator to zero, but instead used a condition to see if the absolute value of denominator was less than a small value. Similarly, if we wanted to know if y was close to the value 10.5, we should use a condition such as fabs(y-10.5) <= 0.0001 instead of y == 10.5. In general, do not use the equality operator with floating-point values.

PRACTICE!

In Problems 1 through 7, draw a flowchart to perform the steps indicated. Then give the corresponding C statements. Assume that the variables have been declared and have reasonable values.

1. If `time` is greater than 15.0, increment `time` by 1.0.
2. When the square root of `poly` is less than 0.5, print the value of `poly`.
3. If the difference between `volt_1` and `volt_2` is larger than 10.0, print the values of `volt_1` and `volt_2`.
4. If the value of `den` is less than 0.05, set `result` to zero; otherwise, set `result` equal to `num` divided by `den`.
5. If the natural logarithm of `x` is greater than or equal to 3, set `time` equal to zero and decrement `count`.
6. If `dist` is less than 50.0 and `time` is greater than 10.0, increment `time` by 2; otherwise, increment `time` by 2.5.
7. If `dist` is greater than or equal to 100.0, increment `time` by 2.0. If `dist` is between 50 and 100, increment `time` by 1. Otherwise, increment `time` by 0.5.

switch Statement

The `switch` statement is used for multiple-selection decision making. In particular, it is often used to replace nested `if/else` statements. Before giving the general discussion of the `switch` statement, we present a simple example that uses nested `if/else` statements and then an equivalent solution that uses the `switch` statement.

Suppose that we have a temperature reading from a sensor inside a large piece of machinery. We want to print a message on the control screen to inform the operator of the temperature status. If the status code is 10, the temperature is too hot and the equipment should be turned off; if the status code is 11, the operator should check the temperature every 5 minutes; if the status code is 13, the operator should turn on the circulating fan; for all other status codes, the equipment is operating in a normal mode. The correct message could be printed with the following set of nested `if/else` statements:

```
if (code == 10)
    printf("Too hot - turn equipment off \n");
else
{
    if (code == 11)
        printf("Caution - recheck in 5 minutes \n");
    else
    {
        if (code == 13)
            printf("Turn on circulating fan \n");
        else
            printf("Normal mode of operation \n");
    }
}
```

An equivalent statement is the following `switch` statement:

```
switch (code)
{
    case 10:
        printf("Too hot - turn equipment off \n");
        break;
    case 11:
        printf("Caution - recheck in 5 minutes \n");
        break;
    case 13:
        printf("Turn on circulating fan \n");
        break;
    default:
        printf("Normal temperature range \n");
        break;
}
Statement 1;
```

The `break` statement causes the execution of the program to continue with the statement following the `switch` statement (Statement 1), thus skipping the rest of the statements in the braces.

Nested `if/else` statements do not always easily translate to a `switch` statement. However,
when the conversion works, the `switch` statement is usually easier to read. It is also easier to
determine the punctuation needed for the `switch` statement. In fact, if the punctuation is not
correct in the `if/else` statements, the compiler may not execute the statements as expected.

Controlling
expression

Case labels

Case structure

Default label

The `switch` statement selects the statements to perform based on a **controlling expression**, which must be an expression of type integer or character. In the general form that follows, **case labels** (`label_1, label_2, . . .`) determine which statements are executed,
and thus, in some languages, this structure is called a **case structure**. The statements executed are the ones that correspond to the case for which the label is equal to the controlling expression. The case labels must be unique constants; an error occurs if two or more of the case
labels have the same value. The **default label** provides a statement to execute if no other
statement is executed; the default label is optional. Here is the code:

```
switch (controlling expression)
{
    case label_1:
        statements;
    case label_2:
        statements;

    . . .
    default:
        statements;
}
```

The statements in the `switch` structure usually contain the `break` statement. When the
`break` statement is executed, the execution of the program breaks out of the `switch` structure,
and continues executing with the statement following the `switch` structure. Without the
`break` statement, the program will execute all statements that follow the ones selected with
the case label.

Although the default clause in the switch statement is optional, we recommend that it be included so that the steps are clearly specified for the situation in which none of the case labels is equal to the controlling expression. We also use the break statement in the default clause to emphasize that the program continues with the statement following the switch statement.

It is valid to use several case labels with the same statement, as in the following:

```c
switch (op_code)
{
    case 'N': case 'R':
      printf("Normal operating range \n");
      break;
    case 'M':
      printf("Maintenance needed \n");
      break;
    default:
      printf("Error in code value \n");
      break;
}
```

When more than one case label is used for the same statement, the evaluation is performed as if the logical operator || joined the cases. For this example, the first statement is executed if op_code is equal to 'N' or if op_code is equal to 'R'.

PRACTICE!

Convert the following nested if/else statements to a switch statement:

```c
if (rank==1 || rank==2)
   printf("Lower division \n");
else
{
    if (rank==3 || rank==4)
      printf("Upper division \n");
    else
    {
      if (rank==5)
          printf("Graduate student \n");
      else
          printf("Invalid rank \n");
    }
}
```

3.4   Problem Solving Applied: Face Recognition

In this section, we use the new statements presented in this chapter to solve a problem related to facial recognition.

One technique for comparing faces uses ratios of distances between key points on a face, as indicated in Figure 3.7. These ratios might include the distance between the eyes divided by the distance between the nose and the chin. Because these measurements are ratios, they

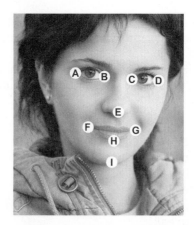

Figure 3.7 *Key points for face recognition.*

can be computed from images of different sizes and should still be similar for the same face. The computer programs that compute these measurements must be able to locate a face in an image and then also locate the eyes and other key points on the face. There are additional challenges if the head is turned in a different direction in one of the images.

For this problem, assume that we have three images of a person looking at the camera. We would like to determine if the two images are likely to be of the same person. The technique that we will use is one that compares ratios of the distances between the outer edges of the eyes to the distances between the tip of the chin and the tip of the nose. Write a C program to read the two distances for each face, compute the ratios, and then determine which two images have the closest ratios.

1. PROBLEM STATEMENT

Given information on three faces, use ratios to determine the two faces that are the most similar.

2. INPUT/OUTPUT DESCRIPTION

The following diagram shows that the inputs to the program are the distance between the outer edges of the eyes and the distance between the tip of the chin and the tip of the nose, for three different images. The output is the image numbers for the two images that are most similar based on ratios of these distances.

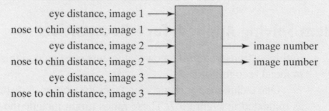

3. HAND EXAMPLE

Assume that the following distances are measured from the three images, in cm:

	Image 1	Image 2	Image 3
Outer eye distance	5.7	6.0	6.0
Nose to chin distance	5.3	5.0	5.6

We can then compute the ratio of the outer eye distance to the nose to chin distance for each image:

$$ratio\_1 = 5.7/5.3 = 1.08$$
$$ratio\_2 = 6.0/5.0 = 1.20$$
$$ratio\_3 = 6.0/5.6 = 1.07$$

We then compute the differences between each pair of ratios, using an absolute value so that each difference is a positive number:

$$diff\_1\_2 = |\,ratio\_1 - ratio\_2\,| = |\,1.08 - 1.20\,| = 0.12$$
$$diff\_1\_3 = |\,ratio\_1 - ratio\_3\,| = |\,1.08 - 1.07\,| = 0.01$$
$$diff\_2\_3 = |\,ratio\_2 - ratio\_3\,| = |\,1.20 - 1.07\,| = 0.13$$

The difference with the smallest value then determines the two images that are the most similar, using these two distances. In this case, image 1 and image 3 are the closest. Note that it is possible that all these differences are large, and thus the smallest of the three may still not represent a good match. Commercial face recognition systems that use distances use a large number of distances in order to improve their accuracy.

4. ALGORITHM DEVELOPMENT

The first step in the development of an algorithm is the decomposition of the problem solution into a set of sequentially executed steps.

Decomposition Outline

1. Read the distances for each image.
2. Compute the ratios for each image.
3. Compute the differences between each pair of ratios.
4. Find the minimum difference.
5. Print the corresponding image numbers as the best match.

This program has a simple structure, so we can convert the decomposition directly into C.

```
/*-------------------------------------------------------------*/
/*  Program chapter3_1                                         */
/*                                                             */
/*  This program selects the two images that are most similar  */
/*  using the distance between the eyes and the distance       */
/*  between the nose and the chin.                             */
```

```c
#include <stdio.h>
#include <math.h>

int main(void)
{
   /*  Declare variables.  */
   double eyes_1, eyes_2, eyes_3, nose_chin_1, nose_chin_2,
           nose_chin_3, ratio_1, ratio_2, ratio_3, diff_1_2,
           diff_2_3, diff_1_3;

   /*  Get user input from the keyboard.  */
   printf("Enter values in cm. \n");
   printf("Enter eye distance and nose-chin distance for image 1: \n");
   scanf("%lf %lf",&eyes_1,&nose_chin_1);
   printf("Enter eye distance and nose-chin distance for image 2: \n");
   scanf("%lf %lf",&eyes_2,&nose_chin_2);
   printf("Enter eye distance and nose-chin distance for image 3: \n");
   scanf("%lf %lf",&eyes_3,&nose_chin_3);

   /*  Compute ratios.  */
   ratio_1 = eyes_1/nose_chin_1;
   ratio_2 = eyes_2/nose_chin_2;
   ratio_3 = eyes_3/nose_chin_3;

   /*  Compute differences.  */
   diff_1_2 = fabs(ratio_1 - ratio_2);
   diff_1_3 = fabs(ratio_1 - ratio_3);
   diff_2_3 = fabs(ratio_2 - ratio_3);

   /*  Find minimum difference and print image numbers.  */
   if ((diff_1_2 <= diff_1_3)) && (diff_1_2 <= diff_2_3)
      printf("Best match is between images 1 and 2 \n");
   if ((diff_1_3 <= diff_1_2)) && (diff_1_3 <= diff_2_3)
      printf("Best match is between images 1 and 3 \n");
   if ((diff_2_3 <= diff_1_3)) && (diff_2_3 <= diff_1_2)
      printf("Best match is between images 2 and 3 \n");

   /*  Exit program.  */
   return 0;
}
/*---------------------------------------------------------------*/
```

5. TESTING

We first test the program with the data from the hand example. This generates the following interaction:

```
Enter values in cm.
Enter eye distance and nose-chin distance for image 1:
5.7 5.3
```

```
Enter eye distance and nose-chin distance for image 2:
6.0 5.0
Enter eye distance and nose-chin distance for image 3:
6.0 5.6
Best match is between images 1 and 3
```

The answer matches the hand example, so we can then test the program with additional lengths. Print an image of each of three different friends and see if you can predict which pair would be the best match for this very simple face recognition system. Run the program with information from two different images of the same person and a third person to see if this technique selects the two images that are from the same person.

MODIFY!

These problems relate to the program developed in this section for finding the best matching pair from three images.

1. Modify the program so that it also prints out all three differences. This gives additional information about the quality of the match; if the differences are small, the quality of the match is better.
2. What would the output be if two of the ratios were the same minimum value? What would the output be if all three of the ratios were the same value? Generate data to make these cases happen and check your answers.
3. Modify the program to use a nested if statement in the solution instead of three independent if statements. Does the answer to Problem 2 change with this solution?
4. Modify the program to print the image numbers for the two images that are the most different.
5. Modify the program to print the best match between four images. (Note that this causes the program to get longer, because with four images there are six possibilities. In Chapter 5 we will learn additional C language functionality that allows us to handle situations such as this one without ever increasing lines of code.)

3.5 Loop Structures

Loops are used to implement repetitive structures. C contains three different loop structures—the while loop, the do/while loop, and the for loop. In addition, C allows us to use two additional statements with loops to modify their performance—the break statement (which we used with the switch statement) and the continue statement.

Before presenting these loop structures, we would like to present two debugging suggestions that are useful when trying to find errors in programs that contain loops. When compiling longer programs, it is not uncommon to have a large number of compiler errors. Rather than trying to find each error separately, we suggest that you recompile your program after correcting several obvious syntax errors. One error will often generate several error messages. Some of these error messages may describe errors that are not in your program, but were printed because the original error confused the compiler.

The second debugging suggestion relates to errors inside a loop. When you want to determine if the steps in a loop are working the way that you want, include `printf` statements in the loop to provide a memory snapshot of key variables each time the loop is executed. Then, if there is an error, you have much of the information that you need to determine the cause of the error.

while Loop

The general form of a `while` loop follows:

```
while (condition)
{
    statements;
}
```

The condition is evaluated before the statements within the loop are executed. If the condition is false, the loop statements are skipped, and execution continues with the statement following the `while` loop. If the condition is true, then the loop statements are executed, and the condition is evaluated again. If it is still true, then the statements are executed again, and the condition is evaluated again. This repetition continues until the condition is false. The statements within the loop must modify variables that are used in the condition; otherwise, the value of the condition will never change, and we will either never execute the statements in the loop or we will never be able to exit the loop. An **infinite loop** is generated if the condition in a `while` loop is always true. Most systems have a system-defined limit on the amount of time that can be used by a program and will generate an execution error when this limit is exceeded. Other systems require that the user enter a special set of characters, such as the control key followed by the character c (abbreviated as $^\wedge$c) to stop or abort the execution of a program. Nearly everyone eventually writes a program that inadvertently contains an infinite loop, so be sure that you know the special characters to abort the execution of a program for your system.

Infinite loop

The following pseudocode and program use a `while` loop to generate a conversion table for converting degrees to radians (note that the degree values start at 0°, increment by 10°, and go through 360°):

Refinement in Pseudocode
main: set degrees to zero
*   while degrees ≤ 360*
*   convert degrees to radians*
*   print degrees, radians*
*   add 10 to degrees*

```
/*-----------------------------------------------------------*/
/*  Program chapter3_2                                       */
/*                                                           */
/*  This program prints a degree-to-radian table            */
/*  using a while loop structure.                           */

#include <stdio.h>
#define PI 3.141593

int main(void)
{
   /*  Declare and initialize variables.  */
   int degrees=0;
   double radians;

   /*  Print radians and degrees in a loop.  */
   printf("Degrees to Radians \n");
   while (degrees <= 360)
   {
      radians = degrees*PI/180;
      printf("%6i %9.6f \n",degrees,radians);
      degrees += 10;
   }

   /*  Exit program.  */
   return 0;
}
/*-----------------------------------------------------------*/
```

The first few lines of output from the program are the following:

```
Degrees to Radians
   0   0.000000
  10   0.174533
  20   0.349066
   . . .
```

do/while Loop

The do/while loop is similar to the while loop except that the condition is tested at the end of the loop instead of at the beginning of the loop. Testing the condition at the end of the loop ensures that the do/while loop is always executed at least once; a while loop may not be executed at all if the condition is initially false. The general form of the do/while loop is as follows:

```
do
{
   statements;
} while (condition);
```

The following pseudocode and program print the degree-to-radian conversion table using a do/while loop instead of a while loop:

Refinement in Pseudocode
main: set degrees to zero
         do
               convert degrees to radians
               print degrees, radians
               add 10 to degrees
         while degrees ≤ 360

```
/*------------------------------------------------------------*/
/*   Program chapter3_3                                       */
/*                                                            */
/*   This program prints a degree-to-radian table            */
/*   using a do-while loop structure.                        */

#include <stdio.h>
#define PI 3.141593

int main(void)
{
   /*   Declare and initialize variables.   */
   int degrees=0;
   double radians;

   /*   Print radians and degrees in a loop.   */
   printf("Degrees to Radians \n");
   do
   {
      radians = degrees*PI/180;
      printf("%6i %9.6f \n",degrees,radians);
      degrees += 10;
   } while (degrees <= 360);

   /*   Exit program.   */
   return 0;
}
/*------------------------------------------------------------*/
```

for LOOP

for loop

Many programs require loops that are based on the value of a variable that increments (or decrements) by the same amount each time through the loop. When the variable reaches a specified value, we will want to exit the loop. This type of loop can be implemented as a while loop, but it can also be easily implemented with the **for loop**. The general form of the for loop is as follows:

```
for (expression_1; expression_2; expression_3)
{
    statements;
}
```

Loop-control
variable

Expression_1 is used to initialize the **loop-control variable**, expression_2 specifies the condition that should be true to continue the loop repetition, and expression_3 specifies the modification to the loop-control variable.

For example, if we want to execute a loop 10 times, with the value of the variable k going from 1 to 10 in increments of 1, we could use the following for loop structure:

```
for (k=1; k<=10; k++)
{
    statements;
}
```

Style

Use consistent placement of braces.

If we want to execute a loop with the value of the variable n going from 20 to 0 in increments of −2, we could use this loop structure:

```
for (n=20; n>=0; n-=2)
{
    statements;
}
```

The for loop could also have been written in this form:

```
for (n=20; n>=0; n=n-2)
{
    statements;
}
```

Both forms are valid, but the abbreviated form is commonly used because it is shorter.

The following expression computes the number of times that a for loop will be executed:

$$\text{floor}\left(\frac{\text{final value} - \text{initial value}}{\text{increment}}\right) + 1.$$

If this value is negative, the loop is not executed. Thus, if a for statement has the structure

```
for (k=5; k<=83; k+=4)
{
    statements;
}
```

it would be executed the following number of times:

$$\text{floor}\left(\frac{83 - 5}{4}\right) + 1 = \text{floor}\left(\frac{78}{4}\right) + 1 = 20.$$

The value of k would be 5, then 9, then 13, and so on, until the final value of 81. The loop would not be executed with the value of 85, because the loop condition is not true when k is equal to 85.

Consider the following set of nested for statements:

```
for (k=1; k<=3; k++)
    for (j=0; j<=1; j++)
        count++;
```

The outer for loop will be executed 3 times. The inner for loop will be executed twice each time the outer for loop is executed. Thus, the variable count will be incremented 6 times.

The following pseudocode and program print the degree-to-radian conversion table shown earlier with a while loop, now modified to use a for loop (note that the pseudocode for the while loop solution to this problem and the pseudocode for the for loop solution to this problem are identical):

Refinement in Pseudocode
main: set degrees to zero
            while degrees ≤ 360
                    convert degrees to radians
                    print degrees, radians
                    add 10 to degrees

```
/*------------------------------------------------------------*/
/*  Program chapter3_4                                        */
/*                                                            */
/*  This program prints a degree-to-radian table             */
/*  using a for loop structure.                               */

#include <stdio.h>
#define PI 3.141593

int main(void)
{
    /*  Declare variables.  */
    int degrees;
    double radians;

    /*  Print radians and degrees in a loop.  */
    printf("Degrees to Radians \n");
    for (degrees=0; degrees<=360; degrees+=10)
    {
        radians = degrees*PI/180;
        printf("%6i %9.6f \n",degrees,radians);
    }

    /*  Exit program.  */
    return 0;
}
/*------------------------------------------------------------*/
```

Observe that the value of degrees did not need to be initialized in the declaration because it is initialized in the for loop statement.

The initialization and modification expressions in a for loop can contain more than one statement, as shown in this statement that initializes and updates two variables in the loop:

```
for (k=1, j=5; k<=10; k++, j++)
{
    sum_1 += k;
    sum_2 += j;
}
```

When more than one statement is used, they are separated by commas and are executed from left to right. This comma operator is executed last in operator precedence.

PRACTICE!

Determine the number of times that the following for loops are executed.

1.
```
for (k=3; k<=20; k++)
{
    statements;
}
```

2.
```
for (k=3; k<=20; ++k)
{
    statements;
}
```

3.
```
for (count=-2; count<=14; count++)
{
    statements;
}
```

4.
```
for (k=2; k>=-10; k--)
{
    statements;
}
```

5.
```
for (time=10; time>=5; time++)
{
    statements;
}
```

6. What is the value of count after the nested for loops are executed?
```
int k, j, count=0;
for (k=-1; k<=3; k++)
    for (j=3; j>=1; j--)
        count++;
```

break AND continue STATEMENTS

We used the break statement in a previous section with the switch statement. The break statement can also be used with any of the loop structures presented in this section to immediately exit from the loop in which it is contained. In contrast, the continue statement is used to

Iteration

skip the remaining statements in the current pass, or **iteration**, of the loop and then continue with the next iteration of the loop. Thus, in a while loop or a do/while loop, the condition is evaluated after the continue statement is executed to determine if the statements in the loop are to be executed again. In a for loop, the loop-control variable is modified, and then the repetition-continuation condition is evaluated to determine whether the statements in the loop are to be executed again. Both the break and continue statements are useful in exiting either the current iteration or the entire loop when error conditions are encountered.

To illustrate the difference between the break and the continue statements, consider the following loop that reads values from the keyboard:

```
sum = 0;
for (k=1; k<=20; k++)
{
    scanf("%lf",&x);
    if (x > 10.0)
        break;
    sum += x;
}
printf("Sum = %f \n",sum);
```

This loop reads up to 20 values from the keyboard. If all 20 values are less than or equal to 10.0, then the statements compute the sum of the values and print the sum. But if a value is greater than 10.0, then the break statement causes control to break out of the loop and execute the printf statement. Thus, the sum printed is only the sum of the values lower than 10.0 and excludes the value greater than 10.0.

Now, consider this variation of the previous loop:

```
sum = 0;
for (k=1; k<=20; k++)
{
    scanf("%lf",&x);
    if (x > 10.0)
        continue;
    sum += x;
}
printf("Sum = %f \n",sum);
```

In this loop, the sum of all 20 values is printed if all values are less than or equal to 10.0. However, if a value is greater than 10.0, then the continue statement causes control to skip the rest of the statements in that iteration of the loop, and it will continue with the next iteration of the loop. Hence, the sum printed is the sum of all values in the 20 values that are less than or equal to 10.0.

3.6   Problem Solving Applied: Wave Interaction

Sea state

The state of the sea, commonly called the **sea state,** is defined in terms of the wind speed and the corresponding waves. As shown in Table 3.3, sea states are defined from 0 to 12, and they cover wind speeds up to and over 70 mph (hurricane speeds). For example, in sea state 1, there is a light breeze (from 1 to 3 miles per hour) and small ripples on the water. In sea state 4, there is a moderate breeze (between 13 and 18 mph), and the sea has small waves with numerous whitecaps. In sea state 6, there is a strong breeze (between 25 and 31 mph), large waves have formed, whitecaps are everywhere, and there is sea spray in the air.

As discussed in the beginning of the chapter, a simple individual wave has crests (high points) and troughs (low points). The vertical distance between a crest and trough is the wave height, and the horizontal distance from crest to crest is the **wavelength.** Deepwater waves occur where the water depth is greater than one-half the wavelength, and are often generated by winds at the ocean surface. The water depth does not affect the speed of deepwater waves. Shallow-water waves occur where the ocean depth is less than 1/20 of the wavelength, and include tide

Wavelength

Table 3.3 Sea State Number

Number	Description	Wind (mph)	Sea Surface Appearance
0	Calm	none	mirror-like
1	Light air	1-3	ripples
2	Light breeze	4-7	small wavelets
3	Gentle breeze	8-12	scattered whitecaps
4	Moderate breeze	13-18	small waves
5	Fresh Breeze	19-24	many whitecaps
6	Strong breeze	25-31	large waves
7	Near gale	32-38	white foam from breaking waves
8	Gale	39-46	moderately high waves
9	Strong gale	47-54	high waves
10	Storm	55-63	very high waves
11	Violent storm	64-72	exceptionally high waves
12	Hurricane	>73	covered with foam and spray

Source: Harold V. Thurman and Elizabeth A Burton. *Introductory Oceanography,* 9th ed. Prentice-Hall, Upper Saddle River, NJ, 2001.

waves. The speed of shallow-water waves is determined by the water depth—the greater the depth, the higher is the wave speed. The speed of transitional waves (those in depths between one-half and 1/20 of the wavelength) are determined by wavelength and water depth. As the velocity of the waves decreases due to friction with the ocean bottom, the wave height increases. When the water depth is 1.3 times the wave height, the wave breaks upon the shore.

In the ocean, waves are generated by many sources, and they come together from many directions. The interaction of individual waves creates a complex set of waves with varying peaks and troughs. If we consider two waves at a time, the combination can be constructive interference in which crests occur at the same time and the troughs occur at the same time, and thus the result has higher crests and lower troughs. The combination is a destructive interference if the crest of one wave occurs at the same time as the trough of another wave, because the sum can cancel the highs and lows. In mixed interference, the two waves have different lengths and heights; thus, the sum is more complicated because it can have components of both constructive and destructive interference.

Interference patterns

To investigate the **interference patterns** of two waves, we now develop a program that will allow the user to enter the wave period and wave height from two different waves. The program will then determine and print the wavelengths of the two waves, and then compute the maximum wave height (assuming that the two waves do not have a phase shift between them). Before we can develop the solution, we first need to discuss sine waves. We will use these sine waves to model the wave and to compute the wavelength of a wave from its period.

Sinusoid

Recall that a sine function is a function of an angle that varies between $+1$ and -1 and has a period of 2π, as shown in Figure 3.8. A **sinusoid** is a form of a sine function that is expressed as a function of time, instead of as a function of angle—for example,

$$s(t) = A \sin(2\pi f t + \phi)$$

where A is the amplitude,
   f is the frequency in cycles per second (or hertz), and
   ϕ is the phase shift in radians.

The period of a sinusoid is $1/f$ in seconds.

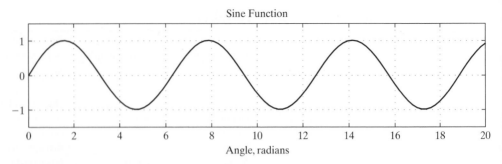

Figure 3.8 *Plot of a sine function.*

Consider the following three functions:

$$s_1(t) = 3\sin(2\pi t),$$
$$s_2(t) = 5\sin(0.4\pi t),\ \text{and}$$
$$s_3(t) = 5\sin(\pi t).$$

Function s_1 has an amplitude of 3, and functions s_2 and s_3 have amplitudes of 5. Function s_1 has a frequency of 1 Hz and thus a period of 1 s; function s_2 has a frequency of 0.2 Hz and thus a period of 5 s; and function s_3 has a frequency of 0.5 Hz and thus a period of 2 s. The phase shift for all three functions is zero. Figure 3.9 contains plots of these three sinusoids.

A property that we will not prove, but that will be useful in analyzing the interaction of waves, is that a sum of two sinusoids is also a periodic signal with a period that is equal to the **least common multiple** (LCM) of the periods of the two individual sinusoids. Thus, if the two periods are 3 s and 6 s, the period of the combined sinusoids is 6 s. If the two periods are 3 s and 5 s, the period of the combined signal is 15 s. If the two periods are $\frac{1}{2}$ s and $\frac{1}{3}$ s, the period of the combined signal is 1 s. If the two periods are $\frac{2}{3}$ s and 2 s, the period of the combined signal is 6 s. Developing an algorithm to determine the least common multiple for two numbers is a nontrivial exercise. Since we want to focus on the characteristics of the wave interaction, we will assume that the wave periods entered into the program will be integers (in seconds). With this assumption, the least common multiple can be shown to be less than or equal to the product of the two integers. Thus, if we use the product as the period for the combined wave, we know that the maximum peak will be in this period. Figure 3.10 contains the sums of each pair of sinusoids in Figure 3.9. Note that the period of $s_1 + s_2$ is 5 (the LCM of 1 and 5), the period of $s_2 + s_3$ is 10 (the LCM of 5 and 2), and the period of $s_1 + s_3$ is 2 (the LCM of 1 and 2).

We now need to consider the relationship between the wavelength L and the period T of a wave. For a wave in deep water [5], this relationship is

$$L = 5.13\,T^2,$$

where T is in seconds and L is in feet. This relationship does not hold for shallow-water waves. We now have all the pieces to describe a solution to the problem presented at the beginning of this section.

Write a program that will allow the user to enter the period and wave height for two waves, where the period (in seconds) is an integer and the wave height is measured in feet. The program should compute and print the wavelength for each wave. It should also compute 200 points of the sum of the two waves over a period equal to the product of the two individual periods. It should then find the maximum of the 200 points of the sum, and print that as an estimate of the maximum wave height of the combined waves, assuming that the phase difference between the two waves is zero.

Least common
multiple

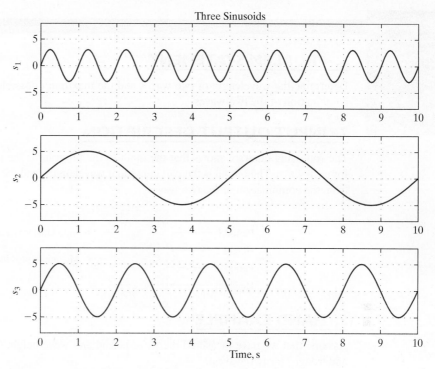

Figure 3.9 *Plot of three sinusoids.*

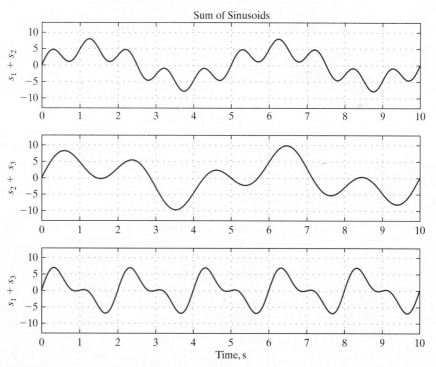

Figure 3.10 *Plot of sums of pairs of sinusoids.*

1. PROBLEM STATEMENT

Determine the wavelength of two waves and the maximum possible wave height from the combination of the two waves.

2. INPUT/OUTPUT DESCRIPTION

The following diagram shows that the input to the program is the period and wave height for each wave. The output is the two individual wavelengths and the maximum wave height from the combination of the two waves.

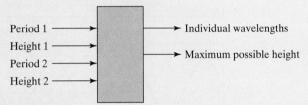

3. HAND EXAMPLE

Assume that we have measured the wave period and wave height (distance between crest and trough) for two waves:

	Period (s)	Height (ft)
Wave 1	4	0.5
Wave 2	10	1.0

We first compute the wavelength for each wave using the equation given earlier in this section:

$$\text{wavelength } 1 = 5.13 \, T_1^2 = 82.08 \text{ ft, and}$$
$$\text{wavelength } 2 = 5.13 \, T_2^2 = 513 \text{ ft.}$$

For the combined wave period, we use the product of the periods of the two waves, or 40 s. Thus, the increment in time between the 200 points is 40/200, or 0.2 s. Using 10 points for the hand example (instead of 200), we have the following set of information:

t	$w_1(t)$	$w_2(t)$	$w_1(t) + w_2(t)$
0	0	0	0
0.2000	0.0773	0.0627	0.1399
0.4000	0.1469	0.1243	0.2713
0.6000	0.2023	0.1841	0.3863
0.8000	0.2378	0.2409	0.4786
1.0000	0.2500	0.2939	0.5439
1.2000	0.2378	0.3423	0.5800
1.4000	0.2023	0.3853	0.5875
1.6000	0.1469	0.4222	0.5691
1.8000	0.0773	0.4524	0.5297

The maximum height for the combined wave (in the first 10 points) is 2(0.5875) ft, or 1.175 ft.

4. ALGORITHM DEVELOPMENT

The first step in the development of an algorithm is the decomposition of the problem solution into a set of sequentially executed steps:

Decomposition Outline

1. Read the periods and wave heights of the two waves to be combined.
2. Compute and print the wavelengths of the individual waves.
3. Compute 200 values of the sum of the two waves and determine the maximum value.
4. Print the maximum value.

The third step in the decomposition outline involves a loop in which we compute 200 values of the new combined wave. Before the loop, we need to compute the period of the combined wave that will be used for the analysis. Since we assume that the wave periods are integers, we can use the product of these periods as the period for the sum of the two waves. (The actual period may be smaller.) We need to think carefully about finding the maximum wave height. Recall the hand example. Once the table has been printed, it is easy to read it and select the maximum height. However, when the program is computing the values for the two waves and adding them, it does not have all the data at one time; it only has the information for the values in the current pass through the loop. Therefore, to keep track of the maximum, we must specify a separate variable to store the maximum value. Each time that we compute a new wave height, we will compare that value to the maximum value. If the new value is larger, we replace the maximum with this new value. Therefore, the refinement in pseudocode is as follows:

Refinement in Pseudocode
main: *read the period and wave height for wave 1*
          read the period and wave height for wave 2
          compute and print the wavelength for each wave
          set new period to the product of the wave periods
          set time increment to new period/200
          set wavemax to 0
          set time to 0
          set steps to 0
          while steps <= 199
                set sum to wave 1 + wave 2
                if sum > wavemax
                      set wavemax to sum
                add 1 to steps
          print wavemax

The steps in the pseudocode are now detailed enough to convert into C:

```
/*-------------------------------------------------------------*/
/*  Program chapter3_5                                         */
/*                                                             */
/*  This program determines the maximum height                */
/*  of a wave that is the sum of two specified waves.         */
```

```c
#include <stdio.h>
#include <math.h>
#define PI 3.141593

int main(void)
{
    /*  Declare variables.  */
    int k;
    double A1, A2, freq1, freq2, height1, height2, length1, length2;
    double T1, T2, w1, w2, sum, new_period, new_height, time_incr, t;
    double maxwave=0;

    /*  Get user input from the keyboard.  */
    printf("Enter integer wave period (s) and wave height (ft) \n");
    printf("for wave 1: \n");
    scanf("%lf %lf",&T1,&height1);
    printf("Enter integer wave period (s) and wave height (ft) \n");
    printf("for wave 2: \n");
    scanf("%lf %lf",&T2,&height2);
    /*  Determine and print wavelengths.  */
    length1 = 5.13*T1*T1;
    length2 = 5.13*T2*T2;
    printf("Wavelengths (in ft) are: %.2f %.2f \n",length1,length2);

    /*  Determine period of combined waves.  */
    new_period = T1*T2;

    /*  Compute 200 points of the combined waves over the   */
    /*  period specified, then find the maximum height.     */
    time_incr = new_period/200;
    A1 = height1/2;
    A2 = height2/2;
    freq1 = 1/T1;
    freq2 = 1/T2;
    for (k=0; k<=199; k++)
    {
        t = k*time_incr;
        w1 = A1*sin(2*PI*freq1*t);
        w2 = A2*sin(2*PI*freq2*t);
        sum = w1 + w2;
        if (sum > maxwave)
            maxwave = sum;
    }
    new_height = maxwave*2;

    /*  Print new wave maximum.  */
    printf("Maximum combined wave height is %.2f ft \n",new_height);

    /*   Exit program.  */
    return 0;
}
/*-----------------------------------------------------------------*/
```

5. TESTING

To test the program using the data from the hand example, we need to modify the program so that it only computes ten points. Thus, we replace the first line of the for loop with this line:

```
for (k=0; k<=9; k++)
```

This generates the following interaction:

```
Enter integer wave period (s) and wave height (ft)
for wave 1:
4 0.5
Enter integer wave period (s) and wave height (ft)
for wave 2:
10 1
Wavelengths (in ft) are: 82.08 513.00
Maximum combined wave height is 1.18 ft
```

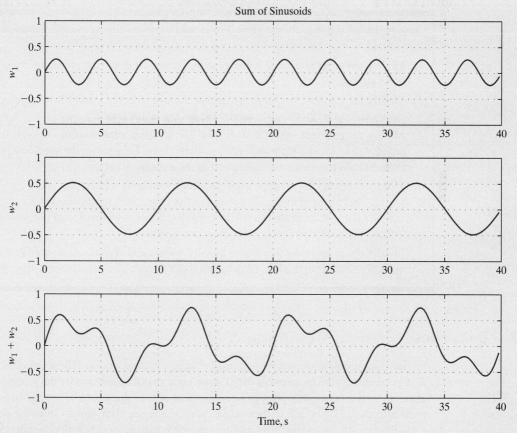

Figure 3.11 *Plot of combined waves.*

The value computed matches the hand example, so we can then test the program. First, we must change the for statement, so it once again specifies that the program should complete the loop two hundred times. The output from this program is

```
Enter integer wave period (s) and wave height (ft)
for wave 1:
4 0.5
Enter integer wave period (s) and wave height (ft)
for wave 2:
10 1
Wavelengths (in ft) are: 82.08 513.00
Maximum combined wave height is 1.46 ft
```

Figure 3.11 contains the two waves and their sums for this example.

MODIFY!

These problems relate to the program developed in this section for computing maximum wave height.

1. Modify the program so that it finds the maximum crest and the minimum trough for the combined waves.

2. Modify the program so that it also allows the user to enter a time, then the program computes and prints the combined wave height at that time.

3. Modify the program so that it allows the user to specify the number of points to compute for determining the maximum combined wave height over the specified period of time. Should this make any difference in the answer? Explain.

4. Modify the program so that it allows the user to specify a phase shift between the two waves. This phase shift should be used as the phase shift for the second wave; the phase shift for the first wave should still be zero. Experiment with different phase shifts to see if they change the maximum possible wave height.

5. Modify the program so that it uses units of meters and seconds, instead of feet and seconds. Remember to modify the program that computes the wavelengths appropriately.

3.7   Data Files

Engineering problem solutions often involve large amounts of data. These data can be output data generated by the program, or they can be input data that are used by the program. It is not generally feasible to either print large amounts of data to the screen or to read large amounts of data from the keyboard. In these cases, we usually use data files to store the data. These data files are similar to the program files that we create to store our C program. In fact, a C program file is an input data file to the C compiler, and the object program is an output file from the C compiler. In this section, we discuss the C statements for interacting with data files and give examples that generate and read information from data files.

When debugging programs read information from data files, echo (or print) the information read from the file to be sure that the data are being read properly. If the data values are all zero, or are unusual numbers, it may be possible that the program cannot find the file because it is in a directory that the program cannot access. The solution is either to move the file to a directory that the program can access or to change some of the operating system parameters so that the program can find the file.

In the examples that follow, we use data files containing sensor data. For this discussion, we assume that the sensor is a seismometer. Seismometers are usually buried near the surface of the earth and record earth motion. These sensors are very sensitive and can record tidal motion, even though they may be located hundreds of miles from the ocean. Seismometer data are collected from sensors all over the earth, and are sent by satellite to central locations for collection and analysis. By studying this motion, scientists and engineers may someday be able to predict earthquakes from seismometer data.

I/O Statements

File pointer

Each data file used in a program must have a **file pointer** associated with it. If a program uses two files, then each file requires a different file pointer. A file pointer is defined with a FILE declaration, as in

```
FILE *sensor1;
```

The FILE data type is defined in the header file stdio.h, and thus the word FILE is capitalized to match the definition in the header file. The asterisk before the identifier specifies that the identifier is a pointer. Later chapters include more information on pointers; in this section, we present only the statements needed to work with file pointers.

After a file pointer is defined, it must then be associated with a specific file. The fopen function obtains the information needed to assign a file pointer to a specific file. The two arguments for this function are the file name and a character that indicates the file status, which

File open mode

is also called the **file open mode**; both the file name and the character need to be enclosed in double quotes. If we are going to read information from a file with a program, the file open mode is r for read. If we are going to write information to a file with a program, the file open mode is w for write. Thus, the following statement specifies that the file pointer sensor is going to be used with a file named sensor.txt, from which we will read information:

```
sensor = fopen("sensor1.txt","r");
```

If a data file cannot be opened, fopen returns a value of NULL. (NULL is a symbolic constant defined in <stdio.h> and has the value of a character zero.) A common reason that a program might not be able to open a data file is because the file cannot be found by the program. Therefore, to be sure that our programs find their data files, it is good practice to check the

Style

value returned by fopen to ensure that the file was successfully opened. The following statements open a file referenced by file1; if the file is not successfully found, an error message is printed and the rest of the if statement is skipped:

```
file1 = fopen(FILENAME,"r");
if (file1 == NULL)
    printf("Error opening input file \n");
else
{
    ...
}
```

One way to be sure that your program can find a data file is to store the data file in the same folder as your program file.

Once an input file and its pointer have been specified, we can read information from the file as we would read information from the keyboard. However, instead of using the scanf function, we use the fscanf function. If each line in the sensor1.txt file contains a time and sensor reading, we can read one line of this information and store the values in the variables t and motion with this statement:

```
fscanf(sensor,"%lf %lf",&t,&motion);
```

Note that the difference between the scanf function and the fscanf function is that the first argument in the fscanf function is the file pointer. Otherwise, both statements are the same. The scanf statement converts the characters received from the keyboard to values, and the fscanf statement converts the characters received from the lines in the data file to values.

If the file is an output file, we can write information to the file with the fprintf function. The first argument of the fprintf statement is the file pointer, and the rest of the arguments define the variables and the form in which the corresponding values are to be written in the file. For example, consider the program developed earlier in this chapter that computed the sum of two waves. If we want to modify this program so that it generates a data file containing this set of data, we could use a pointer waves that would be associated with an output file named waves1.txt using these statements:

```
FILE *waves;
...
waves = fopen("waves1.txt","w");
```

Then, as we compute the information, we can write it to the file with this statement:

```
fprintf(waves,"%.2f %.2f %.2f %.2f\n","%f %f %f\n",
        t,w1,w2,sum);
```

The newline indicator causes a skip to a new line after each group of four values is written to the file.

The fclose function is used to close a file after we are finished with it; the function argument is the file pointer. To close the two files used in these sample statements, we use the following statements:

```
fclose(sensor);
fclose(waves);
```

There is no distinction between closing an input file and closing an output file. If a file has not been closed when the return statement is executed, it will automatically be closed.

A preprocessor directive is often used to specify the data file name because we frequently use the same program with different data files. It is easier to modify the preprocessor directive than it is to search through the statements for the fopen function. An example of a preprocessor directive and a corresponding fopen function are shown in the following code:

```
#define FILENAME "sensor1.txt"
...
sensor = fopen(FILENAME,"r");
```

For the remainder of this textbook, this combination of statements is used in all the example programs that use files.

Reading Data Files

In order to read information from a data file, we must first know some details about the file. Obviously, we must know the file name, so that we can use the `fopen` statement to associate the file with its pointer. We must also know the order and data type of the values stored in the file, so that we can declare corresponding identifiers correctly. Finally, we need to know if there is any special information in the file, so we can help determine the length of the file. If we attempt to execute an `fscanf` statement after we have read all the data in the file, an error occurs. In order to avoid this error, we need to know when we have read all the data.

Data files generally have one of three common structures. Some files have been generated so that the first line in the file contains the number of lines (also called records) with information that follow. For example, suppose that a file containing sensor data has 150 sets of time and sensor information. The data file could be constructed so that the first line contains only the value 150, and that line would then be followed by 150 lines containing the sensor data. To read the data from this file, we read the value from the first line in the file, and then use a `for` loop to read the rest of the information. This type of loop is also called a counter-controlled loop.

Trailer signal
Sentinel signal

Another form of file structure uses a **trailer signal** or **sentinel signal**. These signals are special data values that are used to indicate or signal the last record of a file. For example, the sensor data file constructed with a sentinel signal would contain the 150 lines of information followed by a line with special values, such as −999.0 for the time and sensor value. In order to avoid confusion, these sentinel signals must be values that could not appear as regular data. To read data from this type of file, we use a `while` loop with a condition that is true as long as the data value is not the sentinel signal. This type of loop is also called a sentinel-controlled loop.

The third data file structure does not contain an initial line with the number of valid data records that follow, and it does not contain a trailer or sentinel signal. For this type of data file, we use the value returned by the `fscanf` function to help us determine when we have reached the end of the file. To read data from this type of file, we use a `while` loop with a condition that is true as long as we are not at the end of the file.

Since some operating systems are case sensitive, we will use all lowercase letters in file names to avoid any potential problems. The data file used with a program often changes, so it is helpful if the file name is easy to locate and change. Therefore, use a preprocessor directive to define the filename; otherwise, the filename becomes embedded in the program and cannot be easily changed.

Style

We now present programs for reading sensor information and printing a summary report that contains the number of sensor readings, the average value, the maximum value, and the minimum value. Each of the three common file formats discussed will be used in the programs that follow.

Specified Number of Records.   Assume that the first record in the sensor data file contains an integer that specifies the number of records of sensor information that follow. Each of the following lines contains a time and sensor reading in a file named `sensor1.txt`:

```
10
0.0   132.5
0.1   147.2
0.2   148.3
0.3   157.3
```

```
0.4   163.2
0.5   158.2
0.6   169.3
0.7   148.2
0.8   137.6
0.9   135.9
```

The process of first reading the number of data points and then using that to specify the number of times to read data and accumulate information is easily described using a variable-controlled loop. In the following pseudocode and program, the first actual data value is used to initialize the min and max values. If we set the min value initially to zero and all the sensor values are greater than zero, the program will print the erroneous value of zero for the minimum sensor reading.

The pseudocode and program for this solution are as follows:

Refinement in Pseudocode
main: *set sum to zero*
          if file cannot be opened
                  print error message
          else
                  read number of data points
                  set k to 1
                  while k ≤ number of data points
                          read time, motion
                          if k=1
                                  set max to motion
                                  set min to motion
                          add motion to sum
                          if motion > max
                                  set max to motion
                          if motion < min
                                  set min to motion
                          increment k by 1
                  set average to sum/number of data points
                  print average, max, min

```
/*--------------------------------------------------------------*/
/*  Program chapter3_6                                           */
/*                                                              */
/*  This program generates a summary report from               */
/*  a data file that has the number of data points             */
/*  in the first record.                                       */

#include <stdio.h>
#define FILENAME "sensor1.txt"
```

```
int main(void)
{
   /*  Declare and initialize variables.  */
   int num_data_pts, k;
   double time, motion, sum=0, max, min;
   FILE *sensor;

   /*  Open file and read the number of data points.  */
   sensor = fopen(FILENAME,"r");
   if (sensor == NULL)
      printf("Error opening input file.  \n");
   else
   {
      fscanf(sensor,"%d",&num_data_pts);

      /*  Read data and compute summary information.  */
      for (k=1; k<=num_data_pts; k++)
      {
         fscanf(sensor,"%lf %lf",&time,&motion);
         if (k == 1)
            max = min = motion;
         sum += motion;
         if (motion > max)
            max = motion;
         if (motion < min)
            min = motion;
      }

      /*  Print summary information.  */
      printf("Number of sensor readings: %d \n",
             num_data_pts);
      printf("Average reading:           %.2f \n",
             sum/num_data_pts);
      printf("Maximum reading:           %.2f \n",max);
      printf("Minimum reading:           %.2f \n",min);

      /*  Close file and exit program.  */
      fclose(sensor);
   }

   /*  Exit program.  */
   return 0;
}
/*-----------------------------------------------------------*/
```

The following report will be printed by this program using the sensor1.txt file:

```
Number of sensor readings: 10
Average reading:           149.77
Maximum reading:           169.30
Minimum reading:           132.50
```

Trailer or Sentinel Signals.   Assume that the data file `sensor2.txt` contains the same information as the `sensor1.txt` file, but instead of giving the number of valid data records at the beginning of the file, a final record contains a trailer signal. The time value on the last line in the file will contain a negative value so that we know that it is not a valid line of information. A second number must be included on the trailer line, since the statement that reads each line expects two values; otherwise, an error occurs. The contents of the data file named `sensor2.txt` are as follows:

```
0.0   132.5
0.1   147.2
0.2   148.3
0.3   157.3
0.4   163.2
0.5   158.2
0.6   169.3
0.7   148.2
0.8   137.6
0.9   135.9
-99   -99
```

The process of reading and accumulating information until we read the trailer signal is easily described using a `do/while` loop structure, as shown in the following pseudocode and program:

Refinement in Pseudocode
main: *set sum to zero*
             set number of points to 0
             if file cannot be opened
                     print error message
             else
                     read time, motion
                     set max to motion
                     set min to motion
                     do
                             add motion to sum
                             if motion > max
                                     set max to motion
                             if motion < min
                                     set min to motion
                             increment number of points by 1
                             read time, motion
                     while time ≥ 0
                     set average to sum/number of data points
                     print average, max, min

```
/*------------------------------------------------------------*/
/*  Program chapter3_7                                        */
/*                                                            */
/*  This program generates a summary report from             */
/*  a data file that has a trailer record with               */
/*  negative values.                                         */
#include <stdio.h>
#define FILENAME "sensor2.txt"

int main(void)
{
   /*  Declare and initialize variables.  */
   int num_data_pts=0;
   double time, motion, sum=0, max, min;
   FILE *sensor;

   /*  Open file and read the first data point.  */
   sensor = fopen(FILENAME,"r");
   if (sensor == NULL)
      printf("Error opening input file.  \n");
   else
   {
      fscanf(sensor,"%lf %lf",&time,&motion);

      /*  Initialize variables using first data point.  */
      max = min = motion;

      /*  Update summary data until trailer record read.  */
      do
      {
         sum += motion;
         if (motion > max)
            max = motion;
         if (motion < min)
            min = motion;
         num_data_pts++;
         fscanf(sensor,"%lf %lf",&time,&motion);
      } while (time >= 0);

      /*  Print summary information.  */
      printf("Number of sensor readings: %d \n",
            num_data_pts);
      printf("Average reading:           %.2f \n",
            sum/num_data_pts);
      printf("Maximum reading:           %.2f \n",max);
      printf("Minimum reading:           %.2f \n",min);

      /*  Close file.  */
      fclose(sensor);
   }

   /*  Exit program.  */
   return 0;
}
/*------------------------------------------------------------*/
```

The report printed by this program using the `sensor2.txt` file is exactly the same as the report printed using the `sensor1.txt` file.

End-of-File.   A special **end-of-file indicator** is inserted at the end of every data file; the `feof` function in the Standard C library can be used to detect when this indicator has been reached in a data file. The `fscanf` function can also be used to detect when the end of the data has been reached in a file. Recall that the `fscanf` function returns the number of values successfully read each time that it is executed. Thus, if the function returns a value that is different from the number of values that it was supposed to read, then the end of the data file has been reached, or there are errors in the information in the data file. If the information in the data file is valid, then the `fscanf` function can be used to determine when the end of the data file is reached. Consider the following statements:

```
while ((fscanf(data,"%lf",&x)) == 1)
{
    count++;
    sum += x;
}
ave = sum/count;
```

The `fscanf` function attempts to read a value for x from a data file. If a value is read, the function returns a value of 1, and the statements within the loop are executed. If the end of the data file is reached, there is no more data; thus, the function does not return a value of 1, and control passes to the statement following the `while` loop.

We now assume that the data file `sensor3.txt` contains the same information as the `sensor2.txt` file, except it does not include the trailer signal. In the following pseudocode and program, we read and accumulate information until we reach the end of the data file:

Refinement in Pseudocode
main: *set sum to zero*
           set number of points to 0
           if file cannot be opened
                   print error message
           else
                   while not at the end of the file
                           read time, motion
                           add 1 to the number of points
                           if k=1
                                   set max to motion
                                   set min to motion
                           add motion to sum
                           if motion > max
                                   set max to motion
                           if motion < min
                                   set min to motion
                   set average to sum/number of data points
                   print average, max, min

```
/*------------------------------------------------------------------*/
/*   Program chapter3_8                                             */
/*                                                                  */
/*   This program generates a summary report from                  */
/*   a data file that does not have a header record                */
/*   or a trailer record.                                          */
#include <stdio.h>
#define FILENAME "sensor3.txt"

int main(void)
{
   /*  Declare and initialize variables.  */
   int num_data_pts=0;
   double time, motion, sum=0, max, min;
   FILE *sensor;

   /*  Open file.  */
   sensor = fopen(FILENAME,"r");
   if (sensor == NULL)
      printf("Error opening input file.  \n");
   else
   {
      /*  While not at the end of the file,  */
      /*   read and accumulate information.   */
      while ((fscanf(sensor,"%lf %lf",&time,&motion)) == 2)
      {
         num_data_pts++;

         /*  Initialize variables using first data point.  */
         if (num_data_pts == 1)
            max = min = motion;

         /*  Update summary data.  */
         sum += motion;
         if (motion > max)
            max = motion;
         if (motion < min)
            min = motion;
      }

      /*  Print summary information.  */
      printf("Number of sensor readings: %d \n",
            num_data_pts);
      printf("Average reading:          %.2f \n",
            sum/num_data_pts);
      printf("Maximum reading:          %.2f \n",max);
      printf("Minimum reading:          %.2f \n",min);

      /*  Close file.  */
      fclose(sensor);
   }

   /*  Exit program.  */
   return 0;
}
/*------------------------------------------------------------------*/
```

The report printed using the `sensor3.txt` file is exactly the same as the report printed using `sensor1.dat` or `sensor2.txt`.

All three file structures are commonly used in engineering and scientific applications. Therefore, it is important to know which type of structure is used when you work with a data file. If you make the wrong assumption, you may get incorrect answers instead of an error message. Sometimes the only way to be sure of the file structure is to print the first few lines and the last few lines of the file.

MODIFY!

In two of the programs developed in this chapter, the loop contained a condition that tested for the first time that the loop was executed. When the condition was true, the max and min values were initialized to the first `motion` value. If the data files used with these programs were long, the time required to execute this selection statement could be substantial. One way to avoid this test is to read the first set of data and initialize the variables before entering the loop. This change may also require other changes in the program.

1. Modify program `chapter3_5` to remove the condition that tests for the first time that the loop is executed.
2. Modify program `chapter3_7` to remove the condition that tests for the first time that the loop is executed.

Generating a Data File

Generating a data file is similar to printing a report; however, instead of writing the line to the terminal screen, we write it to a data file. Before we generate the data file, though, we must decide what file structure we want to use. In the previous discussion, we presented the three most common file structures—files with an initial record giving the number of valid records that follow, files with a trailer or sentinel record to indicate the end of the valid data, and files with only valid data records and no special beginning or ending records.

There are advantages and disadvantages to each of the three file structures discussed. If the initial record in the data file contains the number of lines of actual data, we must know how many lines of data will be in the file before we generate the file. It may not always be easy to determine this number. A file with a trailer signal is simple to use, but choosing a value for the trailer signal must be done carefully so that it does not contain values that could occur in the valid data. The simplest file to generate is the one that contains only the valid information, with no special information at the beginning or end of the file. If the information in the file is going to be used with a plotting package, it is usually best to use this third file structure, which includes only valid information.

We now set forth a modification of the program presented earlier in the chapter. The previous program determined the wavelength of two waves and then computed the maximum wave height of the sum of the two waves if the phase angles were zero. In this program, we also write to a data file the time and wave amplitudes for the two individual waves and for the sum of the two waves. Compare this program with the one in Section 3.6.

```
/*-----------------------------------------------------------------*/
/*  Program chapter3_9                                             */
/*                                                                 */
/*  This program determines the maximum height                    */
/*  of a wave that is the sum of two specified waves.             */
```

```c
#include <stdio.h>
#include <math.h>
#define PI 3.141593
#define FILENAME "waves1.txt"

int main(void)
{
   /*  Declare variables.  */
   int k;
   double A1, A2, freq1, freq2, height1, height2, length1, length2;
   double T1, T2, w1, w2, sum, new_period, new_height, time_incr, t;
   double maxwave=0;
   FILE *waves;

   /*  Open output file.  */
   waves = fopen(FILENAME,"w");

   /*  Get user input from the keyboard.  */
   printf("Enter integer wave period (s) and wave height (ft) \n");
   printf("for wave 1: \n");
   scanf("%lf %lf",&T1,&height1);
   printf("Enter integer wave period (s) and wave height (ft) \n");
   printf("for wave 2: \n");
   scanf("%lf %lf",&T2,&height2);

   /*  Determine and print wavelengths.  */
   length1 = 5.13*T1*T1;
   length2 = 5.13*T2*T2;
   printf("Wavelengths (in ft) are: %.2f %.2f \n",length1,length2);

   /*  Determine period of combined waves.  */
   new_period = T1*T2;

   /*  Compute 200 points of the combined waves over the  */
   /*  period specified, and find the maximum height.     */
   time_incr = new_period/200;
   A1 = height1/2;
   A2 = height2/2;
   freq1 = 1/T1;
   freq2 = 1/T2;
   for (k=0; k<=199; k++)
   {
      t = k*time_incr;
      w1 = A1*sin(2*PI*freq1*t);
      w2 = A2*sin(2*PI*freq2*t);
      sum = w1 + w2;
      fprintf(waves,"%.4f %.4f %.4f %.4f \n",t,w1,w2,sum);
      if (sum > maxwave)
         maxwave = sum;
   }
   new_height = maxwave*2;
```

```
    /*  Print new wave maximum.   */
    printf("Maximum combined wave height is %.2f ft \n",new_height);

    /*   Close file and exit program.   */
    fclose(waves);
    return 0;
}
/*----------------------------------------------------------------*/
```

The first few lines of a data file generated by this program using the input values for the hand example are shown below:

```
    0.0000      0.0000      0.0000      0.0000
    0.2000      0.0773      0.0627      0.1399
    0.4000      0.1469      0.1243      0.2713
    0.6000      0.2023      0.1841      0.3863
```

This file is in a form that can be easily plotted using a package such as MATLAB (discussed in Appendix C); a plot of this specific file was shown in Figure 3.11.

MODIFY!

1. Modify program `chapter3_8` so that it generates a file in which the last line of the data file contains negative values for all four values.

2. Modify program `chapter3_8` so that it generates a file in which the first line contains a number that specifies the number of valid lines of data that follow in the data file.

3.8   Numerical Technique: Linear Modeling*

Linear modeling

Linear regression

Linear modeling is the name given to the process that determines the linear equation that is the best fit to a set of data points in terms of minimizing the sum of the squared distances between the line and the data points. (This process is also called **linear regression**.) To understand this process, we first consider the set of temperature values presented in Section 2.6 that were collected from the cylinder head of a new engine:

Time, s	Temperature, °F
0.0	0.0
1.0	20.0
2.0	60.0
3.0	68.0
4.0	77.0
5.0	110.0

---

*Optional sections.

If we plot these data points, we find that they appear to be close to a straight line. In fact, we could determine a good estimate of a straight line through these points by drawing it on a graph and then computing the slope and y-intercept. Figure 3.12 contains a plot of the points (with time on the x-axis and temperature on the y-axis) along with the straight line with the equation

$$y = 20x.$$

To measure the quality of the fit of this linear estimate to the data, we first determine the vertical distance from each point to the linear estimate; these distances are shown in Figure 3.13. The first two points fall exactly on the line, so d_1 and d_2 are zero. The value of d_3 is equal to $60.0 - 40.0$, or 20.0; the rest of the distances can be computed in a similar way. If we compute the sum of the distances, some of the positive and negative values would cancel each other and give a sum that is smaller than it should be. To avoid this problem, we could add absolute values or squared values; linear regression uses squared values. Therefore, the measure of the quality of the fit of this linear estimate is the sum of the squared distances between the points and the linear estimates. This sum can be easily computed; it is 573.

If we drew another line through the points, we could compute the sum of squares that corresponds to this new line. Of the two lines, the better fit is provided by the line with the smaller sum of squared distances. To find the line with the smallest sum of squared distances, we begin with a general linear equation:

$$y = mx + b.$$

We then write an equation that computes the sum of the squared distances between the given data points and this general equation. Using techniques from calculus, we can then compute the derivatives of the equation with respect to m and b, and set the derivatives equal to zero. The values of m and b that are determined in this way represent the straight line with the minimum sum of squared distances, or the **least-squares** distance. Before giving these equations for m and b, we define **summation notation**.

Least-squares
Summation notation

The set of data points given at the beginning of this section can be represented by the points (x_1, y_1), (x_2, y_2) ..., (x_6, y_6). The symbol Σ represents a summation; thus, the sum of the x-coordinates can be expressed in the following notation:

$$\sum_{k=1}^{6} x_k.$$

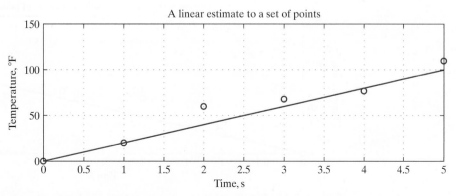

Figure 3.12 A linear estimate to model a set of points.

This summation is read as *the sum of x_k as* k *goes from 1 to 6*. The value of this summation for the example data points is $(0 + 1 + 2 + 3 + 4 + 5)$, or 15. Other sums that could be computed by using the example data points are as follows:

$$\sum_{k=1}^{6} y_k = 0 + 20 + 60 + 68 + 77 + 110 = 335,$$

$$\sum_{k=1}^{6} y_k^2 = 0^2 + 20^2 + 60^2 + 68^2 + 77^2 + 110^2 = 26,653,$$

$$\sum_{k=1}^{6} x_k y_k = 0 \cdot 0 + 1 \cdot 20 + 2 \cdot 60 + 3 \cdot 68 + 4 \cdot 77 + 5 \cdot 110 = 1,202.$$

We now return to the problem of finding the best linear fit to a set of points. Using the preceding procedure, which is based on results from calculus, we find that the slope and y-intercept for the best linear fit to a set of n data points in a least-squares sense, are the following:

$$m = \frac{\sum_{k=1}^{n} x_k \cdot \sum_{k=1}^{n} y_k - n \cdot \sum_{k=1}^{n} x_k y_k}{\left(\sum_{k=1}^{n} x_k\right)^2 - n \cdot \sum_{k=1}^{n} x_k^2}, \tag{3.1}$$

$$b = \frac{\sum_{k=1}^{n} x_k \cdot \sum_{k=1}^{n} x_k y_k - \sum_{k=1}^{n} x_k^2 \cdot \sum_{k=1}^{n} y_k}{\left(\sum_{k=1}^{n} x_k\right)^2 - n \cdot \sum_{k=1}^{n} x_k^2}. \tag{3.2}$$

For the sample set of data, the optimum value for m is 20.83 and the optimum value for b is 3.76. The set of data points and this best fit linear equation are shown in Figure 3.14. The sum of squares for this best fit is 356.82, as compared with 573 for the straight line in Figure 3.13.

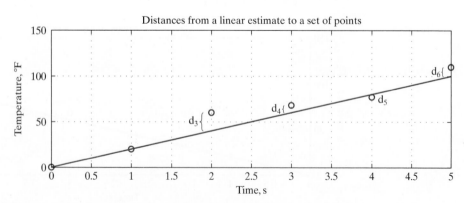

Figure 3.13 *Distances between points and the linear estimate.*

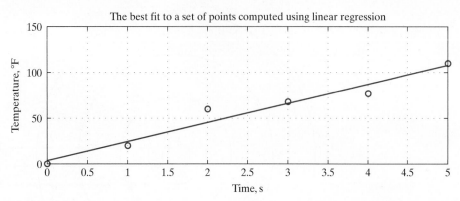

Figure 3.14 *Least-squares linear regression model.*

One of the advantages of performing a linear regression for a set of data points that is nearly linear in nature is that we can then estimate or predict points for which we had no data. For example, in the cylinder-head temperature example, suppose that we want to estimate the temperature for the cylinder head at 3.3 seconds. By using the equation computed with linear regression, the estimated temperature is

$$
\begin{aligned}
y &= mx + b \\
&= (20.83)(3.3) + 3.76 \\
&= 72.5.
\end{aligned}
$$

With an equation model, we can compute estimates that we could not compute with linear interpolation. For example, using the linear model, we can compute an estimate of the temperature for 8 seconds, but we could not compute an estimate at 8 seconds using linear interpolation, because we do not have a point with a time greater than 8 seconds. (This would be
Extrapolation   **extrapolation**, not interpolation.)

It is also important to remember that linear models do not provide a good fit to all sets of data. Therefore, it is important to first determine if a linear model is a good model for the data before using it to predict new data points.

In the next section, we develop a problem solution that determines the best fit for a set of sensor data collected from a satellite, and then we use that model to estimate or predict other sensor values.

3.9   Problem Solving Applied: Ozone Measurements*

Satellite sensors can be used to measure many different pieces of information that help us understand more about the atmosphere, which is composed of a number of layers around the earth. Starting at the earth's surface, the layers are the troposphere, stratosphere, mesosphere, thermosphere, and exosphere, as shown in Figure 3.15. Each layer of the
Troposphere   atmosphere can be characterized by its temperature profile. The **troposphere** is the inner

---

*Optional sections.

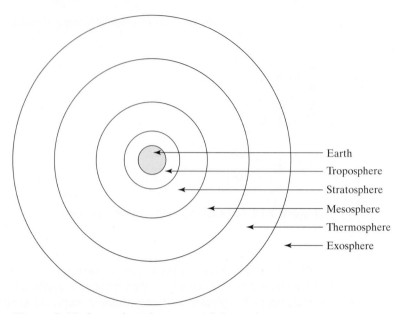

Figure 3.15 *Atmospheric layers around the earth.*

layer of the atmosphere, varying in height from approximately 5 km at the poles to 18 km at the equator. Most cloud formations occur in the troposphere, where there is a steady fall of temperature with increasing altitude. The **stratosphere** is characterized by relatively uniform temperatures over considerable differences in altitude. It extends from the troposphere to approximately 50 km (about 31 miles) above the earth. Pollutants that drift into the stratosphere may remain there for many years before they drift back to the troposphere, where they can be diluted and removed by the weather. The **mesosphere** extends from 50 km to approximately 85 km (about 53 miles) above the earth's surface. In this layer, the air mixes fairly readily. Above the mesosphere is the **thermosphere**, which extends from 85 km to about 140 km (about 87 miles) above the earth. This region is heated by the absorption of solar energy by atomic oxygen. The **ionosphere** is a relatively dense band of charged particles within the thermosphere. Some types of communications use the reflection of radio waves off the ionosphere. Finally, the **exosphere** is the highest region of the atmosphere. In the exosphere, the air density is so low that an air molecule moving upward is more likely to escape the atmosphere than it is to hit another molecule.

Consider a set of data measuring the ozone-mixing ratio in parts per million volume (ppmv). Over small regions, these data are nearly linear, and thus we can use a linear model to estimate the ozone at altitudes other than ones for which we have specific data. Write a program that reads a data file named `zone1.txt` containing the altitude in kilometers and the corresponding ozone-mixing ratios in parts per million volume for a region over which we want to determine a linear model. The data file contains only valid data, and thus does not have a special header line or trailer line. Use the least-squares technique presented in the previous section to determine and print the model. In addition, print the beginning and ending altitudes to indicate the region over which the model is accurate.

Stratosphere

Mesosphere

Thermosphere

Ionosphere

Exosphere

1. PROBLEM STATEMENT

Use the least-squares technique to determine a linear model for estimating the ozone-mixing ratio at a specified altitude.

2. INPUT/OUTPUT DESCRIPTION

The following I/O diagram shows that the input is the data file `zone1.txt`, and the output is the range of altitudes and the linear model:

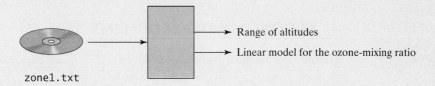

Range of altitudes

Linear model for the ozone-mixing ratio

`zone1.txt`

3. HAND EXAMPLE

Assume that the data consist of the four data points in the following table.

Altitude (km)	Ozone Mixing Ratio (ppmv)
20	3
24	4
26	5
28	6

We now need to evaluate Equations (3.1) and (3.2), which are repeated here for convenience:

$$m = \frac{\sum\limits_{k=1}^{n} x_k \cdot \sum\limits_{k=1}^{n} y_k - n \cdot \sum\limits_{k=1}^{n} x_k y_k}{\left(\sum\limits_{k=1}^{n} x_k\right)^2 - n \cdot \sum\limits_{k=1}^{n} x_k^2};$$

$$b = \frac{\sum\limits_{k=1}^{n} x_k \cdot \sum\limits_{k=1}^{n} x_k y_k - \sum\limits_{k=1}^{n} x_k^2 \cdot \sum\limits_{k=1}^{n} y_k}{\left(\sum\limits_{k=1}^{n} x_k\right)^2 - n \cdot \sum\limits_{k=1}^{n} x_k^2}.$$

To evaluate these equations using the data from the hand example, we need to compute the following group of sums:

$$\sum_{k=1}^{4} x_k = 20 + 24 + 26 + 28 = 98,$$

$$\sum_{k=1}^{4} y_k = 3 + 4 + 5 + 6 = 18,$$

$$\sum_{k=1}^{4} x_k^2 = (20)^2 + (24)^2 + (26)^2 + (28)^2 = 2436,$$

$$\sum_{k=1}^{4} x_k y_k = 20 \cdot 3 + 24 \cdot 4 + 26 \cdot 5 + 28 \cdot 6 = 454.$$

Using these sums, we can now compute the values of m and b:

$$m = 0.37, \quad b = -4.6.$$

4. ALGORITHM DEVELOPMENT

We first develop the decomposition outline.

Decomposition Outline

1. Read data file values and compute corresponding sums and ranges.
2. Compute slope and y-intercept.
3. Print range of altitudes and linear model.

The first step in the decomposition outline involves a loop in which we read the data from the file and, at the same time, add the corresponding values to the sums needed for computing the linear model. We will also need to determine the number of data points as we read the file. The condition to exit the loop will be a test for the end of the file because there is no header or trailer information. Because we want to keep track of the altitude ranges, we also need to save the first altitude value and the last altitude value. Therefore, the refinement in pseudocode is as follows:

Refinement in Pseudocode
Main: set count to zero
   set sumx, sumy, sumxy, sumx2 to zero
   if file cannot be opened
     print error message
  else
     while not at the end of the file
       read x, y
       increment count by 1
       if count = 1
         set first to x
       add x to sumx
       add y to sumy
       add x^2 to sumx2
       add xy to sumxy
    set last to x
    compute slope and y intercept
    print first, last, slope, y intercept

The steps in the pseudocode are now detailed enough to convert into C:

```
/*-------------------------------------------------------------*/
/*  Program chapter3_10                                        */
/*                                                             */
/*  This program computes a linear model for a set             */
/*  of altitude and ozone mixing ratio values.                */

#include <stdio.h>
#define FILENAME "zone1.txt"

int main(void)
{
   /*  Declare and initialize variables.  */
   int count=0;
   double x, y, first, last, sumx=0, sumy=0, sumx2=0,
          sumxy=0, denominator, m, b;
   FILE *zone;

   /*  Open input file.  */
   zone = fopen(FILENAME,"r");
   if (zone == NULL)
      printf("Error opening input file.  \n");
   else
   {
      /*  While not at the end of the file,  */
      /*  read and accumulate information.   */
      while ((fscanf(zone,"%lf %lf",&x,&y)) == 2)
      {
         ++count;
         if (count == 1)
            first = x;
         sumx += x;
         sumy += y;
         sumx2 += x*x;
         sumxy += x*y;
      }
      last = x;

      /*  Compute slope and y-intercept.  */
      denominator = sumx*sumx - count*sumx2;
      m = (sumx*sumy - count*sumxy)/denominator;
      b = (sumx*sumxy - sumx2*sumy)/denominator;
```

```
          /*  Print summary information.  */
          printf("Range of altitudes in km: \n");
          printf("%.2f to %.2f \n\n",first,last);
          printf("Linear model: \n");
          printf("ozone-mix-ratio = %.2f altitude + %.2f \n",
                 m,b);

          /*  Close file.  */
          fclose(zone);
   }

   /*  Exit program.  */
   return 0;
}
/*-------------------------------------------------------------*/
```

5. TESTING

Using the data from the hand example as the contents of the data file `zone1.txt`, we get the following program output:

```
Range of altitudes in km:
20.00 to 28.00

Linear model:
ozone-mix-ratio = 0.37 altitude + -4.60
```

This matches the values computed from the hand example.

MODIFY!

These problems relate to the program developed in this section. You may need to use the following relationship in some of the problems:

$$1 \text{ km} = 0.621 \text{ mi.}$$

1. Add statements to the program so that it allows you to enter an altitude in kilometers. The program should use the model to estimate the corresponding ozone mix ratio.

2. Modify the program in Problem 1 so that it checks the altitude that you enter to determine whether it is appropriate for this model.

3. Modify the program in Problem 2 so that it allows you to enter the altitude in miles. (The program should convert miles to kilometers.)

4. Modify the original program so that it also prints a linear model so that it can be used with altitudes that are in miles instead of kilometers. Assume that the data file still contains altitudes in kilometers.

SUMMARY In this chapter, we covered the use of conditions and if statements to select the proper statements to be executed. We also presented techniques for repeating sets of statements in loops. These loops can be implemented as while loops or for loops. Selection and repetition structures are used in most programs. In addition, we included the statements necessary to read information from a data file so that we could use the information in the program. We also presented the statements to generate a data file from a program. Data files are commonly used in solving engineering problems; therefore, this concept was presented early in the text so that we could use it in many of the later problem solutions. Finally, we covered the concept of generating a linear model for a set of data points and included the equations for determining the best fit in terms of least squares.

KEY TERMS

case label	linear regression
case structure	logical operator
compound statement	loop
condition	loop control variable
conditional operator	program walkthrough
controlling expression	pseudocode
data file	relational operator
default label	repetition
divide and conquer	selection
empty statement	sequence
end-of-file indicator	sentinel signal
error condition	stepwise refinement
file open mode	structured program
file pointer	summation notation
flowchart	test data
for loop	top-down design
iteration	trailer signal
least squares	validation and verification
linear modeling	while loop

C STATEMENT SUMMARY

Declaration for file pointer:

```
FILE *sensor;
```

if statement:

```
if (temp > 100)
    printf("Temperature exceeds limit \n");

temp>100 ? printf("Caution \n"): printf("Normal \n");
```

if/else statement:

```
if (d <= 30)
   velocity = 4.25 + 0.00175*d*d;
else
   velocity = 0.65 + 0.12*d - 0.0025*d*d;
```

switch statement:

```
switch (op_code)
{
   case 'n': case 'r':
      printf("Normal operating range \n");
      break;
   case 'm':
      printf("Maintenance needed \n");
      break;
   default:
      printf("Error in code value \n");
      break;
}
```

while loop:

```
while (degrees <= 360)
{
   radians = degrees*PI/180;
   printf("%6.0f %9.6f \n",degrees,radians);
   degrees += 10;
}
```

do/while loop:

```
do
{
   radians = degrees*PI/180;
   printf("%6.0f %9.6f \n",degrees,radians);
   degrees += 10;
} while (degrees <= 360);
```

for loop:

```
for (degrees=0; degrees<=360; degrees+=10)
{
   radians = degrees*PI/180;
   printf("%6.0f %9.6f \n",degrees,radians);
}
```

break statement:

```
break;
```

continue statement:

```
continue;
```

File open function:

```
sensor = fopen("sensor1.txt","r");
waves = fopen(FILENAME,"w");
```

File input function:

```
fscanf(sensor,"%lf %lf",&t,&motion);
```

File output function:

```
fprintf(waves,"%.2f %.2f %.2f %.2f \n",t,w1,w2,sum);
```

File close function:

```
fclose(sensor);
```

NOTES

1. Use spaces around the relational operator in a logical expression in a simple condition; use spaces around the logical operator and not around the relational operators in a complicated condition.
2. Indent the statements within a compound statement or inside a loop. If loops or compound statements are nested, indent each nested set of statements from the previous statement.
3. Even when they are not required, use braces to clearly identify the structure of a complicated statement.
4. Use the default case within the `switch` statement to emphasize the action to take when none of the case labels matches the controlling expression.
5. Put each brace on a line by itself so that the body of the loop is easily identified.
6. Define file names with preprocessor directives so that they can easily be changed.

DEBUGGING NOTES

1. When you discover and correct an error in a program, start the testing step over again. In particular, rerun the program with all the test data sets.
2. Be sure to use the relational operator == instead of = in a condition for equality.
3. Put the braces surrounding a block of statements on lines by themselves; this will help you avoid omitting them.
4. Do not use the equality operator with floating-point values; instead, test for values "close to" a desired value.
5. Recompile your program frequently when correcting syntax errors; correcting one error may remove many error messages.
6. When debugging loops, use the `printf` statement to give memory snapshots of the values of key variables.
7. Be sure you know the special characters needed to abort the execution of a program on your system if it goes into an infinite loop. It is easier than you think to generate an infinite loop.
8. When debugging a program that reads data from a data file, print the values as soon as they are read. This will help you check for errors in reading the information.
9. When debugging a program that reads a data file, be sure that your program can access the directory that contains the data file.
10. To avoid problems with operating systems that are not case sensitive, use file names with lowercase letters.

PROBLEMS

SHORT ANSWER PROBLEMS

True–False Problems

Indicate whether the following statements are true (T) or false (F):

1. If a condition's value is zero, then the condition is evaluated as false.   T   F
2. The contents within the bracket following the if condition should always be a relational expression.   T   F
3. A nested if structure can be always substituted with a switch case structure. T   F
4. The logical operators && and || have the same precedence level.   T   F
5. Unlike the while loop, a do/while loop forces execution of the statements inside the loop atleast once.   T   F
6. A for loop may contain multiple initialization expressions and multiple increment expressions.   T   F

Syntax Problems

Identify any syntax errors in the following statements (assume that the variables have all been defined as integers):

7. `for (b=1, b=<25, b++)`

8. `while(m-2==3);`

9.
```
int m=65;
switch(m)
{
  case 'A':
   printf("hi");
   break;
  case 65:
     printf("hello");
     break;
  default :
     ;
}
```

Multiple Choice Problems

Circle the letter for the best answer to complete each statement or for the correct answer to each question.

10. Consider the following statement:

    `int i=100, j=0;`

    Which of the following statements are true?
    (a) `i<3`
    (b) `!(j<1)`
    (c) `(i>0) || (j>50)`
    (d) `(j<i) && (i<=10)`

11. Which of the following expressions evaluate to 1, if a = -1, b = 0, and c = 1?
    (a) b||!(a&&c)
    (b) !a||b||!c
    (c) a&&b||c
    (d) !(a||b&&c)

12. Which of the following are unary operators?
    (a) !
    (b) ||
    (c) &&

13. What will be the output of the following problem?

    ```
    int m=43, n=-9,p=0,x,y;

    x=m+p&&n;

    y=p||n/m;

    printf("x = %d   y = %d", x, y);
    ```
    (a) x = 1 y = 0
    (b) x = 43 y = 0
    (c) x = 44 y = 1
    (d) syntax error

Problems 14–16 refer to the following section of code:

```
int s=0,t=1;
for(int i=1;;i++)
{ if(i>4) break;
   t*=i;
   s+=t;}
printf("\ns=%d",s);
```

14. What will be the value of s after the code is executed?
    (a) s = 10   (b) s = 33
    (c) s = 24   (d) syntax error

15. The if condition within the loop has the same impact as
    (a) an if condition if(i<=4) continue; within the for loop block
    (b) a for loop statement for (int i=1;i<4;i++)
    (c) a while loop block while (i<4) break; within the for loop block
    (d) a for loop statement for (int i=1;i<=4;i+=1)

16. The values of s computed is
    (a) sum of first four natural numbers, i.e. $1 + 2 + 3 + 4$
    (b) product of first four natural numbers, i.e. $1 \times 2 \times 3 \times 4$
    (c) sum of squares of first four natural numbers, i.e. $1^2 + 2^2 + 3^2 + 4^2$
    (d) sum of factorials of first four natural numbers, i.e. $1! + 2! + 3! + 4!$

Memory Snapshot Problems

Give the corresponding snapshots of memory after the following set of statements is executed.

17.
```
int x=100,y;
if(x>0)
if(x<1000)
```

```
        x*=5;
            else
        x+=100;
            else
                x*=(-1);
            y=x%55;
```

PROGRAMMING PROBLEMS

Unit Conversions. The following problems generate tables of unit conversions. Include a table heading and column headings for the tables. Choose the number of decimal places based on the values to be printed.

18. Generate a table of conversions from radians to degrees. Start the radian column at 0.0, and increment by $\pi/10$, until the radian amount is 2π.

19. Generate a table of conversions from degrees to radians. The first line should contain the value for $0°$ and the last line should contain the value for $360°$. Allow the user to enter the increment to use between lines in the table.

20. Generate a table of conversions from feet-inches to millimeters. Start the feet-inch column at 0'-0" and increment by 3". The last line should contain the value 2'-9". (Recall that 1' = 12" and 1" = 25.4 mm.)

21. Generate a table of conversions from kilometers per hours to meters per second. Start the km/hr column with 10km/hr and increment by 1 km/hr. Show exactly ten conversion data. (Recall that 1km = 1000 m and 1hr = 60 X 60 sec.)

22. Generate a table of conversions from CGS unit to SI unit of magneto-motive force. The CGS unit is Gilbert and the SI unit is Ampere. Start the Gilbert column with 1Gi and increment by 10 Gi. The last line should contain the value 91 Gi. Use a do while loop for your purpose. (Recall that 1 Gilbert = 10/4 π amperes)

Currency Conversions. The following problems generate tables of currency conversions. Use title and column headings. Assume the following conversion rates:

1 dollar ($) = 0.737938 Euro (Europe)

1 yen (Y) = $0.013005

1 dollar ($) = 0.632293 pounds (£) (UK)

23. Generate a table of conversions from Euros to dollars. Start the Euros column at 5 Euros and increment by 5 Euros. Print 25 lines in the table.

24. Generate a table of conversions from pounds (£) to dollars. Start the pounds column at 1 £ and increment by 2 £. Print 30 lines in the table.

25. Generate a table of conversions from yen to pounds. Start the yen column at 100 Y and print 25 lines, with the final line containing the value 10,000 Y.

26. Generate a table of conversions from dollars to Euros, yen, and pounds. Start the column with $1 and increment by $1. Print 50 lines in the table.

Temperature Conversions. The following problems generate temperature-conversion tables. Use the following equations that give relationships between temperatures in degrees Fahrenheit (T_F), degrees Celsius (T_C), degrees Kelvin (T_K), and degrees Rankin (T_R):

$$T_F = T_R - 459.67° \text{ R}$$
$$T_F = (9/5)T_C + 32° \text{ F}$$
$$T_R = (9/5)T_K$$

27. Write a program to generate a table of conversions from Fahrenheit to Celsius for values from 0°F to 100°F. Print a line in the table for each 5° change. Use a `while` loop in your solution.

28. Write a program to generate a table of conversions from Fahrenheit to Kelvin for values from 0°F to 200°F. Allow the user to enter the increment in degrees Fahrenheit between lines. Use a `do while` loop in your solution.

29. Write a program to generate conversions from Celsius to Fahrenheit, Rankin, and Kelvin for 0°C to 100°C with an increment of 5°C. Use for loop for your solution.

Sounding Rocket Trajectory. Sounding rockets are used to probe different levels of the atmosphere to collect information (such as that used to monitor the levels of ozone in the atmosphere). In addition to carrying the scientific package for collecting data in the upper atmosphere, the rocket also carries a telemetry system to transmit scientific data to a receiver at the launch site. Performance measurements on the rocket itself are also transmitted, so they can be monitored by range safety personnel and later analyzed by engineers. These performance data include altitude, velocity, and acceleration data. Assume that this information is stored in a file and that each line contains four values—time, altitude, velocity, and acceleration. Assume that the units are seconds, meters, meters/second, and meters/second^2, respectively.

30. Assume that the file `rocket1.txt` contains an initial line that contains the number of actual data lines that follows. Write a program that reads these data and determines the time at which the rocket begins falling back to earth. (*Hint*: Determine the time at which the altitude begins to decrease.)

31. The number of stages in the rocket can be determined by the number of times that the velocity increases to some peak and then begins decreasing. Write a program that reads these data and determines the number of stages on the rocket. Use the data file `rocket2.txt`. It contains a trailer line with the value −99 for all four values.

32. Modify the program in Problem 31 so that it prints the times that correspond to the firing of each stage. Assume that the firing corresponds to the point at which the velocity begins to increase.

33. After each stage of the rocket is fired, the acceleration will initially increase and then decrease to −9.8 m/s^2, which is the downward acceleration due to gravity. Find the time periods of the rocket flight during which the acceleration is due only to gravity. Allow the acceleration to range up to 65% of the theoretical value for these time periods. Use the data file `rocket3.txt`, which does not contain a header line or a trailer line.

Suture Packaging. Sutures are strands or fibers used to sew living tissue together after an injury or an operation. Packages of sutures must be sealed carefully before they are shipped to hospitals so that contaminants cannot enter the packages. The object that seals the package is referred to as a sealing die. Generally, sealing dies are heated with an electric heater. For the sealing process to be a success, the sealing die is maintained at an established temperature and must contact the package with a predetermined pressure for an established time period. The time period in which the sealing die contacts the package is called the dwell time. Assume that the acceptable range of parameters for an acceptable seal are the following:

Temperature:	150–170°C,
Pressure:	60–70 psi,
Dwell time:	2–2.5 s.

34. A data file named `suture1.txt` contains information on batches of sutures that have been rejected during a one-week period. Each line in the data file contains the batch number, temperature, pressure, and dwell time for a rejected batch. The quality control engineer must analyze this information, and needs a report that computes the percent of the batches rejected due to temperature, the percent rejected due to pressure, and the percent rejected due to dwell time. It is possible that a specific batch may have been rejected for more than one reason, and it should be counted in all applicable totals. Write a program to compute and print these three percentages.

35. Modify the program developed in Problem 34 such that it also prints the number of batches in each rejection category and the total number of batches rejected. (Remember that a rejected batch should appear only once in the total, but could appear in more that one rejection category.)

36. Write a program to read the data file `suture1.txt` and make sure that the information relates only to batches that should have been rejected. If any batch should not be in the data file, print an appropriate message with the batch information.

Timber Regrowth. One problem in timber management is to determine how much of an area to leave uncut so that the harvested area is reforested in a certain period of time. It is assumed that reforestation takes place at a known rate per year, depending on climate and soil conditions. A reforestation equation expresses this growth as a function of the amount of timber standing and the reforestation rate. For example, if 100 acres are left standing after harvesting and the reforestation rate is 0.05, then 100 + (0.05 · 100), or 105 acres, are forested at the end of the first year. At the end of the second year, the number of acres forested is 105 + (0.05 · 105), or 110.25 acres.

37. Assume that the area has a total of 14,000 acres, with 2500 acres uncut and a reforestation rate of 0.02. Print a table showing the number of acres forested at the end of each year, for a total of 20 years.

38. Modify the program developed in Problem 37 so that the user can enter the number of years to be used for the table.

39. Modify the program developed in Problem 37 so that the user can enter a number of acres and the program will determine how many years are required for the number of acres to be completely reforested.

Cloud Computing. Cloud computing is a technology that uses the internet and central remote servers to maintain data and applications. It is basically a resource delivery and usage model to get resources (infrastructure, platform, and software) through the internet as services in a multi-tenant environment. All resources in the cloud are scalable infinitely and used whenever required as utility. A cloud provider supplies a pool of resources to the consumer. A cloud provider may provide on-demand resources and reserved resources. On-demand instances are paid by the hour with no long-term commitments. Reserved instances allow making a low, one-time payment for each instance that is reserved and in turn receive a significant discount on the hourly usage charge for that instance. Instance types may be classified into a number of types having different memory, computing capacity, platform etc.

Assume that the pricing structure of a cloud provider that provides only on-demand instances is stored in a data file. Each line of the data file contains instance ID, memory, computing capacity, and prices in LINUX and Windows environments. The contents of the data file price.txt are as below:

Instance ID	Memory(GB)	Computing Capacity(GHz)	Price($/hr) LINUX usage	Price($/hr) Windows Usage
1	1.7	1	0.09	0.11
2	7.5	4	0.42	0.48
3	15	8	0.70	0.94
4	0.5	2	0.05	0.03
5	1.7	5	0.48	0.62
6	6.5	10	0.98	1.22
7	17.1	6	0.50	0.64
8	34.6	12	1.02	1.20
9	69.5	24	2.26	2.56
10	20	32	1.95	1.98

40. Write a program to read the pricing structure from the data file and print the maximum, minimum and average price in $/hr of instances that provide a memory 1GB to 10GB for LINUX usage.

41. Write a program to read the pricing structure from the data file and show the pricing details of the instance with maximum capacity.

42. Write a program to read the pricing structure from the data file and prepare a report showing the total number of instance types and a chart showing the average price of each instance.

Weather Balloons. Weather balloons are used to gather temperature and pressure data at various altitudes in the atmosphere. The balloon rises because the density of the helium inside the balloon is less than the density of the surrounding air outside the balloon. As the balloon rises, the surrounding air becomes less dense, and thus the balloon's ascent slows until it reaches a point of equilibrium. During the day, sunlight warms the helium trapped inside the balloon, which causes the helium to expand and become less dense; thus, the balloon will rise higher. During the night, however, the helium in the balloon cools and becomes more dense; thus, the balloon will descend to a lower altitude. The next day, the sun heats the helium again and the balloon rises. Over time, this process generates a set of altitude measurements that can be approximated with a polynomial equation. Assume that the following polynomial represents the altitude or height in meters during the first 48 hours following the launch of a weather balloon:

$$\text{alt}(t) = -0.12t^4 + 12t^3 - 380t^2 + 4100t + 220,$$

where the units of t are hours. The corresponding polynomial model for the velocity in meters per hour of the weather balloon is

$$v(t) = -0.48t^3 + 36t^2 - 760t + 4100.$$

Figure 3.16 contains a plot of the altitude and velocity of the balloon for a period of 48 hours. From the plots, we can see the periods during which the balloon rises or falls.

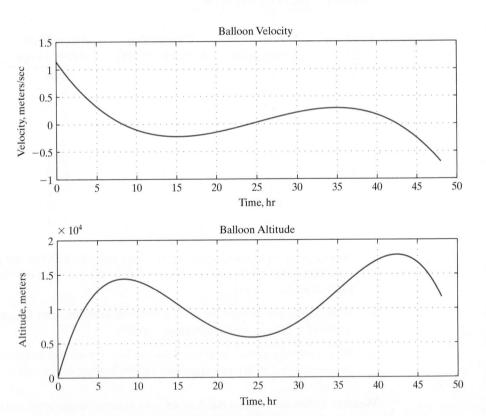

Figure 3.16 Velocity and altitude data for a weather balloon.

43. Write a program that will print a table of the altitude and the velocity for this weather balloon using units of meters and meters per second. Let the user enter the start time, the increment in time between lines of the table, and the ending time, where all the time values must be less than 48 hours. Use the program to generate a table showing the weather balloon information every 10 minutes over a 2-hour period, starting 4 hours after the balloon was launched.

44. Modify the program in Problem 43 so that it also prints the peak altitude and its corresponding time.

45. Modify the program in Problem 43 so that it will check to be sure that the final time is greater than the initial time. If it is not, ask the user to reenter the complete set of report information.

46. The preceding equations are accurate only for the time from 0 to 48 hours, so modify the program generated in Problem 43 so that it prints a message to the user that specifies an upper bound of 48 hours. Also, check the user input to make sure that it stays within the proper bounds. If there are errors, ask the user to reenter the complete set of report information.

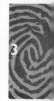

47. Modify the program in Problem 43 so that it stores the time, altitude, and velocity information in a data file named `balloon1.txt`.

CHAPTER FOUR

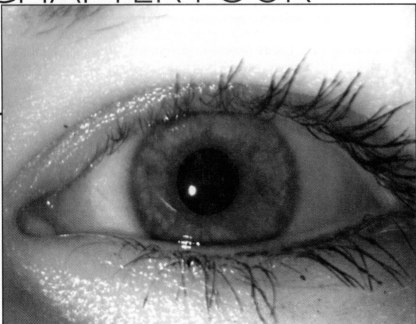

4

Crime Scene Investigation: Iris Recognition

Iris recognition is based on matching patterns in the iris, the donut-shaped colored part of the eye. Iris recognition is performed with an image collected by an IR (infrared) camera that takes a picture of the eyeball. IR images are in black and white, so the color information is not part of the image. However, IR images can capture the patterns in very dark brown eyes—something not possible with visible light images. Iris recognition is one of the most accurate biometrics because of the complex structure in the iris. The patterns, or striations, in the iris are formed by tearing of the membrane when a baby is still in the womb. Your left iris is very different from your right iris, and the irises of identical twins are different. The iris is also one of the few biometrics that does not change as a person ages. For example, faces change over time, and bones change as a child grows to be an adult. Iris recognition has been showcased in a number of movies, but the technology is not always represented accurately. For example, most iris recognition systems contain "liveness" testing. This means that they include tests to determine if an eyeball is part of a living person. These liveness tests can include checks such as of the jitter of the eye. All eyes have a very small of amount of jitter, and if an eye is encountered with none, then it would not be a live eyeball. Another liveness test can be done by changing the light illumination of the collection system. Such a change should cause the pupil to change size; if it does not, then, again, it is not a live eyeball. Thus, cutting out eyeballs to spoof an iris recognition system—as was done in the movie *Minority Report*—would not work in real life!

MODULAR PROGRAMMING WITH FUNCTIONS

CHAPTER OUTLINE

OBJECTIVES *In this chapter, we develop problem solutions containing:*

- modules from the Standard C library,
- programmer-defined modules,
- functions that generate random numbers,
- macro functions,
- recursive functions, and
- techniques for finding real roots to polynomials.

4.1 Modularity

Functions
Modules

The execution of a C program begins with the statements in the `main` function. A program may also contain other functions, and it may refer to functions in another file or in a library. These **functions**, or **modules**, are sets of statements that typically perform an operation or that compute a value. For example, the `printf` function prints a line of information on the terminal screen, and the `sqrt` function computes the square root of a value.

To maintain simplicity and readability in longer and more complex problem solutions, we develop programs that use a `main` function and additional functions, instead of using one long `main` function. By separating a solution into a group of modules, each module is easier to understand, thus adhering to the basic guidelines of structured programming presented in Chapter 3.

---

*Optional section.

The process of developing a problem solution is often one of "divide and conquer," as was discussed in Chapter 3 when we first discussed the decomposition outline. The decomposition outline is a set of sequentially executed steps that solves the problem, so it provides a good starting point for selecting potential functions. In fact, it is not uncommon for each step in the decomposition outline to correspond to one or more function references in the `main` function.

Breaking a problem solution into a set of modules has many advantages. Because a module has a specific purpose, it can be written and tested separately from the rest of the problem solution. An individual module is smaller than the complete solution, so testing it is easier. Also, once a module has been carefully tested, it can be used in new problem solutions without being retested. For example, suppose that a module is developed to find the average of a group of values. Once this module is written and tested, it can be used in other programs that

Reusablity

need to compute an average. This **reusability** is a very important issue in the development of large software systems, because it can save development time. In fact, libraries of commonly used modules (such as the Standard C library) are often available on computer systems.

Modularity

The use of modules (called **modularity**) often reduces the overall length of a program because many problem solutions include steps that are repeated several places in the program. By incorporating these steps that are repeated in a function, the steps can be referenced with a single statement each time that they are needed.

Several programmers can work on the same project if it is separated into modules, because the individual modules can be developed and tested independently of each other. This allows the development schedule to be accelerated, because some of the work can be done in parallel.

Abstraction

The use of modules that have been written to accomplish specific tasks supports the concept of **abstraction**. The modules contain the details of the tasks, and we can reference the modules without worrying about these details. The I/O diagrams that we use in developing a problem solution are an example of abstraction—we specify the input information and the output information without giving the details of how the output information is determined. In a similar way, we can think of modules as "black boxes" that have a specified input and that compute specified information; we can use these modules to help develop a solution. Thus, we are able to operate at a higher level of abstraction to solve problems. For example, the Standard C library contains functions that compute the logarithms of values. We can reference these functions without being concerned about the specific details, such as whether the functions are using infinite series approximations or lookup tables to compute the specified logarithms. By using abstraction, we can reduce the software development time while we increase its quality.

To summarize, there are several advantages to using modules in a problem solution:

- A module can be written and tested separately from other parts of the solution, and thus module development can be done in parallel for large projects.
- A module is a small part of the solution, and thus testing it separately is easier.
- Once a module is tested carefully, it does not need to be retested before it can be used in new problem solutions.
- The use of modules usually reduces the length of a program, making it more readable.
- The use of modules promotes the concept of abstraction, which allows us to "hide" the details in modules; this allows us to use modules in a functional sense without being concerned about the specific details.

Additional benefits of modules will be pointed out as we progress through this chapter.

Structure charts
Module charts

Structure charts, or **module charts**, show the module structure of a program. The `main` function references additional functions, which may also reference other functions themselves. Figure 4.1 contains some of the structure charts for the programs developed in the Problem Solving Applied sections in this chapter and the next chapter. Note that a structure

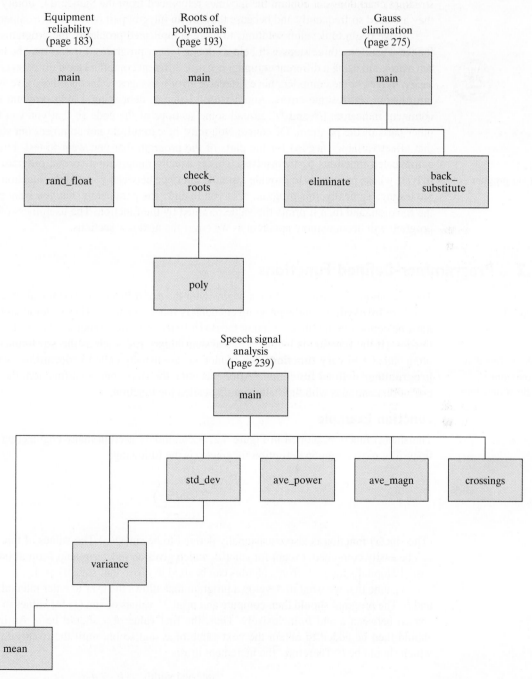

Figure 4.1 *Examples of structure charts.*

chart does not indicate the sequence of steps that are contained in the decomposition outline. The structure chart shows the separation of the program tasks into modules and indicates which modules reference other modules. Therefore, both the decomposition outline and the structure chart provide different but useful views of a problem solution. Also, note that the structure chart does not contain the modules referenced from the Standard C library because they are used so frequently and because they are an integral part of the C environment.

As we begin to develop solutions to more complicated problems, the programs become longer. Therefore, three steps will help us debug longer programs. First, it may be helpful to run a program using a different compiler because different compilers have different error messages; in fact, some compilers have extensive error messages, whereas others give very little information about some errors. Another useful step in debugging a long program is to add comment indicators (/* and */) around some sections of the code so that you can focus on other parts of the program. Of course, you must be careful; do not comment out statements that affect variables needed for the parts of the program that you want to test. Finally, test complicated functions by themselves. This is usually done with a special program called a **driver**, whose purpose is to provide a simple interface between you and the function that you are testing. Typically, this program asks you to enter the parameters that you want passed to the function, and then it prints the value returned by the function. The usefulness of a driver program will become more apparent as we cover the next few sections.

Driver program

4.2   Programmer-Defined Functions

The execution of a program always begins with the `main` function. Additional functions are called, or **invoked**, when the program encounters function names. These additional functions must be defined in the file containing the `main` function or in another available file or library of files. (If the function is included in a system library file, such as the `sqrt` function, it is often called a **library function**; other functions are usually called **programmer-written** or **programmer-defined functions**.) After executing the statements in a function, the program execution continues with the statement that called the function.

Invoked

Library function
Programmer–
defined functions

Function Example

The sinc(x) function, plotted in Figure 4.2, is commonly used in many engineering applications. The most common definition for sinc(x) is the following:

$$f(x) = \text{sinc}(x)$$
$$= \frac{\sin(\pi x)}{\pi x}.$$

(The sinc(x) function is also occasionally defined to be sin(x)/x.) The values of this function can be easily computed, except for sinc(0), which gives an indeterminant form of 0/0. In this case, l'Hôpital's theorem from calculus can be used to prove that sinc(0) = 1.

Assume that we want to develop a program that allows the user to enter interval limits, a and b. The program should then compute and print 21 values of sinc(x) for values of x evenly spaced between a and b, inclusively. Thus, the first value of x should be a. An increment should then be added to obtain the next value of x, and so on, until the twenty-first value, which should be b. Therefore, the increment in x is

$$\text{x\_increment} = \frac{\text{interval width}}{20} = \frac{b - a}{20}.$$

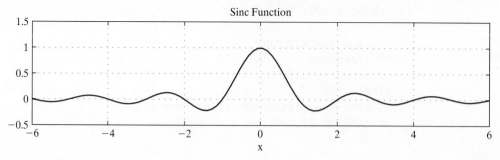

Figure 4.2 *Sinc function in* $[-20, 20]$.

Select values for a and b, and convince yourself that, with this increment and with a as the first value, the twenty-first value will be b.

Because sinc(x) is not part of the mathematical functions provided by the Standard C library, we implement this problem solution in two ways. In one solution, we include the statements to perform the computations of sinc(x) in the main function; in the other solution, we write a programmer-defined function to compute sinc(x), and then reference the programmer-defined function each time that the computations are needed. Both solutions are now presented so that you can compare them.

Solution I

```
/*------------------------------------------------------------*/
/*  Program chapter4_1                                        */
/*                                                            */
/*  This program prints 21 values of the sinc                 */
/*  function in the interval [a,b] using                      */
/*  computations within the main function.                    */

#include <stdio.h>
#include <math.h>
#define PI 3.141593

int main(void)
{
    /*  Declare variables.  */
    int k;
    double a, b, x_incr, new_x, sinc_x;

    /*  Get interval endpoints from the user.  */
    printf("Enter endpoints a and b (a<b): \n");
    scanf("%lf %lf",&a,&b);
    x_incr = (b - a)/20;

    /*  Compute and print table of sinc(x) values.  */
    printf("x and sinc(x) \n");
```

```
        for (k=0; k<=20; k++)
        {
           new_x = a + k*x_incr;
           if (fabs(new_x) < 0.0001)
              sinc_x = 1.0;
           else
              sinc_x = sin(PI*new_x)/(PI*new_x);
           printf("%f %f \n",new_x,sinc_x);
        }

        /* Exit program.  */
        return 0;
}
/*------------------------------------------------------------*/
```

Solution 2

```
/*------------------------------------------------------------*/
/*  Program chapter4_2                                        */
/*                                                            */
/*  This program prints 21 values of the sinc function        */
/*  using a programmer-defined function.                      */

#include <stdio.h>
#include <math.h>
#define PI 3.141593

int main(void)
{
   /* Declare variables.  */
   int k;
   double a, b, x_incr, new_x;
   double sinc(double x);

   /* Get interval endpoints from the user.  */
   printf("Enter endpoints a and b (a<b): \n");
   scanf("%lf %lf",&a,&b);
   x_incr = (b - a)/20;

   /* Compute and print table of sinc(x) values.  */
   printf("x and sinc(x) \n");
   for (k=0; k<=20; k++)
   {
      new_x = a + k*x_incr;
      printf("%f %f \n",new_x,sinc(new_x));
   }

   /* Exit program.  */
   return 0;
}
```

```
/*------------------------------------------------------------*/
/*  This function evaluates the sinc function.              */
double sinc(double x)
{
   if (fabs(x) < 0.0001)
      return 1.0;
   else
      return sin(PI*x)/(PI*x);
}
/*------------------------------------------------------------*/
```

The following output represents an example interaction that could occur with either program:

```
Enter endpoints a and b (a<b):
-2 2
x and sinc(x)
-2.000000 0.000000
-1.800000 -0.103943
-1.600000 -0.189207
-1.400000 -0.216236
-1.200000 -0.155915
-1.000000 0.000000
-0.800000 0.233872
-0.600000 0.504551
-0.400000 0.756827
-0.200000 0.935489
0.000000 1.000000
0.200000 0.935489
0.400000 0.756827
0.600000 0.504551
0.800000 0.233872
1.000000 0.000000
1.200000 -0.155915
1.400000 -0.216236
1.600000 -0.189207
1.800000 -0.103943
2.000000 0.000000
```

Figure 4.3 contains plots of the 21 values computed for four different intervals [a, b]. The program computes only 21 values, so the resolution in the plots is affected by the size of the interval—a smaller interval has better resolution than a larger interval. Note that the main function of Solution 2 is easier to read because it is shorter than the main function in the first solution. Now that you have an example of a program with a programmer-defined function, we present a more general discussion of the statements in a function.

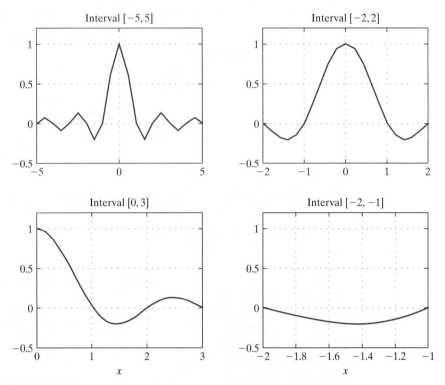

Figure 4.3 *Program output for four intervals.*

Function Definition

A function consists of a definition statement followed by declarations and statements. The first part of the definition statement defines the type of value that is computed by the function (double, in our example); if the function does not compute a value, the type is void. The function name and parameter list follow the return_type. Thus, the general form of a function is

```
return_type function_name (parameter_declarations)
{
    declarations;
    statements;
}
```

The parameter declarations represent the information passed to the function; if there are no input parameters (also called arguments), then the parameter declarations should be void. Additional variables used by a function are defined in the declarations. The declarations and the statements within a function are enclosed in braces. The function name should be selected to help document the purpose of the function. Comments should also be included within the function to further describe the purpose of the function and to document the steps. We also use a comment line with dashes to separate a programmer-defined function from the main function and from other programmer-defined functions.

All functions should include a `return` statement, which has the following general form:

`return expression;`

The expression specifies the value to be returned to the statement that referenced the function. The expression type should match the return_type indicated in the function definition to avoid potential errors. The cast operator (discussed in Chapter 2) can be used to explicitly specify the type of the expression if necessary. A `void` function does not return a value, and thus has this general definition statement:

`void function_name(parameter declarations)`

The `return` statement in a `void` function does not contain an expression and has this form:

`return;`

Functions can be defined before or after the `main` function. (Remember that a right brace specifies the end of the `main` function.) However, one function must be completely defined before another function begins; function definitions cannot be nested within each other. In our programs, we include the `main` function first, and then additional functions are included in the order in which they are referenced in the program.

We now look closer at the interaction between a statement that references a function and the function itself.

Function Prototype

The `main` function presented in program `chapter4_2` contained the following statement in its declarations:

`double sinc(double x);`

Function prototype

This statement is a **function prototype** statement. It informs the compiler that the `main` function will reference a function named `sinc`, that the `sinc` function expects a `double` parameter, and that the `sinc` function returns a `double` value. The identifier x is not being defined as a variable; it is just used to indicate that a value is expected as an argument by the `sinc` function. In fact, it is valid to include only the argument types in the function prototype statement:

`double sinc(double);`

Both of these prototype statements give the same information to the compiler. We recommend using parameter identifiers in prototype statements because the identifiers help document the order and definition of the parameters.

A function prototype can be included with preprocessor directives, or, because a function prototype is defining the type of value being returned by the function, it can also be included with other variable declarations. For example, the declarations of program `chapter4_2` are

```
/*  Declare variables and function prototypes.  */
int k;
double a, b, x_incr, new_x;
double sinc(double x);
```

These statements could also have been written in the following form:

```
/*  Declare variables and function prototypes.      */
int k;
double a, b, x_incr, new_x, sinc(double x);
```

In our programs, we list function prototypes on separate declaration statements to make it easier to identify them.

Function prototype statements should be included for all functions referenced in a program. Header files, such as stdio.h and math.h, contain the prototype statements for many of the functions in the Standard C library; otherwise, we would need to include individual prototype statements for functions such as printf and sqrt in our programs. If a programmer-defined function references other programmer-defined functions, then it will also need additional prototype statements.

If a program references a large number of programmer-defined functions, it becomes cumbersome to include all the function prototype statements. In these cases, a custom header file can be defined that contains the function prototypes and any related symbolic constants. A header file should have a file name that ends with a suffix of .h. The file is then referenced with an include statement, using double quotes around the file name. In Chapter 5, we develop a set of functions for computing common statistics from a set of values. If a header file containing the corresponding function prototypes is named stat_lib.h, then the prototypes are all included in a program with this statement:

```
#include "stat_lib.h"
```

Custom header files are often used to accompany routines that are shared by programmers.

Parameter List

The definition statement of a function defines the parameters that are required by the function; these are called **formal parameters**. Any statement that references the function must include values that correspond to the parameters; these are called **actual parameters**. For example, consider the sinc function developed earlier in this section. The definition statement of this function is

Formal parameters
Actual parameters

```
double sinc(double x)
```

and the statement from the main program that references the function is

```
printf("%f %f \n",new_x,sinc(new_x));
```

Thus, the variable x is the formal parameter, and the variable new_x is the actual parameter. When the reference to the sinc function in the printf statement is executed, the value in the actual parameter is copied to the formal parameter, and the steps in the sinc function are executed using the new value in x. The value returned by the sinc function is then printed. It is important to note that the value in the formal parameter is not moved back to the actual parameter when the function is completed. We illustrate these steps with a memory snapshot that shows the transfer of the value from the actual parameter to the formal parameter, assuming that the value of new_x is 5.0:

actual parameter

new_x | 5.0 | ⟶ x | 5.0 |

formal parameter

After the value in the actual parameter is copied to the formal parameter, the steps in the sinc function are executed. When debugging a function, it is a good idea to use printf statements to provide a memory snapshot of the actual parameters before the function is referenced, as well as a memory snapshot of the formal parameters at the beginning of the function.

Valid references to the sinc function can also include expressions or other function references, as shown in these example references to the sinc function:

```
printf("%f \n",sinc(x+2.5));

scanf("%lf",&y);
printf("%f \n",sinc(y));

z = x*x + sinc(2*x);

w = sinc(fabs(y));
```

In all these example references, the formal parameter is still x, but the actual parameter is x+2.5, or y, or 2*x, or fabs(y), depending on the reference selected.

If a function has more than one parameter, the formal parameters and the actual parameters must match in number, type, and order. A mismatch between the number of formal parameters and actual parameters can be detected by the compiler using the function prototype statement. If the type of an actual parameter is not the same as the corresponding formal parameter, then the value of the actual parameter will be converted to the appropriate type; this conversion is also called **coercion of arguments** and may or may not cause errors. The coercion occurs according to the discussion given in Chapter 2, which discussed moving values stored as one type to a variable with a different type. Converting values to a higher type (such as from float to double) generally works correctly; converting values to a lower type (such as from float to int) often introduces errors.

Coercion of
arguments

To illustrate the coercion of arguments, consider the following function that returns the maximum of two values:

```
/*-----------------------------------------------------------*/
/*  This function returns the maximum of two                 */
/*  integer values.                                          */

int max(int a,int b)
{
   if (a > b)
      return a;
   else
      return b;
}
/*-----------------------------------------------------------*/
```

Assume that a reference to this function is max(x_sum,y_sum) and that x_sum and y_sum are integers containing the values 3 and 8, respectively. Then, the following memory snapshot shows the transfer of values from the actual parameters to the formal parameters when the reference max(x_sum,y_sum) is made:

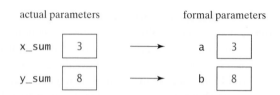

actual parameters   formal parameters

x_sum | 3 | ⟶   a | 3 |

y_sum | 8 | ⟶   b | 8 |

The statements in the function will then return the value 8 as the value of the reference max(x_sum,y_sum).

Now suppose that a reference to the function max is made using float variables t_1 and t_2. If t_1 and t_2 contain the values 2.8 and 4.6, then the following transfer of parameters occurs when the reference max(t_1,t_2) is executed:

actual parameters   formal parameters

t_1 | 2.8 | ⟶   a | 2 |

t_2 | 4.6 | ⟶   b | 4 |

The statements in the function will then return the value 4 to the statement containing the reference max(t_1,t_2). Obviously, the wrong value has been returned by the function. However, the problem is not in the function; the problem is that the function was referenced with the wrong types of actual parameters.

Additional errors can be introduced if the actual parameters are out of order. These errors may not be detected by the compiler, and they can be difficult to locate; therefore, be especially careful that the order of the formal parameters and the actual parameters match.

Call-by-value
Reference-by value

The function reference in the sinc example is a **call-by-value** reference, or a **reference-by value**. When a function reference is made, the value of the actual parameter is passed to the function and is used as the value of the corresponding formal parameter. In general, a C function cannot change the value of an actual parameter. Exceptions occur when the actual parameters are arrays (discussed in Chapter 5) or are pointers (discussed in Chapter 6); these exceptions generate a **call-by-reference** or a **reference-by-address**, which will be discussed in these later chapters.

Call-by-reference
Reference-by-
address

Storage Class and Scope

In the sample programs presented thus far, we have declared variables within a main function and within programmer-defined functions. It is also valid to define a variable before the main function. Therefore, it is important to be able to determine the **scope** of a function or a variable, where scope refers to the portion of the program in which it is valid to reference the function or variable; scope is also sometimes defined in terms of the portion of the program in

Scope

PRACTICE!

Consider the following function:

```
/*------------------------------------------------------------*/
/*  This function counts positive parameters.              */
int positive(double a,double b,double c)
{
   int count;

   count = 0;
   if (a >= 0)
      count++;
   if (b >= 0)
      count++;
   if (c >= 0)
      count++;
   return count;
}
/*------------------------------------------------------------*/
```

Assume that the function is referenced with the following statements:

```
x = 25;
total = positive(x,sqrt(x),x-30);
```

1. Show the memory snapshot of the actual parameters and the formal parameters.
2. What is the new value of total?

Storage class

Local variables

Global variables

Automatic storage class

External storage class

which the function or variable is visible or accessible. Because the scope of a variable is directly related to its **storage class**, we also discuss the four storage classes—automatic, external, static, and register.

First, we define the difference between local variables and global variables. **Local variables** are defined within a function, and thus include the formal parameters and any other variables declared in the function. A local variable can be accessed only in the function that defines it. A local variable has a value when its function is being executed, but its value is not retained when the function is completed. **Global variables** are defined outside the main function or other programmer-defined functions. The definition of a global variable is outside of all functions, so it can be accessed by any function within the program. However, to reference a global variable, some compilers require that the declaration within the function include the keyword extern before the type designation to tell the computer to look outside the function for the variable. The **automatic storage class** is used to represent local variables; this is the default storage class, but it can also be specified with the keyword auto before the type designation. The **external storage class** is used to represent global variables; the extern designation must be used within functions, and it is optional in the original definition of a global variable.

Consider a program that contains the following statements:

```c
#include <stdio.h>
int count=0;
...
int main(void)
{
   int x, y, z;
   ...
}
int calc(int a,int b)
{
   int x;
   extern int count;
   ...
}
void check(int sum)
{
   extern int count;
   ...
}
```

The variable count is a global variable that can be referenced by the functions calc and check. The variables x, y, and z are local variables that can be referenced only in the main program; similarly, the variables a, b, and x are local variables that can be referenced only in the function calc, and sum is a local variable that can be referenced only in the function check. Note that there are two local variables x—these are two different variables with different scopes.

Style

The memory assigned to an external variable is retained for the duration of the program. Although an external variable can be referenced from a function using the proper declaration, using global variables is generally discouraged. In general, parameters are preferred for transferring information to a function because the parameter is evident in the function prototype, whereas the external variable is not visible in the function prototype. The use of global variables should be avoided whenever possible.

Function names also have an external storage class, and thus can be referenced from other functions. Function prototypes included outside of any function are also external references, and thus are available to all other functions in the program; this explains why we do not need to include math.h in every function that references a mathematical function. However, the parameter variables in the function prototype are known only in the function prototype statement.

Static

The **static** storage class is used to specify that the memory for a local variable should be retained during the entire program execution. Therefore, if a local variable in a function is given a static storage class assignment by using the keyword static before its type specification, the variable will not lose its value when the program exits the function in which it is defined. A static variable could be used to count the number of times that a function was invoked, because the value of the count would be preserved from one function call to another.

Register

The keyword register is used before the type designation of a variable to specify that it should be placed in a **register**, as opposed to a memory location. Accessing registers is faster than accessing memory, so this storage class is used for frequently accessed values. Because the number of registers available is system dependent, and because the time required for a memory access is steadily being reduced, this type of storage class is seldom used.

PRACTICE!

Using the program on page 183 in Section 4.6, give the following information (you do not need to understand the program to determine the requested information):

1. List the external identifiers.
2. List the local variables and identify their scope.

Using the program on page 193 in Section 4.8, give the following information (you do not need to understand the program to determine the requested information):

3. List the external identifiers.
4. List the local variables and identify their scope.

4.3   Problem Solving Applied: Computing the Boundaries of the Iris

In this section, we use the new statements presented in this chapter to solve a problem related to iris recognition. Most techniques for performing iris recognition first start with a segmentation operation. This operation identifies the iris/pupil boundary and the boundary between the iris and the sclera (or the white of the eye), as shown in Figure 4.4. This is an easy step for us to accomplish visually, but segmentation is a very complex process to carry out automatically with a computer algorithm. For some analysis, a manual segmentation program is used in which a user is presented with an image of an eye on the screen, and the user then clicks on three points on the pupil boundary and three points on the boundary between the iris and the sclera. With three points, the computer can compute the equation of the circle. Hence, we can then compute the equations of the circles that form the boundaries of the iris. With this information, we can extract the iris and continue the process of iris recognition.

The problem that we will solve in this section is the one of taking three points on a plane and then computing the equation for the circle through these points. We will also compute the location of the center of the circle. There are a number of techniques for finding the equation of a circle from three points on the circle. The one that we use is based on finding the equation of a line through points P_1 and P_2, and the equation of the line through points P_2 and P_3. If we assume that

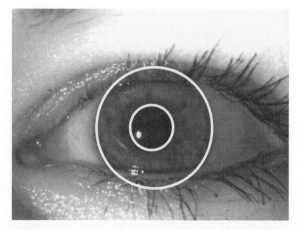

Figure 4.4 *Image of an eye with iris boundaries identified.*

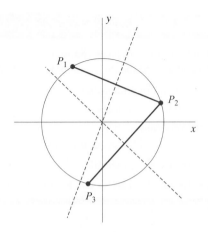

Figure 4.5 *Relationship between points on a circle and the center.*

these two lines are not parallel, then the lines perpendicular to these two line segment (P_1P_2 and P_2P_3), that also go through the midpoints of the line segments, intersect in the center of the circle. (See Figure 4.5). We are not going to go through the details of the derivation of these equations, but a number of derivations are available on the Internet. We now present the equations for computing the equations of the lines P_1P_2 and P_2P_3 and for computing the equations of the lines perpendicular to these line segments through their midpoints. We then present the equation for finding the intersection point of these perpendicular lines; this point is also the center of the circle represented by the three original points. Once we have the center point, we can make use of the points on the circle to compute the radius. Here are the equations needed for this function, assuming that the coordinates of P_1, P_2, and P_3 are (x_1, y_1), (x_2, y_2), and (x_3, y_3), respectively:

- Slope of line P_1P_2:

$$m_{12} = \frac{y_2 - y_1}{x_2 - x_1} \tag{4.1}$$

- Slope of line P_2P_3:

$$m_{23} = \frac{y_3 - y_2}{x_3 - x_2} \tag{4.2}$$

- Equation of line P_1P_2:

$$y_{12} = m_{12}(x - x_1) + y_1 \tag{4.3}$$

- Equation of line P_2P_3:

$$y_{23} = m_{23}(x - x_2) + y_2 \tag{4.4}$$

- Equation of line perpendicular to line P_1P_2 that bisects the line segment:

$$y_{p12} = -\frac{1}{m_{12}}\left(x - \frac{x_1 + x_2}{2}\right) + \left(\frac{y_1 + y_2}{2}\right) \tag{4.5}$$

- Equation of line perpendicular to line P_2P_3 that bisects the line segment:

$$y_{p23} = -\frac{1}{m_{23}}\left(x - \frac{x_2 + x_3}{2}\right) + \left(\frac{y_2 + y_3}{2}\right) \tag{4.6}$$

- Equation for x coordinate of the center of the circle:

$$x_c = \frac{m_{12}m_{23}(y_1 - y_3) + m_{23}(x_1 + x_2) - m_{12}(x_2 + x_3)}{2(m_{23} - m_{12})} \quad (4.7)$$

- Equation for y coordinate of the center of the circle:

$$y_c = -\frac{1}{m_{12}}\left(x_c - \frac{x_1 + x_2}{2}\right) + \left(\frac{y_1 + y_2}{2}\right) \quad (4.8)$$

- Equation for radius of the circle:

$$r = \sqrt{(x_1 - x_c)^2 + (y_1 - y_c)^2} \quad (4.9)$$

We are now ready to develop the solution in C.

1. PROBLEM STATEMENT

Given three points in a plane, determine the coordinates of the center of the circle and the radius of the circle that contains the three points.

2. INPUT/OUTPUT DESCRIPTION

The following diagram shows that the inputs to the program are the x, y coordinates of three points. The outputs are the coordinates of the center and the radius of the circle.

point 1: x_1, y_1 ⟶ [ ] ⟶ circle center: x_c, y_c
point 2: x_2, y_2 ⟶   ⟶ circle radius
point 3: x_3, y_3 ⟶

3. HAND EXAMPLE

Using the equations provided earlier in this section, we now compute the center of the corresponding circle and its radius. Assume that we start with the following three points on a circle:

$$P_1 = (-1, -1)$$
$$P_2 = (1, 1)$$
$$P_3 = (3, -1)$$

Using equations (4.1) and (4.2), we can compute the slopes m_{12} and m_{23} for the lines P_1P_2 and P_2P_3:

$$m_{12} = 2/2 = 1$$
$$m_{23} = -2/2 = -1$$

Using equations (4.3) and (4.4), we can compute the equations for the line P_1P_2 and P_2P_3:

$$y_{12} = x$$
$$y_{23} = -x + 2$$

Using Equations (4.5) and (4.6), we can compute the equations for the lines perpendicular to P_1P_2 and P_2P_3 that bisect the line segment:

$$y_{p12} = -x$$
$$y_{p23} = x - 2$$

Using equations (4.7) and (4.8), we can compute the coordinates of the center point:

$$x_c = 1$$
$$y_c = -1$$

Using equation (4.9), we can compute the radius of the circle:

$$r = 2$$

Thus, the equation for the circle is:

$$(x - 1)^2 + (y + 1)^2 = 4$$

(Recall that the equation for a circle with center coordinates of (x_c, y_c) and radius r is:

$$(x - x_c)^2 + (y - y_c)^2 = r^2$$

where the center coordinates are (x_c, y_c) and the radius is r.)

When we substitute the coordinates for points P_1, P_2, and P_3 in this equation, we confirm that these points are on the circle.

4. ALGORITHM DEVELOPMENT

Because there are several equations to evaluate to compute the coordinates of the center of the circle, this is a good candidate for a function. If we look back at the hand example, we see that most of the equations are computing intermediate values needed for the computation of the x coordinate of the center of the circle. Let's put these computations into a function, and then the main program will be shorter and thus easier to understand. We now develop the decomposition outline for the main program and for the function.

Decomposition Outline of the main program:

1. Read the coordinates of the three points.
2. Use a function to determine the x coordinate of the center of the circle.
3. Compute the y coordinate of the center of the circle.
4. Compute the radius of the circle.
5. Print the coordinates of the center of the circle and the radius.

Decomposition Outline of the function (assuming the coordinates of the three points are input parameters to the function and the output of the function is the x coordinate of the center of the circle):

1. Compute the equations for the two lines connecting the three points.
2. Compute the equations for the two lines perpendicular to the lines connecting the three points that also bisect the line segments.
3. Compute the x coordinate of the center of the circle.

Both the main program and the function have simple structures, so we can convert the decomposition directly into C.

```
/*-------------------------------------------------------------*/
/*  Program chapter4_3                                         */
/*                                                            */
/*  This program reads the coordinates of three points and then */
/*  references a function to determine the coordinates of the  */
/*  center of a circle through the points and the radius of the */
/*  circle.                                                    */

#include <stdio.h>
#include <math.h>

int main(void)
{
   /*  Declare variables.                                     */
   double x1, x2, x3, y1, y2, y3, m12, xc, yc, r;
   double circle_x_coord(double x1,double y1,double x2,double y2,
                         double x3,double y3);

   /*  Get user input from the keyboard.                      */
   printf("Enter x and y coordinates for first point: \n");
   scanf("%lf %lf",&x1,&y1);
   printf("Enter x and y coordinates for second point: \n");
   scanf("%lf %lf",&x2,&y2);
   printf("Enter x and y coordinates for third point: \n");
   scanf("%lf %lf",&x3,&y3);

   /*  Use a function to determine the x coordinate of the center. */
   xc = circle_x_coord(x1,y1,x2,y2,x3,y3);

   /*  Compute the y coordinate of the center.                */
   m12 = (y2 - y1)/(x2 - x1);
   yc = -(1/m12)*(xc - (x1 + x2)/2) + (y1 + y2)/2);

   /*  Compute the radius of the circle.                      */
   r = sqrt((x1 - xc)*(x1 - xc) + (y1 - yc)*(y1 - yc));

   /*  Print circle parameters.  */
   printf("\nCenter of Circle: (%.1f,%.1f) \n",xc,yc);
   printf("Radius of Circle: %.1f \n",r);
```

```
        /*  Exit program.  */
        return 0;
}
/*----------------------------------------------------------------*/
/*  This function computes the x coordinate of the center of a    */
/*  circle given three points on the circle.                     */

double circle_x_coord(double x1,double y1,double x2,double y2,
                      double x3,double y3)
(
        /*  Declare variables.                                   */
        double m12, m23, xc_num, xc_den, xc;
        /*  Compute slopes of the two lines between points.      */
        m12 = (y2 - y1)/(x2 - x1);
        m23 = (y3 - y2)/(x3 - x2);

        /*  Compute the x coordinate of the center of the circle.   */
        xc_num = m12*m23*(y1 - y3) + m23*(x1 + x2) - m12*(x2 + x3);
        xc_den = 2*(m23 - m12);
        xc = xc_num/xc_den;

        /*  Return the x coordinate of the center.  */
        return xc;
)
/*----------------------------------------------------------------*/
```

5. TESTING

We first test the program with the data from the hand example. This generates the following interaction:

```
Enter x and y coordinates for first point:
-1 -1
Enter x and y coordinates for second point:
1  1
Enter x and y coordinates for third point:
3 -1
Center of Circle: (1.0,-1.0)
Radius of Circle: 2.0
```

The answer matches the hand example, so we can then test the program with additional points.

MODIFY!

These problems relate to the program developed in this section for finding the center and the radius of a circle given three points on the circle.

1. Modify the program so that it also prints the equation of the circle.
2. If either line is vertical, then the corresponding slope is infinite. Determine if this occurs, and print an error message before exiting the program.
3. If line P_1P_2 is vertical, exchange the values of P_1 with P_3. Then check to see if this now yields two non-vertical lines. If so, continue processing.
4. If line P_2P_3 is vertical, exchange the values of P_2 with P_1. Then check to see if this now yields two non-vertical lines. If so, continue processing.
5. Combine the solutions to problems 3 and 4 above. This solution will now work properly unless all three points are vertical, but that indicates an error in the points, because they cannot all three be on the same circle.

4.4   Problem Solving Applied: Iceberg Tracking

Large icebergs are tracked by satellites, and their location is specified using latitude and longitude. It is important to be able to determine the distance between the iceberg and nearby ships. In this section, we will develop a program that determines the distance between two objects when we are given their latitudes and longitudes. Before developing the program, we need to briefly discuss latitudes and longitudes and develop the equation that allows us to determine the distance between two points using latitudes and longitudes.

Assume that the earth is represented by a sphere with a radius of 3960 miles. We can then define a position on the earth's surface in terms of a grid determined by a latitude and a longitude measurement. To understand these measurements, we first need to review the definition

Great circle

of a **great circle**—a circle formed by the intersection of a sphere and a plane that passes through the center of the sphere. If the plane does not pass through the center of the sphere, it

Prime meridian

will be a circle with a smaller circumference and hence is not a great circle. The **prime meridian** is a north–south great circle that passes through Greenwich, just outside London,

Equator

and through the North Pole. The **equator** is an east–west great circle that is equidistant from the North Pole and South Pole. Thus, we can define a rectangular coordinate system such that the origin is the center of the earth, the z-axis goes from the center of the earth through the North Pole, and the x-axis goes from the center of the earth through the point where the prime

Latitude

meridian intersects the equator (see Figure 4.6). The **latitude** is an angular distance, is measured in degrees, and extends northward or southward from the equator (as in 25°N); and the

Longitude

longitude is an angular distance, is measured in degrees, and extends westward or eastward from the prime meridian (as in 120° W).

Global Positioning
System

The **Global Positioning System** (GPS), originally developed for military use, uses 24 satellites circling the earth to pinpoint a location on the surface. Each satellite broadcasts a coded radio signal indicating the time and the satellite's exact position 11,000 miles above the earth. The satellites are equipped with atomic clocks that are accurate to within one second every 70,000 years. A GPS receiver picks up the satellite signal and measures the time between the signal's transmission and its reception. By comparing signals from at least three satellites, the receiver can determine the latitude, longitude, and altitude of its position.

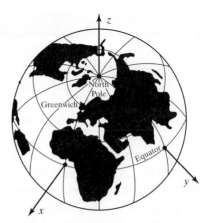

Figure 4.6 *Rectangular Coordinate System for the Earth.*

The shortest distance between two points on a sphere is shown to be on the arc of the great circle containing them. If we know the angle between vectors from the center of the earth to the two points defining the arc, we can then estimate the distance as a proportion of the earth's circumference. To illustrate, suppose that the angle between two vectors from the center of the earth is 45°. Then the angle is 45/360, or 1/8 of a complete revolution. Hence, the distance between the two points is 1/8 of the earth's circumference (π times twice the radius) or 3,110 miles.

The best way to compute the shortest distance between two points that are specified in latitude and longitude is through a series of coordinate transformations. Recall that the spherical coordinates (ρ, ϕ, θ) of a point P in a rectangular coordinate system represent the length ρ (rho) of the vector connecting the point to the origin, the angle ϕ (phi) between the positive z-axis and the vector, and the angle θ (theta) between the x-axis and the projection of the vector in the xy-plane (see Figure 4.7). We then convert the spherical coordinates to rectangular coordinates (x, y, z). Finally, a simple trigonometric equation computes the angle between two points (or vectors) in rectangular coordinates. Once we know the angle between the two points, we can then use the technique described in the previous paragraph to find the distance between the two points.

We need to use equations that relate latitude and longitude to spherical coordinates, that convert spherical coordinates to rectangular coordinates, and that compute the angle between two vectors. Figure 4.7 is useful in relating the notation to the following equations:

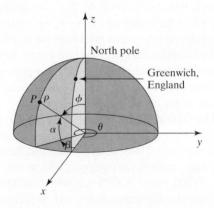

Figure 4.7 *Spherical Coordinate System.*

- Latitude/longitude and spherical coordinates:

$$\alpha = 90° - \phi, \quad \beta = 360° - \theta$$

- Spherical and rectangular coordinates:

$$x = \rho \sin \phi \cos \theta, \quad y = \rho \sin \phi \sin \theta, \quad z = \rho \cos \theta$$

- Angle γ between two vectors **a** and **b**:

$$\cos \gamma = \mathbf{a} \cdot \mathbf{b}/(|\mathbf{a}||\mathbf{b}|)$$

where $\mathbf{a} \cdot \mathbf{b}$ is the dot product of **a** and **b** and $|\mathbf{a}|$ is the length of the vector **a**.

- Dot product of two vectors (x_a, y_a, z_a) and (x_b, y_b, z_b) in rectangular coordinates:

$$\mathbf{a} \cdot \mathbf{b} = x_a x_b + y_a y_b + z_a z_b$$

- Length of a vector (x_a, y_a, z_a) in rectangular coordinates:

$$|\mathbf{a}| = \sqrt{(x_a^2 + y_a^2 + z_a^2)}$$

- Great circle distance:

$$\text{distance} = (\gamma/(2\pi))(\text{earth's circumference}) = (\gamma/(2\pi))(\pi \cdot 2 \text{ radius})$$
$$= \gamma \cdot 3960.$$

Pay close attention to the angle units in these equations. Unless otherwise specified, it is assumed that the angles are measured in radians. (Recall that an angle in degrees can be converted to radians by multiplying it by the factor $\pi/180$.)

Write a program that asks the user to enter the latitude and longitude coordinates for two points in the Northern Hemisphere. Then compute and print the shortest distance between the two points.

1. PROBLEM STATEMENT

Compute the shortest distance between two points in the Northern Hemisphere.

2. INPUT/OUTPUT DESCRIPTION

For this program, the input is the latitude and longitude of two points in the Northern Hemisphere. The output is the shortest distance between the points, as shown in the following diagram:

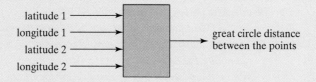

3. HAND EXAMPLE

For a hand example, we will compute the great circle distance between New York and London. The latitude and longitude of New York is 40.75° N and 74° W, respectively, and the latitude and longitude of London is 51.5° N and 0° W, respectively.

The spherical coordinates for New York are

$$\phi = (90 - 40.75)° = 49.25(\pi/180) = 0.8596$$
$$\theta = (360 - 74)° = 286(\pi/180) = 4.9916$$
$$\rho = 3960.$$

The rectangular coordinates for New York (to two decimal places) are

$$x = \rho \sin \phi \cos \theta = 826.90$$
$$y = \rho \sin \phi \sin \theta = -2883.74$$
$$z = \rho \cos \theta = 2584.93.$$

Similarly, the rectangular coordinates for London can be computed to be

$$x = 2465.16, y = 0, z = 3099.13.$$

The cosine of the angle between the two vectors is equal to the dot product of the two vectors divided by the product of their lengths, or 0.6408. Using an inverse cosine function, γ can be determined to be 0.875 radians. Finally, the distance between New York and London is

$$0.875 \cdot 3960 = 3466 \text{ miles.}$$

4. ALGORITHM DEVELOPMENT

To develop the algorithm, we must first decompose the problem solution into a set of sequentially executed steps, as follows:

Decomposition Outline

1. Read latitude and longitude for the two locations.
2. Compute the distance between the two locations.
3. Print the distance computed.

After completing the hand example, converting the decomposition to pseudocode is straightforward. Since other programs may also need to determine the distance between two points that are specified with latitude and longitude, we will develop the computations using a function.

Refinement in Pseudocode
main: *read latitude and longitude for the two points*
          compute great circle distance using a function, gc_distance
          print distance
gc_distance(lat1N,long1W,lat2N,long2W):
          convert latitude and longitude to spherical coordinates
          convert spherical coordinates to rectangular coordinates
          compute the angle between the two vectors
          compute the great circle distance for the arc between the two vectors

The steps in the pseudocode are now detailed enough to convert into C:

```
/*------------------------------------------------------------*/
/*   Program chapter4_4                                       */
/*                                                            */
/*   This program determines the distance between two points  */
/*   that are specified with latitude and longitude values    */
/*   that are in the Northern Hemisphere.                     */

#include <stdio.h>
#include <math.h>
#define PI 3.141593

int main(void)
{
    /*  Declare variables and function prototype.  */
    double lat1, long1, lat2, long2;
    double gc_distance(double lat1,double long1,
                       double lat2,double long2);

    /*  Get locations of two points.  */
    printf("Enter latitude north and longitude west ");
    printf("for location 1: \n");
    scanf("%lf %lf",&lat1,&long1);
    printf("Enter latitude north and longitude west ");
    printf("for location 2: \n");
    scanf("%lf %lf",&lat2,&long2);

    /*  Print great circle distance.  */
    printf("Great Circle Distance: %.0f miles \n",
           gc_distance(lat1,long1,lat2,long2));

    /*  Exit program.  */
    return 0;
}
/*------------------------------------------------------------*/
/*   This function computes the distance between two          */
/*   points using great circle distances.                    */

double gc_distance(double lat1,double long1,
                   double lat2,double long2)
{
    /*  Declare variables.  */
    double rho, phi, theta, gamma, dot, dist1, dist2,
           x1, y1, z1, x2, y2, z2;

    /*  Convert latitude,longitude to rectangular coordinates.  */
    rho = 3960;
    phi = (90 - lat1)*(PI/180.0);
    theta = (360 - long1)*(PI/180.0);
    x1 = rho*sin(phi)*cos(theta);
    y1 = rho*sin(phi)*sin(theta);
    z1 = rho*cos(phi);
    phi = (90 - lat2)*(PI/180.0);
    theta = (360 - long2)*(PI/180.0);
```

```
    x2 = rho*sin(phi)*cos(theta);
    y2 = rho*sin(phi)*sin(theta);
    z2 = rho*cos(phi);

    /*  Compute angle between vectors.  */
    dot = x1*x2 + y1*y2 + z1*z2;
    dist1 = sqrt(x1*x1 + y1*y1 + z1*z1);
    dist2 = sqrt(x2*x2 + y2*y2 + z2*z2);
    gamma = acos(dot/(dist1*dist2));

    /*  Compute and return great circle distance.  */
    return gamma*rho;
}
/*------------------------------------------------------------*/
```

5. TESTING

We start testing with the hand example, which gives the following interaction:

```
Enter latitude north and longitude west for location 1:
40.75 74
Enter latitude north and longitude west for location 2:
51.5 0
Great Circle Distance: 3466 miles
```

MODIFY!

These problems relate to the preceding program, which computes the great circle distance between two points in the Northern Hemisphere.

1. Write a function that computes the angle (in radians) between two vectors. Use the following prototype statement:

```
double angle(double x1, double y1, double z1,
             double x2, double y2, double z2)
```

2. Modify the program developed in this section so that it uses the function in Problem 1.

3. Modify the program so that it will allow the user to enter a latitude in the Northern Hemisphere or the Southern Hemisphere. The program should ask the user to enter the latitude value; then it should ask the user to enter either N or S to specify the hemisphere. If the latitude is in the Southern Hemisphere, change the sign of the value to a negative number to give an N reference. Note that you do not need to change the gc_distance function—but you do need to make the adjustment to a Southern Hemisphere reference (so that the calculations start with a northern reference).

4. Modify the program so that it will allow the user to enter a longitude in either the west or the east. The program should ask the user to enter the longitude value; then it should ask the user to enter either W or E to specify the hemisphere. If the longitude is a reference to the east, subtract the number from 360 to give a western reference. Note that

you do not need to change the gc_distance function—but you do need to make the adjustment to an eastern reference (so that the calculations start with a western reference).

5. Modify the program so that it combines the modifications in Problems 3 and 4. Thus, the user can enter a latitude with a northern or southern reference, and a longitude with a western or eastern reference. Again, the function does not need to be modified.

4.5   Random Numbers

Random numbers

A sequence of **random numbers** is not defined by an equation; instead, it has certain characteristics that define it. These characteristics include the minimum and maximum values and the average. They also indicate whether the possible values are equally likely to occur or whether some values are more likely to occur than others. Sequences of random numbers can be generated from experiments, such as tossing a coin, rolling a die, or selecting numbered balls. Sequences of random numbers can also be generated using the computer.

Many engineering problems require the use of random numbers in the development of a solution. In some cases, the numbers are used to develop a simulation of a complicated problem. The simulation can be run over and over to analyze the results; each repetition represents a repetition of the experiment. We also use random numbers to approximate noise sequences. For example, the static that we hear on a radio is a noise sequence. If our test program uses an input data file that represents a radio signal, we may want to generate noise and add it to a speech signal or a music signal to provide a more realistic signal.

Engineering applications often require random numbers distributed between specified values. For example, we may want to generate random integers between 1 and 500, or we may want to generate random floating-point values between 5 and −5. We now present discussions on generating random numbers between two specified values. The random numbers generated are equally likely to occur; that is, if the random number is supposed to be an integer between 1 and 5, each of the integers in the set $\{1, 2, 3, 4, 5\}$ is equally likely to occur. Another way of saying this is that each integer should occur approximately 20% of the time.

Uniform random numbers,

Random numbers that are equally likely to be any value in a specified set are also called **uniform random numbers,** or uniformly distributed random numbers.

Integer Sequences

The Standard C library contains a function rand that generates a random integer between 0 and RAND_MAX, where RAND_MAX is a system-dependent integer defined in stdlib.h. (A common value for RAND_MAX is 32,767.) The rand function has no input arguments and is referenced by the expression rand(). Thus, to generate and print a sequence of two random numbers, we could use this statement:

```
printf("random numbers: %i %i \n",rand(),rand());
```

Pseudo-random

Each time that a program containing this statement is executed, the same two values are printed, because the rand function generates integers in a specified sequence. (Because this sequence eventually begins to repeat, it is sometimes called a **pseudo-random** sequence instead of a random sequence.) However, if we generate additional random numbers in the same program, they will be different. Thus, this pair of statements generates four random numbers:

```
printf("random numbers: %i %i \n",rand(),rand());
printf("random numbers: %i %i \n",rand(),rand());
```

Each time that the rand function is referenced in a program, it generates a new value; however, each time that the program is run, it generates the same sequence of values.

In order to cause a program to generate a new sequence of random values each time that

Random-number
seed

it is executed, we need to give a new **random-number seed** to the random-number generator. The function srand (from stdlib.h) specifies the seed for the random-number generator; for each seed value, a new sequence of random numbers is generated by rand. The argument of the srand function is an unsigned integer that is used in computations that initialize the sequence; the seed value is not the first value in the sequence. If an srand function is not used before the rand function is referenced, the computer assumes that the seed value is 1. Therefore, if you specify a seed value of 1, you will get the same sequence of values from the rand function that you will get without specifying a seed value.

In the next program, the user is asked to enter a seed value, and then the program generates 10 random numbers. Each time the user executes the program and enters the same seed, the same set of 10 random integers is generated. Each time a different seed is entered, a different set of 10 random integers is generated. The function prototype statements for rand and srand are included in stdlib.h. Here is the program:

```
/*---------------------------------------------------------------*/
/*  Program chapter4_5                                           */
/*                                                               */
/*  This program generates and prints ten                       */
/*  random integers between 1 and RAND_MAX.                     */

#include <stdio.h>
#include <stdlib.h>

int main(void)
{
    /*  Declare variables.   */
    unsigned int seed;
    int k;

    /*  Get seed value from the user.   */
    printf("Enter a positive integer seed value: \n");
    scanf("%u",&seed);
    srand(seed);

    /*  Generate and print ten random numbers.   */
    printf("Random Numbers: \n");
    for (k=1; k<=10; k++)
       printf("%i ",rand());
    printf("\n");

    /*  Exit program.   */
    return 0;
}
/*---------------------------------------------------------------*/
```

A sample output follows, using the Microsoft Visual C++ 2010 Express compiler:

```
Enter a positive integer seed value:
123
Random Numbers:
440 19053 23075 13104 32363 3265 30749 32678 9760 28064
```

Experiment with the program on your computer system; use the same seed to generate the same numbers, and use different seeds to generate different numbers.

Because the prototype statements for rand and srand are included in stdlib.h, we do not need to include them separately in a program. However, it is instructive to analyze these prototype statements. Because the rand function returns an integer and has no input, its prototype statement is

```
int rand(void);
```

Because the srand function returns no value and has an unsigned integer as an argument, its prototype statement is

```
void srand(unsigned int);
```

Generating random integers over a specified range is simple with the rand function. For example, suppose that we want to generate random integers between 0 and 7. The following statement first generates a random number that is between 0 and RAND_MAX; then it uses the modulus operator to compute the modulus of the random number and the integer 8:

```
x = rand()%8;
```

The result of the modulus operation is the remainder after rand() is divided by 8, so the value of x can assume integer values between 0 and 7.

Suppose that we want to generate a random integer between -25 and 25. The total number of possible integers is 51, and a single random number in this range can be computed with this statement:

```
y = rand()%51 - 25;
```

This statement first generates a value between 0 and 50. Then it subtracts 25 from the value, yielding a new value between -25 and 25.

We can now write a function that generates an integer between two specified integers, a and b. The function first computes n, which is the number of all integers between a and b, inclusive; this value is equal to $b - a + 1$. The function then uses the modulus operation with the rand function to generate a new integer between 0 and $n - 1$. Finally, the lower limit, a, is added to the new integer to give a value between a and b. All three steps can be combined in one expression in the return statement in the function:

```
/*------------------------------------------------------------*/
/*  This function generates a random integer                  */
/*  between specified limits a and b (a<b).                   */
int rand_int(int a,int b)
{
    return rand()%(b-a+1) + a;
}
/*------------------------------------------------------------*/
```

To illustrate the use of this function, the following program generates and prints 10 random integers between user-specified limits (the user also enters the seed to initiate the sequence):

```
/*-----------------------------------------------------------*/
/*  Program chapter4_6                                       */
/*                                                           */
/*  This program generates and prints ten random            */
/*  integers between user-specified limits.                 */

#include <stdio.h>
#include <stdlib.h>

int main(void)
{
    /*  Declare variables and function prototype.  */
    unsigned int seed;
    int a, b, k;
    int rand_int(int a,int b);

    /*  Get seed value and interval limits.  */
    printf("Enter a positive integer seed value: \n");
    scanf("%u",&seed);
    srand(seed);
    printf("Enter integer limits a and b (a<b): \n");
    scanf("%i %i",&a,&b);

    /*  Generate and print ten random numbers.  */
    printf("Random Numbers: \n");
    for (k=1; k<=10; k++)
       printf("%i ",rand_int(a,b));
    printf("\n");

    /*  Exit program.  */
    return 0;
}
/*-----------------------------------------------------------*/
/*  This function generates a random integer                 */
/*  between specified limits a and b (a<b).                  */

int rand_int(int a,int b)
{
    return rand()%(b-a+1) + a;
}
/*-----------------------------------------------------------*/
```

A sample set of values generated from this program is as follows:

```
Enter a positive integer seed value:
13
Enter integer limits a and b (a<b):
-5 5
Random Numbers:
-1 3 1 4 -2 -3 5 0 -2 4
```

Remember that the values generated are system dependent; you should not expect to get this same set of random numbers from a different compiler.

MODIFY!

Using different seed values, modify the preceding program to generate several sets of random integers in each of the following ranges:

1. 0 through 500
2. −10 through 200
3. −50 through −10
4. −5 through 5

Floating-Point Sequences

In many engineering problems, we need to generate random floating-point values in a specified interval $[a, b]$. The computation to convert an integer between 0 and RAND_MAX to a floating-point value between a and b has three steps. The value from the rand function is first divided by RAND_MAX to generate a floating-point value between 0 and 1. The value between 0 and 1 is then multiplied by $(b - a)$ (the width of the interval $[a, b]$) to give a value between 0 and $(b - a)$. The value between 0 and $(b - a)$ is then added to a to adjust it so that it will be between a and b. These three steps are combined in the expression on the return statement in the following function:

```
/*-----------------------------------------------------------*/
/*  This function generates a random                         */
/*  double value between a and b.                            */

double rand_float(double a,double b)
{
    return ((double)rand()/RAND_MAX)*(b-a) + a;
}

/*-----------------------------------------------------------*/
```

Note that a cast operator was needed to convert the integer rand() to a double value so that the result of the division would be a double value.

The program presented earlier in this section can be easily modified to generate and print floating-point values. A sample set of values from such a modification is shown in the following sequence:

```
Enter a positive integer seed value:
82
Enter limits a and b (a<b):
-5 5
Random Numbers:
-4.906613 -3.671834 -1.478164 -2.086093 -4.181494 -3.135624
-4.559923 1.599628 1.382031 0.490280
```

MODIFY!

Modify the program for generating integers so that it generates 10 random floating-point values within a user-specified range. Then, using different seed values, generate several sets of numbers from each of the following ranges:

1. 0.0 through 1.0   2. −0.1 through 1.0
3. −5.0 through −4.5   4. 5.1 through 5.1

4.6   Problem Solving Applied: Instrumentation Reliability

Reliability

Equations for analyzing the reliability of instrumentation can be developed from the study of statistics and probability, where the **reliability** is the proportion of the time that the component works properly. Thus, if a component has a reliability of 0.8, then it should work properly 80% of the time. The reliability of combinations of components can also be determined if the individual component reliabilities are known. Consider the diagrams in Figure 4.8. In order for information to flow from point a to point b in the series design, all three components must work properly. In the parallel design, only one of the three components must work properly for information to flow from point a to point b. If we know the reliability of an individual component, then the reliability of a specific combination of components can be

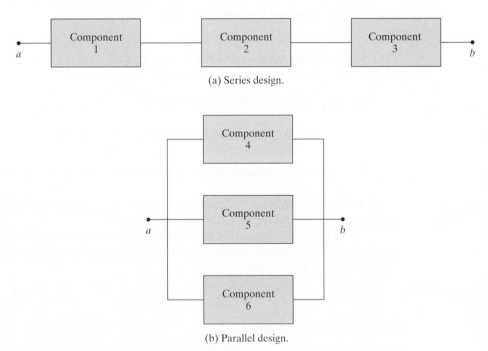

(a) Series design.

(b) Parallel design.

Figure 4.8 *Series and parallel configurations.*

determined in two ways; an analytical reliability can be computed using theorems and results from probability and statistics, and a computer simulation can be developed to give an estimate of the reliability.

Consider the series configuration of Figure 4.8(a). If r is the reliability of a component, and if all three components have the same reliability, then it can be shown that the reliability of the series configuration is r^3. Thus, if the reliability of each component is 0.8 (which means that a component works properly 80% of the time), then the analytical reliability of the series configuration is $(0.8)^3$, or 0.512. Thus, this series configuration should work properly 51.2% of the time.

Consider the parallel configuration of Figure 4.8(b). If r is the reliability of a component, and if all three components have the same reliability, then it can be shown that the reliability of the parallel configuration is $3r - 3r^2 + r^3$. Thus, if the reliability of each component is 0.8, then the analytical reliability of the parallel configuration is $3(0.8) - 3(0.8)^2 + (0.8)^3$, or 0.992. The parallel configuration should work properly 99.2% of the time. Your intuition probably tells you that the parallel configuration is more reliable because only one of the components must be working for the overall configuration to perform properly, whereas all three components must work properly in order for the series configuration to perform properly.

We can also estimate the reliability of these two designs using random numbers from a **computer simulation**. First, we need to simulate the performance of a single component. If the reliability of a component is 0.8, then it works properly 80% of the time. To simulate this performance, we could generate a random value between 0 and 1. If the value is between 0 and 0.8, we can assume that the component worked properly; otherwise, it failed. (We could also have used the values 0 to 0.2 for a failure and 0.2 to 1.0 for a component that worked properly.) To simulate the series design with three components, we would generate three floating-point random numbers between 0 and 1. If all three numbers are less than or equal to 0.8, then the design works for this one trial; if any one of the numbers is greater than 0.8, then the design does not work for this one trial. If we run hundreds or thousands of trials, we can compute the proportion of the time that the overall design works. This simulation estimate is an approximation to the analytically computed reliability.

To estimate the reliability of the parallel design with a component reliability of 0.8, we again generate three random floating-point numbers between 0 and 1. If any one of the three numbers is less than or equal to 0.8, then the design works for this one trial; if all of the numbers are greater than 0.8, then the design does not work for one trial. To estimate the reliability determined by the simulation, we divide the number of trials for which the design works by the total number of trials performed.

As indicated by the previous discussion, we can use computer simulations to provide a validation for the analytical results because the simulated reliability should approach the analytically computed reliability as the number of trials increases. There are also cases in which it is very difficult to analytically compute the reliability of a piece of instrumentation. In these cases, a computer simulation can be used to provide a good estimate of the reliability.

Develop a program to compare the analytical reliabilities of the series and parallel configurations in Figure 4.8 with simulation results. Allow the user to enter the individual component reliability and the number of trials to use in the simulation.

Computer
simulation

1. PROBLEM STATEMENT

Compare the analytical and simulation reliabilities for a series configuration with three components and for a parallel configuration with three components. Assume that all components have the same reliability.

2. INPUT/OUTPUT DESCRIPTION

The I/O diagram shows that the input values are the component reliability, the number of trials, and a random-number seed for initiating the sequence. The output consists of the analytical reliability and the simulation reliability for the series and the parallel configurations.

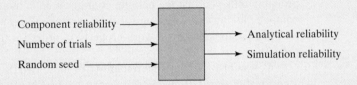

3. HAND EXAMPLE

For the hand example, we use a component reliability of 0.8 and three trials. Since each trial requires three random numbers, assume that the first nine random numbers generated are the following (they were generated from the `rand_float` function using a seed of 47 for values between 0 and 1).

First set of three random values:

```
0.005860 0.652303 0.271187
```

Second set of three random values:

```
0.589007 0.064699 0.992248
```

Third set of three random values:

```
0.719565 0.786615 0.001923
```

From each group of three random numbers, we can determine whether a series configuration would work properly and whether a parallel configuration would work properly. For the first group of three random numbers, all the values are less than 0.8; so both the series and parallel configurations would work properly. One of the values in the second set of numbers is greater than 0.8, so only the parallel configuration would work properly. Both configurations work properly with the third set of random numbers. Thus, the analytical results (computed earlier in this section) and the simulation results for three trials are the following:

```
Analytical Reliability:
Series: 0.512  Parallel: 0.992
Simulation for 3 Trials
Series: 0.667  Parallel: 1.000
```

As we increase the number of trials, the simulation results should approach the analytical results. If we change the random number seed, the simulation results may also change even with only three trials.

4. ALGORITHM DEVELOPMENT

We first develop the decomposition outline because it divides the solution into a series of sequential steps:

Decomposition Outline

1. Read component reliability, number of trials, and random number seed.
2. Compute analytical reliabilities.
3. Compute simulation reliabilities.
4. Print comparison of reliabilities.

Step 1 prompts the user to enter the necessary information and then read it. Step 2 uses the equations given earlier to compute the analytical reliabilities. Because the computations are straightforward, we compute them in the `main` function. Step 3 involves a loop to generate the random numbers and to determine whether the configurations would perform properly for each trial. The `rand_float` function is used to compute the random numbers in the loop. In step 4, we print the results of the computations. The structure chart for this solution was shown in Figure 4.1. The refinement in pseudocode is the following:

> *Refinement in Pseudocode*
> *main: read component reliability, number of trials,*
> *   and random number seed*
> *   compute analytical reliabilities*
> *   set series_success to zero*
> *   set parallel_success to zero*
> *   set k to 1*
> *   while k ≤ number of trials*
> *   generate three random numbers between 0 and 1*
> *   if each number ≤ component reliability,*
> *   increment series_success by 1*
> *   If any number ≤ component reliability,*
> *   increment parallel_success by 1*
> *   increment k by 1*
> *   print analytical reliabilities*
> *   print simulation reliabilities*

The steps in the pseudocode are now detailed enough to convert to C. We also include the `rand_float` function in the program:

```
/*-----------------------------------------------------------*/
/*  Program chapter4_7                                       */
/*                                                           */
/*  This program estimates the reliability of a series and   */
/*  a parallel configuration using a computer simulation.    */
```

```c
#include <stdio.h>
#include <stdlib.h>
#include <math.h>

int main(void)
{
   /*  Declare variables and function prototypes.  */
   unsigned int seed;
   int n, k;
   double component_reliability, a_series, a_parallel,
          series_success=0, parallel_success=0,
          num1, num2, num3;
   double rand_float(double a,double b);

   /*  Get information for the simulation.  */
   printf("Enter individual component reliability: \n");
   scanf("%lf",&component_reliability);
   printf("Enter number of trials: \n");
   scanf("%i",&n);
   printf("Enter unsigned integer seed: \n");
   scanf("%u",&seed);
   srand(seed);
   printf("\n");

   /*  Compute analytical reliabilities.  */
   a_series = pow(component_reliability,3);
   a_parallel = 3*component_reliability
                - 3*pow(component_reliability,2)
                + pow(component_reliability,3);

   /*  Determine simulation reliability estimates.  */
   for (k=1; k<=n; k++)
   {
      num1 = rand_float(0,1);
      num2 = rand_float(0,1);
      num3 = rand_float(0,1);
      if (((num1<=component_reliability) &&
           (num2<=component_reliability)) &&
           (num3<=component_reliability))
              series_success++;
      if (((num1<=component_reliability) ||
           (num2<=component_reliability)) ||
           (num3<=component_reliability))
              parallel_success++;
   }
   /*  Print results.  */
   printf("Analytical Reliability \n");
   printf("Series: %.3f Parallel: %.3f \n",
           a_series,a_parallel);
   printf("Simulation Reliability, %i trials \n",n);
   printf("Series: %.3f Parallel: %.3f \n",
           (double)series_success/n,
           (double)parallel_success/n);
```

```
    /*  Exit program.  */
    return 0;
}
/*---------------------------------------------------------*/
/*  This function generates a random                       */
/*   double value between a and b (a<b).                   */
double rand_float(double a,double b)
{
    return ((double)rand()/RAND_MAX)*(b-a) + a;
}
/*---------------------------------------------------------*/
```

5. TESTING

If we use the data from the hand example, we have the following interaction, wherein output matches the data that we computed by hand:

```
Enter individual component reliability:
0.8
Enter number of trials:
3
Enter unsigned integer seed:
47

Analytical Reliability
Series: 0.512 Parallel: 0.992
Simulation Reliability, 3 trials
Series: 0.667 Parallel: 1.000
```

Here are results from two more simulations which demonstrate that the simulation results approach the analytical results as the number of trials increases:

```
Enter individual component reliability:
0.8
Enter number of trials:
100
Enter unsigned integer seed:
123

Analytical Reliability
Series: 0.512 Parallel: 0.992
Simulation Reliability, 100 trials
Series: 0.470 Parallel: 1.000

Enter individual component reliability:
0.8
Enter number of trials:
1000
Enter unsigned integer seed:
3535

Analytical Reliability
Series: 0.512 Parallel: 0.992
Simulation Reliability, 1000 trials
Series: 0.530 Parallel: 0.990
```

MODIFY!

These problems relate to the program developed in this section, which compares the analytical and simulated reliabilities.

1. Use this program to compute information comparing the simulation results for 10, 100, 1000, and 10000 trials, assuming that the component reliability is 0.85.

2. Use this program to compute information comparing the simulation results for 1000 trials, using five different random-number seeds. Assume that the component reliability is 0.75.

3. What component reliability is necessary to give a series reliability of 0.7? (*Hint*: Use the analytical reliability equation.) Validate your answer using this program.

4. What component reliability is necessary to give a parallel reliability of 0.9? Using the analytical reliability equation is not as easy in this case. If your calculator does not find roots of polynomial equations, just experiment with the program until you are close to the desired reliability.

4.7 Numerical Technique: Roots of Polynomials*

A polynomial is a function of a single variable that can be expressed in the following general form:

$$f(x) = a_0 x^N + a_1 x^{N-1} + a_2 x^{N-2} + \cdots + a_{N-2} x^2 + a_{N-1} x + a_N, \quad (4.10)$$

where the variable is x and the coefficients are represented by $a_0, a_1, \ldots a_N$. The degree of a polynomial is equal to the largest nonzero exponent. Therefore, the general form for a cubic (degree 3) polynomial is

$$g(x) = a_0 x^3 + a_1 x^2 + a_2 x + a_3,$$

and a specific example of a cubic polynomial is

$$h(x) = x^3 - 2x^2 + 0.5x - 6.5.$$

Note that, for each term in the equation, the sum of the coefficient subscript and the variable exponent is equal to the polynomial degree using the notation in Equation (4.10).

Polynomial Roots

The solutions to many engineering problems involve finding the roots of an equation of the form

$$y = f(x),$$

Roots

where the **roots** are the values of x for which y is equal to zero. Examples of applications in which we need to find roots of equations include designing the control system for a robotic arm, designing springs and shock absorbers for an automobile, analyzing the response of a motor, and analyzing the stability of a digital filter.

If a function $f(x)$ is a polynomial of degree N, then $f(x)$ has exactly N roots. These N roots may contain real roots or complex roots, as shown in the following examples. If we

---

*Optional section.

assume that the coefficients $(a_0, a_1, \ldots, a_N)$ of the polynomial are real values, then complex roots will always occur in complex conjugate pairs. (Recall that a complex number can be expressed as $a + ib$, where $i = \sqrt{-1}$. The complex conjugate of $a + ib$ is $a - ib$.)

If a polynomial is factored into linear terms, it is easy to identify the roots of the polynomial by setting each term to zero. For example, consider the following equation:

$$f(x) = x^2 + x - 6$$
$$= (x - 2)(x + 3).$$

If $f(x)$ is equal to zero, we have the following:

$$(x - 2)(x + 3) = 0.$$

The roots of the equation, or the values of x for which $f(x)$ is equal to zero, are then $x = 2$ and $x = -3$. These roots also correspond to the values of x where the polynomial crosses the x-axis, as shown in Figure 4.9.

If a quadratic equation (polynomial of degree 2) cannot easily be factored, we can use the quadratic formula to determine the two roots of the equation. Recall that for a general quadratic equation

$$y = ax^2 + bx + c,$$

the roots can be computed as follows:

$$x_1 = \frac{-b + \sqrt{b^2 - 4ac}}{2a},$$

$$x_2 = \frac{-b - \sqrt{b^2 - 4ac}}{2a}.$$

Thus, for the quadratic equation

$$f(x) = x^2 + 3x + 3,$$

the roots are

$$x_1 = \frac{-3 + \sqrt{-3}}{2} = -1.5 + 0.87\sqrt{-1},$$

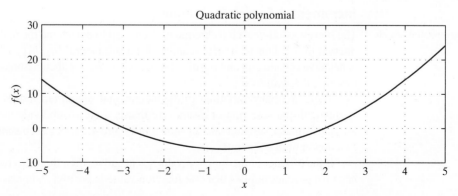

Quadratic polynomial

Figure 4.9 *Polynomial with two real roots.*

and

$$x_2 = \frac{-3 - \sqrt{3}}{2} = -1.5 - 0.87\sqrt{-1}.$$

Because a cubic polynomial has degree 3, it has exactly three roots. If we assume that the coefficients are real, then there are only these four possibilities:

- 3 real roots at different values (distinct roots)
- 3 real roots at the same value (multiple roots)
- 2 real roots at the same value (double root) and 1 real root at a distinct value, or
- 1 real root and a complex conjugate pair of roots.

Examples of functions that illustrate each of these cases are as follows:

$$
\begin{aligned}
f_1(x) &= (x - 3)(x + 1)(x - 1) \\
&= x^3 - 3x^2 - x + 3, \\
f_2(x) &= (x - 2)^3 \\
&= x^3 - 6x^2 + 12x - 8, \\
f_3(x) &= (x + 4)(x - 2)^2 \\
&= x^3 - 12x + 16, \\
f_4(x) &= (x + 2)[x - (2 + i)][x - (2 - i)] \\
&= x^3 - 2x^2 - 3x + 10.
\end{aligned}
$$

Figure 4.10 contains plots of these functions. Note again that the real roots correspond to the points where the function crosses the x axis.

It is relatively easy to determine the roots of polynomials of degree 1 or 2, but it can be difficult to determine the roots of polynomials of degree 3 and higher. A number of numerical techniques exist for determining the roots of polynomials. Techniques such as the incremental search, the bisection method, and the false-position technique identify the real roots by searching for intervals in which the function changes sign because this indicates that the function has crossed the x-axis. Additional techniques, such as the Newton–Raphson method, can be used to find complex roots.

Incremental-Search Technique

Incremental-search   The **incremental-search** technique is often used to determine the real roots of a function in an interval $[a, b]$. This technique searches for a subinterval $[a_k, b_k]$ such that the function value is negative on one end and positive on the other. We are then assured that there is at least one root in this subinterval.

There are many variations of the incremental-search technique. The one that we discuss begins with the selection of a step size that is used to subdivide the original interval into a group of smaller subintervals, as shown in Figure 4.11. For each subinterval, we evaluate the function at both endpoints.

If the product of the function values is negative, then there is a root in this subinterval. (A negative product implies that one function value is positive, whereas the other function value

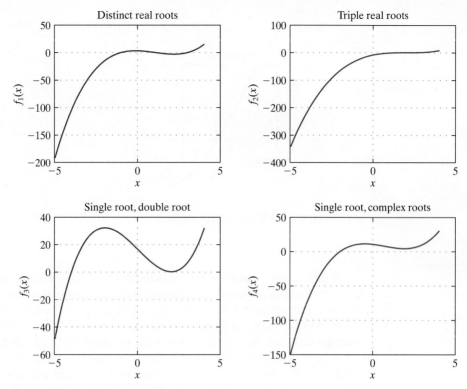

Figure 4.10 *Cubic polynomials.*

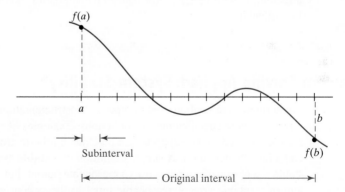

Figure 4.11 *Incremental search.*

is negative; hence, the function must cross the *x* axis in the interval.) At this point, we can estimate the root to be the midpoint of this small segment, as shown in Figure 4.12(a). It is also possible that one of the subinterval endpoints might be a root, or be very close to a root, as shown in Figure 4.12(b). Remember that it is not likely that a floating-point value will be exactly equal to zero, so the test to determine if an endpoint is a root should compare the function value to a very small number, but not to zero.

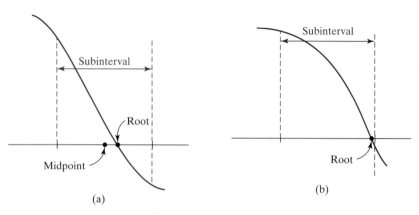

Figure 4.12 *Subinterval analysis.*

It is also important to recognize that there are cases in which this incremental-search technique fails. For example, suppose that there are two roots in one of the subintervals. In this case, because the function values at the endpoints will have the same sign, their product will be positive and the algorithm will skip to the next subinterval. As another example, consider the case with three roots in one of the subintervals. In this case, because the function values at the endpoints have different signs, the estimate of the root is the midpoint of the subinterval. We then continue with the next subinterval, and thus miss the other two roots in the previous subinterval. These examples are used to illustrate the fact that the incremental-search technique has some flaws, although, in general, it works reasonably well. If we need a technique with better performance characteristics, other root-finding methods should be investigated.

4.8 Problem Solving Applied: System Stability*

System

Stable system

The term **system** is often used to represent instrumentation or equipment for which specified inputs generate specified outputs or actions. Examples of systems include the cooling equipment connected to the supports of a pipeline, a robotic arm used in a manufacturing facility, and a fast "bullet" train. A simple definition of a **stable system** is the following: A system is stable if a reasonable input causes a reasonable output. For example, consider the control system of a robotic arm. A reasonable input to the system would specify that the arm should move in a direction that is valid for the robotic arm. If a reasonable input causes the arm to become erratic or to attempt to move in invalid directions, then the system is not stable. The analysis of the stability of the design of a system involves determining dynamic properties of the system. A discussion of the types of analyses involved, or of the functions involved, is beyond the scope of this text, but one component of the analysis requires the determination of the roots of polynomials. Usually, both the real and complex roots are needed, but the techniques for finding complex roots involve using the derivative of the polynomial, and thus become

---

*Optional section.

more mathematically involved. Therefore, we reduce the scope of this problem to finding only the real roots of a polynomial given a specified interval in which to search. We also assume that the polynomial is a cubic polynomial, but the solution can easily be extended to handle higher degree polynomials.

Develop a program to determine the real roots of a cubic polynomial. Allow the user to enter the coefficients of the polynomial, the interval to be searched, and the step size of the subintervals used in the search.

1. PROBLEM STATEMENT

Determine the real roots of a cubic polynomial.

2. INPUT/OUTPUT DESCRIPTION

The I/O diagram shows that the input values are the polynomial coefficients, the interval endpoints, and the step size of the subintervals, while the output values are the roots identified in the specified interval:

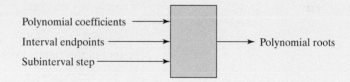

3. HAND EXAMPLE

For the hand example, we use the equation

$$y = 2x - 4.$$

This function can be described as a cubic polynomial with $a_0 = 0$, $a_1 = 0$, $a_2 = 2$, and $a_3 = -4$. If we set the polynomial to zero, we easily observe that the root is equal to 2. To examine the incremental-search technique, we first use a step size such that the root falls on one of the endpoints of a subinterval; then we use a step size such that the root does not fall on one of the endpoints of a subinterval. If the root falls on an endpoint, we can easily identify it because the polynomial value will be very close to zero. If the root falls within a subinterval, the product of the function values at the endpoints will be negative, and we then estimate the root to be the midpoint of the interval.

First, consider the interval [1, 3] with a step size of 0.5. The subintervals and the corresponding information derived from them are as follows:

Subinterval 1:   [1.0, 1.5]
$f(1.0) \cdot f(1.5) = (-2) \cdot (-1) = 2$
No root in this interval.

Subinterval 2:   [1.5, 2.0]
When we evaluate the endpoints, we detect the root at $x = 2.0$.

Subinterval 3:   [2.0, 2.5]
                           When we evaluate the endpoints, we again detect the root
                           at $x = 2.0$. Note that we will need to be careful that we do
                           not identify this root twice in the program.

Subinterval 4:   [2.5, 3.0]
                           $f(2.5) \cdot f(3.0) = (1) \cdot (2) = 2$
                           No root in this interval.

We now consider the interval [1, 3] with a step size of 0.3. The subintervals and the corresponding information derived from them are as follows:

Subinterval 1:   [1.0, 1.3]
                           $f(1.0) \cdot f(1.3) = (-2) \cdot (-1.4) = 2.8$
                           No root in this interval.

Subinterval 2:   [1.3, 1.6]
                           $f(1.3) \cdot f(1.6) = (-1.4) \cdot (-0.8) = 1.12$
                           No root in this interval.

Subinterval 3:   [1.6, 1.9]
                           $f(1.6) \cdot f(1.9) = (-0.8) \cdot (-0.2) = 1.6$
                           No root in this interval.

Subinterval 4:   [1.9, 2.2]
                           $f(1.9) \cdot f(2.2) = (-0.2) \cdot (0.4) = -0.08$
                           The root in this interval is estimated to occur at the midpoint:
                           $$x = \frac{1.9 + 2.2}{2} = 2.05.$$

Subinterval 5:   [2.2, 2.5]
                           $f(2.2) \cdot f(2.5) = (0.4) \cdot (1.0) = 0.4$
                           No root in this interval.

Subinterval 6:   [2.5, 2.8]
                           $f(2.5) \cdot f(2.8) = (1.0) \cdot (1.6) = 1.6$
                           No root in this interval.

Subinterval 7:   [2.8, 3.1]
                           Note that the right endpoint exceeds the overall endpoint. In
                           the program, we will modify such an interval so that it ends on
                           the original right endpoint.
                           $f(2.8) \cdot f(3.0) = (1.6) \cdot (2.0) = 3.2$
                           No root in this interval.

4. ALGORITHM DEVELOPMENT

We first develop the decomposition outline because it breaks the solution into a series of sequential steps:

Decomposition Outline

1. Read polynomial coefficients, interval of interest, and step size.
2. Locate roots using subintervals.

Step 1 involves prompting the user to enter the necessary information and then reading it. Step 2 requires a loop to compute the subinterval endpoints and then to determine if a root occurs on an endpoint or in the subinterval. When a root is located, a corresponding message is printed. There are a number of operations involved in Step 2, so we should consider using functions to keep the main function from getting long. Because we need to evaluate the cubic polynomial several places in the program, it is a good candidate for a function. Within each subinterval, we need to search for a root; this search is also a good candidate for a function. The structure chart for this solution was shown in Figure 4.1. The refinement in pseudocode is the following:

Refinement in Pseudocode
main: *read coefficients, interval endpoints a and b,*
                and step size
             compute the number of subintervals, n
             set k to 0
             while k ≤ n–1
                    compute left subinterval endpoint
                    compute right subinterval endpoint
                    check_roots (left, right, coefficients)
                    increment k by 1
             check_roots (b,b,coefficients)
check_roots (left, right, coefficients):
             set f_left to poly(left,coefficients)
             set f_right to poly(right,coefficients)
             if f_left is near zero
                    print root at left endpoint
             else
                    if f_left · f_right < 0
                    print root at midpoint of subinterval
             return
poly(x,a_0,a_1,a_2,a_3):
             return $a_0x^3 + a_1x^2 + a_2x + a_3$

Note in the pseudocode for the check_roots function that we check to see if the left subinterval endpoint is a root, but we do not check the right subinterval endpoint. This is necessary to avoid identifying the same root twice—the first time, it is a right endpoint for one interval; the second time, it is a left endpoint for the next subinterval. Because we only check the left endpoints, we need to check the final point in the interval because it never becomes a left endpoint.

The steps in the pseudocode are detailed enough to convert to C:

```
/*-----------------------------------------------------------------*/
/*  Program chapter4_8                                             */
/*                                                                 */
/*  This program estimates the real roots of a                    */
/*  polynomial function using incremental search.                 */

#include <stdio.h>
#include <math.h>

int main(void)
{
```

```
      /*  Declare variables and function prototype.  */
      int n, k;
      double a0, a1, a2, a3, a, b, step, left, right;
      void check_roots(double left,double right,double a0,
                       double a1,double a2,double a3);
      /*  Get user input.  */
      printf("Enter coefficients a0, a1, a2, a3: \n");
      scanf("%lf %lf %lf %lf",&a0,&a1,&a2,&a3);
      printf("Enter interval limits a, b (a<b): \n");
      scanf("%lf %lf",&a,&b);
      printf("Enter step size: \n");
      scanf("%lf",&step);
      /*  Check subintervals for roots.  */
      n = ceil((b - a)/step);
      for (k=0; k<=n-1; k++)
      {
         left = a + k*step;
         if (k == n-1)
            right = b;
         else
            right = left + step;
         check_roots(left,right,a0,a1,a2,a3);
      }

      /*  Exit program.  */
      return 0;
}
/*------------------------------------------------------------------*/
/*  This function checks a subinterval for a root.               */
void check_roots(double left,double right,double a0,
                 double a1,double a2,double a3)
{
   /*  Declare variables and function prototype.  */
   double f_left, f_right;
   double poly(double x,double a0,double a1,
               double a2,double a3);
   /*  Evaluate subinterval endpoints and test for roots.  */
   f_left = poly(left,a0,a1,a2,a3);
   f_right = poly(right,a0,a1,a2,a3);
   if (fabs(f_left) < 0.1e-04)
      printf("Root detected at %.3f \n",left);
   else
      if (fabs(f_right) < 0.1e-04)
         ;
      else
         if (f_left*f_right < 0)
            printf("Root detected at %.3f \n",(left+right)/2);
   return;
}
/*------------------------------------------------------------------*/
```

```
/*  This function evaluates a cubic polynomial.              */
double poly(double x,double a0,double a1,double a2,double a3)
{
    return a0*x*x*x + a1*x*x + a2*x + a3;
}
/*---------------------------------------------------------------*/
```

5. TESTING

If we use the data from the hand example, we have the following interaction with the program. The roots match the ones that we computed by hand:

```
Enter coefficients a0, a1, a2, a3:
0 0 2 -4
Enter interval limits a, b (a<b):
1 3
Enter step size:
0.5
Root detected at 2.000

Enter coefficients a0, a1, a2, a3:
0 0 2 -4
Enter interval limits a, b (a<b):
1 3
Enter step size:
0.3
Root detected at 2.050
```

Use the polynomials given on page 188 to test this program. Use intervals and step sizes so that the roots do not always fall on subinterval endpoints.

MODIFY!

1. The size of the interval affects the estimate of the root if the root is not on an endpoint of a subinterval. Using the polynomial from the hand example and the interval [0.5, 3], experiment with several step sizes, including 1.1, 0.75, 0.5, 0.3, and 0.14.

2. Using the first cubic polynomial given on page 188, test the program using intervals in which the roots fall on the endpoints of the interval [a, b] entered as input.

3. Using the first cubic polynomial given on page 188, find a step size that causes the program to miss some of the roots for an initial interval of [−10, 10]. Explain why the roots were missed by the program. Is this an error in the program?

4. Modify the program so that it can accept and locate the real roots of a fourth-degree polynomial.

5. Use this program to help answer Problem 4 of the previous problems on page 186.

4.9 Macros*

Macro

Before compiling a program, the preprocessor performs any actions specified by preprocessing directives, such as the inclusion of header files or the symbolic definition of constants. A simple operation can also be specified by a preprocessing directive called a **macro**, which has the following general form:

#define macro_name(parameters) macro_text

The macro_text replaces references to the macro_name in the program. If the macro does not have parameters, then it is essentially a symbolic constant. If a macro has parameters, then it can represent a simple function. If the description of the macro takes more than one line, a backslash (\) must be used at the end of each line but the last to indicate that the line is continued on the next line.

An advantage of using a macro instead of a function is that the macro does not need to be defined in a separate module; thus, the compilation and linking/loading process is simplified, and the execution time is reduced. During preprocessing, each reference to the macro is replaced with the macro text.

To illustrate, consider the following simple program that converts degrees Fahrenheit to degrees Centigrade:

```
/*------------------------------------------------------------*/
/*  Program chapter4_9                                        */
/*                                                            */
/*  This program converts a temperature in                    */
/*  Fahrenheit to Centigrade.                                 */

#include <stdio.h>
#define degrees_C(x) (((x) - 32)*(5.0/9.0))

int main(void)
{
   /*  Declare variables.  */
   double temp;

   /*  Get temperature in Fahrenheit.  */
   printf("Enter temperature in degrees Fahrenheit: \n");
   scanf("%lf",&temp);

   /*  Convert and print temperature in Centigrade.  */
   printf("%f degrees Centigrade \n",degrees_C(temp));

   /*  Exit program.  */
   return 0;
}
/*------------------------------------------------------------*/
```

When the printf statement from the program is compiled, the macro text replaces the macro reference, giving the following:

```
printf("%f degrees Centigrade \n",(((temp) - 32)*(5.0/9.0)));
```

---

*Optional section.

When this statement is executed, the value in `temp` is correctly converted from degrees Fahrenheit to degrees Centigrade before it is printed.

It is important to include parentheses around each individual argument and around the complete macro_text in the macro definition so that the macro will work properly when it is referenced with an expression as an actual parameter. To illustrate, consider these macros that convert temperatures in degrees Centigrade to degrees Fahrenheit:

```
#define degrees1_F(x) ((x)*(9.0/5.0) + 32)
#define degrees2_F(x) x*(9.0/5.0) + 32
```

When these macros are used with a variable as an actual parameter, they both work properly. For example, the statements

```
max_temp1 = degrees1_F(temp);
max_temp2 = degrees2_F(temp);
```

are correctly compiled as the following equivalent computations:

```
max_temp1 = ((temp)*(9.0/5.0) + 32);
max_temp2 = temp*(9.0/5.0) + 32;
```

However, the statements

```
max_temp1 = degrees1_F(temp+10);
max_temp2 = degrees2_F(temp+10);
```

are compiled as the following statements, which do not yield the same values:

```
max_temp1 = ((temp+10)*(9.0/5.0) + 32);
max_temp2 = temp+10*(9.0/5.0) + 32;
```

Therefore, the parentheses around the macro arguments and around the macro_text are necessary to ensure correct calculations.

The following macro computes the area of a triangle with a specified base and height as shown in Figure 4.13:

```
#define area_tri(base,height) (0.5*(base)*(height))
```

Note that parentheses are included around each argument and around the complete macro_text in the macro definition.

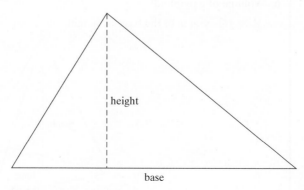

Figure 4.13 *Triangle.*

PRACTICE!

Give macros to compute the following values, and, along with each macro, give an example of a statement that references it:

1. Area of a square:

   $A = \text{side}^2$.

2. Area of a rectangle:

   $A = \text{side}_1 \times \text{side}_2$.

3. Area of a parallelogram:

   $A = \text{base} \times \text{height}$.

4. Area of trapezoid:

   $A = 1/2 \text{ base} \times (\text{height}_1 + \text{height}_2)$.

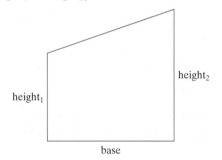

5. Volume of a sphere:

   $V = 4/3 \, \pi \times \text{radius}^3$.

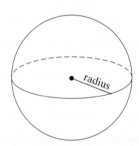

6. Volume of a pyramid:

   $V = 1/3 \times \text{area of the base} \times \text{height}$.

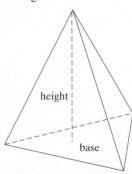

7. Volume of a right circular cone:

   $V = 1/3 \ \pi \times \text{radius}^3 \times \text{height}$.

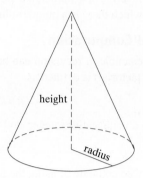

height

radius

8. Volume of a cube:

   $V = \text{side}^3$.

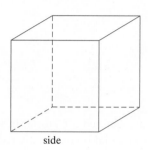

side

9. Volume of a rectangular parallelepiped:

   $V = \text{length} \times \text{width} \times \text{height}$.

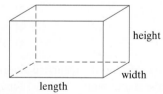

height

width

length

4.10 Recursion*

Recursive function

A function that invokes itself (or calls itself) is a **recursive function**. Recursion can be a powerful tool for solving certain classes of problems in which the solution can be defined in terms of a similar but smaller problem, and then the smaller problem can be defined in terms of a similar but still smaller problem. This redefinition of the problem into smaller problems continues until the smaller problem has a unique solution that is then used to determine the overall solution. There are system-dependent limitations to the number of times that a recursive function can call itself as it is continually redefining a problem into smaller and smaller problems, but these limitations do not usually cause difficulties.

---

*Optional section.

In the next two examples, we illustrate problems that can be solved with a recursive algorithm. In the examples, note the two parts to a recursive solution. First, the solution has to be redefined in terms of a similar, but smaller, problem; second, the smaller problems must reach a point at which there is a unique solution.

Factorial Computation

Factorial

A simple example of recursion can be shown using the **factorial** computation. Recall that $k!$ (read as k factorial) is defined as

$$k! = (k)(k - 1)(k - 2) \ldots (3)(2)(1),$$

where k is a nonnegative integer, and where $0! = 1$. Thus,

$$5! = 5 \cdot 4 \cdot 3 \cdot 2 \cdot 1 = 120.$$

We could also compute 5! using the following steps:

$$5! = 5 \cdot 4!$$
$$4! = 4 \cdot 3!$$
$$3! = 3 \cdot 2!$$
$$2! = 2 \cdot 1!$$
$$1! = 1 \cdot 0!$$
$$0! = 1.$$

Thus, we have defined a factorial in terms of a product that involves smaller factorials. The smaller factorial is continually redefined until we reach 0!. We substitute the value of 0! in the last equation, and then begin going back up the list of equations, substituting values for the factorials:

$$1! = 1 \cdot 1$$
$$2! = 2 \cdot 1! = 2$$
$$3! = 3 \cdot 2! = 6$$
$$4! = 4 \cdot 3! = 24$$
$$5! = 5 \cdot 4! = 120.$$

We have now developed a recursive algorithm for computing a factorial.

We present a program with two functions to compute a factorial. The first function is a nonrecursive (iterative) function, and the second is a recursive function. A factorial value becomes large quickly, so we use long integers for the factorial value. Note that both functions are referenced similarly in the main function:

```
/*-------------------------------------------------------------*/
/*  Program chapter4_10                                         */
/*                                                             */
/*  This program compares a recursive function and             */
/*  a nonrecursive function for computing factorials.          */

#include <stdio.h>

int main(void)
{
   /*  Declare variables and function prototypes.  */
   int n;
```

```c
long factorial(int k);
long factorial_r(int k);

/* Get user input. */
printf("Enter positive integer: \n");
scanf("%i",&n);

/* Compute and print factorials. */
printf("Nonrecursive: %i! = %li \n",n,factorial(n));
printf("Recursive: %i! = %li \n",n,factorial_r(n));

/* Exit program. */
return 0;
}
/*------------------------------------------------------------*/
/* This function computes a factorial with a loop.        */

long factorial(int k)
{
   /* Declare variables. */
   int j;
   long term;

   /* Compute factorial with multiplication. */
   term = 1;
   for (j=2; j<=k; j++)
      term *=j;

   /* Return factorial value. */
   return term;
}
/*------------------------------------------------------------*/
/* This function computes a factorial recursively.        */

long factorial_r(int k)
{
   /* Use recursive reference until k=0. */
   if (k == 0)
      return 1;
   else
      return k*factorial_r(k-1);
}
/*------------------------------------------------------------*/
```

The condition k == 0 keeps the recursive routine from becoming an infinite loop; this routine calls itself recursively with an argument that is continually being decremented by 1, until the argument reaches zero.

For large values of k, the value of $k!$ can exceed even long integers. In these cases, the computations should be done using double or long double values. An interesting approximation to $k!$ is also discussed in the end-of-chapter problems.

Fibonacci Sequence

Fibonacci sequence   A **Fibonacci sequence** is a sequence of numbers $\{f_0, f_1, f_2, f_3, \dots\}$ in which the first two values (f_0 and f_1) are equal to 1, and each succeeding number is the sum of the previous two numbers. Thus, the first few values of the Fibonacci sequence are

$$1 \quad 1 \quad 2 \quad 3 \quad 5 \quad 8 \quad 13 \quad 21 \quad 34 \quad \dots$$

This sequence was first described in the year 1202, and it has applications that range from biology to electrical engineering. For example, Fibonacci sequences are often used in studies of rabbit population growth.

A function to compute the kth value in the Fibonacci sequence is a good candidate for a recursive function because each new value in the sequence is computed from the two previous values. The following functions implement both nonrecursive and recursive algorithms for computing a Fibonacci number:

```c
/*-----------------------------------------------------------------*/
/*   Program chapter4_11                                           */
/*                                                                 */
/*   This program compares a recursive function and a             */
/*   nonrecursive function for computing Fibonacci numbers.       */

#include <stdio.h>

int main(void)
{
   /*  Declare variables and function prototypes.  */
   int n;
   int fibonacci(int k);
   int fibonacci_r(int k);

   /*  Get user input.  */
   printf("Enter positive integer: \n");
   scanf("%i",&n);

   /*  Compute and print factorials.  */
   printf("Nonrecursive: Fibonacci number = %li \n",
           fibonacci(n));
   printf("Recursive:    Fibonacci number = %li \n",
           fibonacci_r(n));

   /*  Exit program.  */
   return 0;

}
/*-----------------------------------------------------------------*/
/*   This function computes the kth Fibonacci                      */
/*   number using a nonrecursive algorithm.                       */

int fibonacci(int k)
{
   /*  Declare variables.  */
   int term, prev1, prev2, n;
```

```
      /*  Compute kth Fibonacci number with a loop.  */
      term = 1;
      if (k > 1)
      {
         prev1 = prev2 = 1;
         for (n=2; n<=k; n++)
         {
            term = prev1 + prev2;
            prev2 = prev1;
            prev1 = term;
         }
      }

      /*  Return kth Fibonacci number.  */
      return term;
}
/*------------------------------------------------------------*/
/*  This function computes the kth Fibonacci                  */
/*  number using a recursive algorithm.                       */

int fibonacci_r(int k)
{
      /*  Declare variables.  */
      int term;

      /*  Compute kth Fibonacci number recursively until k=1.  */
      term = 1;
      if (k > 1)
         term = fibonacci_r(k-1) + fibonacci_r(k-2);

      /*  Return kth Fibonacci number.  */
      return term;
}
/*------------------------------------------------------------*/
```

In the recursive function, the condition k > 1 keeps the function from going into an infinite loop.

MODIFY!

1. Use the program chapter4_12 to compute values of 1!, 2!, and so on, until you reach the limits for long integers. What kind of error message occurred when the value of $k!$ exceeded the limits on your system?

2. Modify chapter4_12 so that it uses double values instead of integers to compute factorials. Explain why the number of digits of precision determines the maximum value of $k!$ that can be correctly computed using double values. What is the maximum value of $k!$ that can be computed using double values on your system?

Most programs in C benefit from using both library and programmer-defined functions. Functions allow us to reuse software and to employ abstraction in our solution and, hence, reduce development time and increase the quality of the software. Numerous examples were developed to illustrate using programmer-defined functions to solve problems, including examples of macros and recursive functions. Specific applications were presented to illustrate generating random numbers (integers or floating-point values) and to implement the incremental search technique for identifying real roots of polynomials.

KEY TERMS

abstraction	local variable
actual parameter	macro
automatic class	modularity
call-by-reference	module
call-by-value	module chart
coercion of arguments	programmer-defined function
computer simulation	pseudorandom
driver program	random number
external class	random number seed
factorial	recursion
Fibonacci sequence	register class
formal parameter	reusability
function	root
function prototype	scope
global variable	static class
incremental search	storage class
invoke	structure chart
library function	

C STATEMENT SUMMARY

Function definition:

```
return_type function_name(parameter types)
{
    declarations;
    statements;
}
```

Return statement:

```
return;
return (a + b)/2;
```

Function prototype:

```
double sinc(double x);
double sinc(double);
void check_roots(double left,double right,double a0,
                 double a1,double a2,double a3)
```

Macro:

```
#define degrees_C(x) (((x) - 32)*(5.0/9.0))
```

NOTES

1. A program with several modules is easier to read and understand than one long `main` function.
2. Select the name of the function to indicate the purpose of the function.
3. Use a special line, such as a line of dashes, to separate programmer-defined functions from the `main` function and other programmer-defined functions.
4. Use a consistent order for functions. For instance, place the `main` function first, followed by additional functions in the order in which they are referenced.
5. Use parameter identifiers in prototype statements to help document the order and definition of the parameters.
6. List the function prototypes on separate lines so that they are easy to identify.
7. Use the parameter list instead of external variables to transmit information to a function.

DEBUGGING NOTES

1. If you are having difficulty understanding the error messages from a compiler, try running the program on another compiler to obtain different error messages.
2. When debugging a long program, add comment indicators (/* and */) around some sections of the code so that you can focus on other parts of the program.
3. Test a complicated function by itself using a driver program.
4. Make sure that the value returned from a function matches the function return type. If necessary, use the cast operator to convert a value to the proper type.
5. Functions can be defined before or after the `main` function, but not within it.
6. Always use function prototype statements to avoid errors in parameter passing.
7. Use `printf` statements to generate memory snapshots of the actual parameters before a function is referenced, and of the formal parameters at the beginning of the function.
8. Carefully match the type, order, and number of actual parameters with the formal parameters of a function.
9. In a macro definition, each argument and the entire body should be enclosed in its own set of parentheses.
10. System-dependent limitations can occasionally cause problems with recursive solutions to a problem.

PROBLEMS

SHORT-ANSWER PROBLEMS

True–False Problems

Indicate whether the following statements are true (T) or false (F):

1.	A function is called recursive if it calls itself through a statement in its function definition.	T	F
2.	A function may return zero, one, or two values.	T	F
3.	In a call-by-value reference, a function cannot change the value of an actual parameter.	T	F
4.	A static variable is declared inside a function, but it retains its value from one reference call to another.	T	F

Multiple Choice Problems

Circle the letter for the best answer to complete each statement or for the correct answer to each question.

5. For the function definition, `int myfunc(int x)`, a valid return statement is
   (a) `return(&x);`
   (b) `return(pow(x,3));`
   (c) `return("x");`
   (d) `return;`
   (e) none of the above are valid

6. In a function call, the actual parameters are separated by
   (a) commas.
   (b) semicolons.
   (c) colons.
   (d) spaces.

7. The definition of the statements in which an identifier is known (or can be used) is
   (a) global.
   (b) local.
   (c) static.
   (d) scope.

Program Analysis

Use the following function for Problems 8–11 :

```
/*-----------------------------------------------------------*/
void change(int *i, int j)
{
    int t;
    if(*i>j)
    {
        t=*i;
        *i=j;
        j=t;
    }
}
/*-----------------------------------------------------------*/
```

8. If the initial values are x=8 and y=9, what will be their values after `change(&x, y)` is executed?

9. If the initial values are x=18 and y=9, what will be their values after `change(&x, y)` is executed?

10. If the initial values are x=012 and y=12, what will be the output of the following program segment?

    ```
    ................
        change(&x,y);
        printf("x=%d    y=%d",x,y);
    ................
    ```

11. Can the function serve to swap two integers? Justify your answer.

PROGRAMMING PROBLEMS

Simple Simulations. In the following problems, develop simple simulations using the functions `rand_int` and `rand_float`:

12. Write a program to simulate tossing a "fair" coin. Allow the user to enter the number of tosses. Print the number of tosses that yield heads and the number of tosses that yield tails. What should be the percentage distribution of heads and tails?

13. Write a program to simulate tossing two coins that have been weighted such that they land with heads up 70% of times. Assume that a total of 50 tosses have been made. Print the number of tosses that yield both tails.

14. Write a program to simulate rolling a six-sided "fair" die with one dot on one side, two dots on another side, three dots on another side, and so on. Allow the user to enter the number of rolls. Print the number of rolls that gave one dot, the number of rolls that gave two dots, and so on. What should be the percentage distribution of the number of dots from the rolls?

15. Write a program to simulate an experiment rolling two six-sided "fair" dice. Allow the user to enter the number of rolls of the dice to simulate. What percentage of the time does the sum of the dots on the dice equal 8 in the simulation?

16. Write a program to simulate drawing 2 balls from a bag that contains red and blue balls. Assume that the bag contains an infinite number of balls and the ratio of the number of red balls to blue balls is 3:2. Assume that the two balls are drawn at random. Allow the user to enter the number of drawings to simulate. What percentage of the time (a) are both balls red? (b) are both balls blue? (c) is one ball red and the other blue?

Component Reliability. The problems that follow specify computer simulations to evaluate the reliability of several component configurations. Use the function `rand_float` developed in this chapter.

17. Write a program that simulates the design shown in Figure 4.14 using a component reliability of 0.8 for components 1 and 2, 0.85 for component 3, 0.75 for component 4 and 0.9 for component 5. Print the estimate of the reliability using 5,000 simulations.

18. Write a program that simulates the design shown in Figure 4.15 using a component reliability of 0.95 for component 1, 0.75 for component 2, 0.8 for component 3, 0.9 for component 4, and 0.85 for component 5. Print the estimate of the reliability using 5000 simulations.

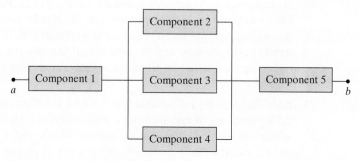

Figure 4.14 *Configuration 1.*

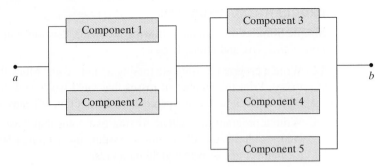

Figure 4.15 *Configuration 2.*

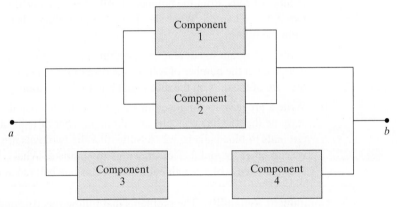

Figure 4.16 *Configuration 3.*

19. Write a program that simulates the design shown in Figure 4.16 using a component relia-bility of 0.95 for all components. Print the estimate of the reliability, using 5000 simula-tions. (The analytical reliability of this system is 0.99976.)

Flight-Simulator Wind Speed. This set of problems relates to a computer simulation of wind speed for a flight simulator. Assume that the wind speed for a particular region can be modeled using an average value and a range of gust values that is added to the average. For example, the wind speed might be 10 miles an hour, with added noise (that represents gusts) that range from -2 miles per hour to 2 miles per hour, as shown in Figure 4.17. Use the func-tion `rand_float` developed in this chapter.

20. Write a program to generate a data file named `wind1.dat` that contains 1 hour of simu-lated wind speeds. Each line of the data file should contain the time in seconds and the corresponding wind speed. The time should start with 0 seconds. The increment in time should be 10 seconds, and the final line of the data file should correspond to 3600 sec-onds. The user should be prompted to enter the average wind speed and the range of val-ues of the gusts.

21. Redo Problem 20, but assume that we want the flight-simulator wind data to include a 0.5% possibility of encountering a small storm at each time step. Therefore, modify the solution to Problem 20 so that the average wind speed is increased by 10 mph for a peri-od of 5 minutes when a storm is encountered. A plot of an example data file with three storms is shown in Figure 4.18.

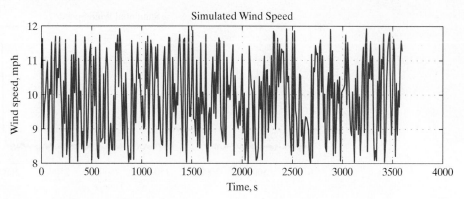

Figure 4.17 *Simulated wind speed.*

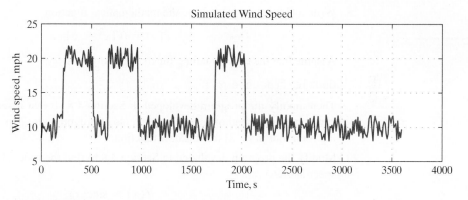

Figure 4.18 *Simulated wind speeds with three storms.*

22. Redo Problem 21, but assume that there is a 1% possibility of encountering a microburst at each time step in a small storm. Therefore, modify the solution to Problem 21 so that the wind speed is increased by 50 mph over the storm values for a period of 1 minute if a microburst is encountered. A plot of an example data file with a microburst within a storm is shown in Figure 4.19.

23. Modify the program in Problem 21 so that the user enters the possibility of encountering a storm.

24. Modify the program in Problem 21 so that the user enters the length in minutes for the duration of a storm.

25. Modify the program in Problem 21 so that the length of a storm is a random number that varies between 3 and 5 minutes.

Roots of Functions. The following problems relate to finding real roots for functions:

26. Write a program to determine the real roots of a quadratic equation, assuming that the user enters the coefficients of the quadratic equation. If the roots are complex, print an appropriate message.

27. Modify Problem 26 so that the program also computes the real and imaginary parts of the roots if they are complex.

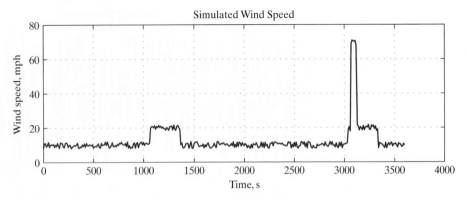

Figure 4.19 *Simulated wind speeds with a microburst.*

28. Write a C function to evaluate the mathematical function

$$f(x) = 0.1x^2 - x \ln x.$$

Assume that the corresponding function prototype is

```
double f(double x);
```

Then modify the program developed in Section 4.8 so that it searches for roots of this new function instead of searching for roots of polynomials. Test the program by searching for a root in [1, 2] for this new function.

29. Modify the program developed in Section 4.8 to find the roots of this function in a user-specified interval:

$$f(x) = \text{sinc}(x).$$

Use the `sinc` function developed in this chapter.

30. In the program developed in Section 4.8 we searched for subintervals for which the function values at the endpoints had different signs; we then estimated the root location to be the midpoint of the subinterval. A more accurate estimate of the root location is usually the intersection of a straight line through the function values with the x axis, as shown in Figure 4.20. Using similar triangles, it can be shown that the intersection point c can be computed using the equation

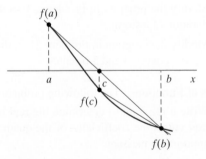

Figure 4.20 *Straight line intersection in (a,b).*

$$c = \frac{a \cdot f(b) - b \cdot f(a)}{f(b) - f(a)}.$$

Modify program chapter4_7 to estimate the root of a subinterval using this approximation.

Factorials. The following set of problems relates to computing factorials. If you did not cover the section on recursion, read the material on pages 200–201 for the definition of a factorial and then review the nonrecursive function for computing the factorial.

31. A convenient approximation for computing the factorial $n!$ for large values of n is given by the Stirling formula

$$n! = \sqrt{2\pi n}\left(\frac{n}{e}\right)^n,$$

where e is the base for natural logarithms, or approximately 2.718282. Write an integer function for computing this approximation to a factorial. Assume that the corresponding prototype is

```
int n_fact(int n);
```

32. Suppose that we have n distinct objects. There are many different orders that we can select to line up the objects in a row. In fact, there are $n!$ orderings, or **permutations**, that can be obtained with n objects. If we have n objects, and select k of the objects, then there are $n!/(n - k)!$ possible orderings of k objects. That is, the number of different permutations of n different objects taken k at a time is $n!/(n - k)!$. Write a function named permute that receives values for n and k, and then returns the number of permutations of the n objects taken k at a time. (If we consider the set of digits $\{1, 2, 3\}$, the different permutations of two digits are $\{1, 2\}, \{2, 1\}, \{1, 3\}, \{3, 1\}, \{2, 3\}$, and $\{3, 2\}$.) Assume that the corresponding prototype is

Permutations

```
int permute(int n,int k);
```

33. Whereas permutations (Problem 32) are concerned with order, combinations are not. Thus, given n distinct objects, there is only one combination of n objects taken n at a time, but there are $n!$ permutations of n distinct objects taken n at a time. The number of **combinations** of n objects, taken k at a time is equal to $n!/((k!)(n - k)!)$. Write a function named combine that receives values for n and k and then returns the number of combinations of the n objects taken k at a time. (If we consider the set of digits $\{1, 2, 3\}$, the different combinations of two digits are $\{1, 2\}, \{1, 3\}$, and $\{2, 3\}$.) Assume that the corresponding prototype is

Combinations

```
int combine(int n, int k);
```

34. The sine of an angle can be computed from the following infinite series:

$$\sin x = x - \frac{x^3}{3!} + \frac{x^5}{5!} - \frac{x^7}{7!} + \cdots.$$

Write a program that reads an angle x (in radians) from the keyboard. Then, in a function, compute the cosine of the angle using the first five terms of this series. Print the value computed along with the value of the cosine computed using the C library function.

35. Modify the program in Problem 34 so that the function accepts the angle x in degrees. Also use terms from the series as long as the absolute value of a term is greater than 0.000001. Then print the output in a tabular form starting from x = 0°, increment by 10° and end at 90°.

CHAPTER FIVE

5

What are some lessons
learned from this situation?

listening...

Crime Scene Investigation: Speech Analysis and Speech Recognition

In the four-color insert in Chapter 1 on biometrics, we discussed the use of speech recognition as a biometric. We also pointed out that speaker recognition (identifying people by their speech signals) is a very challenging problem. Some of the challenges relate to the fact that things can cause a voice to change. For example, when you have a cold, your voice is different; when you are emotional or stressed, your voice is different. Current research has advanced enough that there are some examples of speech recognition (identifying the words in the speech, but not who is speaking) in commercial systems. A new fifth-generation fighter aircraft, the Joint Strike Fighter, is being designed with special speech recognition capabilities. In the initial checkout before a flight, the pilot will load the characteristics of his/her speech, and then during missions, the pilot will be able to verbally ask for information from the on-board computers. Speech analysis is a very common part of both commercial systems and crime scene investigations. For example, speech analysis can include analyzing audio signals to reduce the noise in the signals. It can also isolate background sounds such as sirens or aircraft noise to attempt to determine the location where the signal was collected. If there are a lot of speech data to analyze, automated routines can be written to do a number of things such as convert speech signals to text, determine the language being spoken, determine if the speaker is male or female, and even identify a regional dialect in the speech. In this chapter, we will develop a C program to perform some common speech analysis routines.

ARRAYS AND MATRICES

CHAPTER OUTLINE

OBJECTIVES *In this chapter, we develop problem solutions containing:*

- one-dimensional arrays,
- programmer-defined modules for the statistical analysis of data,
- functions that sort information and search for information,

- two-dimensional arrays,
- vector and matrix computations, and
- techniques for solving a system of simultaneous equations.

5.1 One-Dimensional Arrays

When solving engineering problems, it is important to be able to visualize the data related to the problem. Sometimes the data consist of just a single number, such as the radius of a circle. At other times, the data may be a coordinate in a plane that can be represented as a pair of numbers, with one number representing the *x*-coordinate and the other number representing the *y*-coordinate. There are also times when we want to work with a set of similar data values, but we do not want to give each value a separate name. For example, suppose that we have a set of 100 temperature measurements that we want to use to perform several computations.

*Optional section.

Array

One-Dimensional
Array

Obviously, we do not want to use 100 different names for the temperature measurements, so we need a method for working with a group of values using a single identifier. One solution to this problem uses a data structure called an **array**. A **one-dimensional array** can be visualized as a list of values arranged in either a row or a column, as follows:

5	0	-1	2	15	2
s[0]	s[1]	s[2]	s[3]	s[4]	s[5]

t[0]	0.0
t[1]	0.1
t[2]	0.2
t[3]	0.3

'a'	'e'	'i'	'o'	'u'
v[0]	v[1]	v[2]	v[3]	v[4]

Elements

Subscripts

We assign an identifier to an array, and then we distinguish between **elements** or values in the array using **subscripts**. In C, the subscripts always start with 0 and increment by 1. Thus, by using the example arrays, the first value in the s array is referenced by s[0], the third value in the t array is referenced by t[2], and the last value in the array v is referenced by v[4].

Arrays are convenient for storing and handling large amounts of data, so there is a tendency to use them in algorithms when they are not necessary. Arrays are more complicated to use than simple variables, and thus make programs longer and more difficult to debug. Therefore, use arrays only when it is necessary to have the complete set of data available in memory.

Definition and Initialization

An array is defined using declaration statements. The identifier is followed by an integer expression in brackets that specifies the number of elements in the array. Note that all elements in an array must be the same data type. The declaration statements for the three example arrays are as follows:

```
int s[6];
char v[5];
double t[4];
```

An array can be initialized with declaration statements or with program statements. To initialize the array with a declaration statement, the values are specified in a sequence that is separated by commas and enclosed in braces. To define and initialize the sample arrays s, v, and t, use the following statements:

```
int s[6]={5,0,-1,2,15,2};
char v[5]={'a','e','i','o','u'};
double t[4]={0.0,0.1,0.2,0.3};
```

If the initializing sequence is shorter than the array, then the rest of the values are initialized to zero. Hence, if we want to define an integer array of 100 values, where each value is also initialized to zero, we would use the following statement:

```
int s[100]={0};
```

If an array is specified without a size, but with an initialization sequence, the size is defined to be equal to the number of values in the sequence, as follows:

```
int s[]={5,0,-1,2,15,2};
double t[]={0.0,0.1,0.2,0.3};
```

The size of an array must be specified in the declaration statement by using either a constant within brackets or by an initialization sequence within braces.

Arrays can also be initialized with program statements. For example, suppose that we want to fill a `double` array g with the values 0.0, 0.5, 1.0, 1.5, ..., 10.0. Because there are 21 values, listing the values on the declaration statement would be tedious. Thus, we use the following statements to define and initialize this array:

```
/*  Declare variables.  */
int k;
double g[21];
...
/*  Initialize the array g.  */
for (k=0; k<=20; k++)
   g[k] = k*0.5;
```

It is important to recognize that the condition in this `for` statement must specify a final subscript value of 20, and not 21, since the array elements are g[0] through g[20]. It is a common mistake to specify a subscript that is one value more than the largest valid subscript, and this error can be very difficult to find because it accesses values outside the array. Accessing values outside of an array can produce execution errors such as "segmentation fault" or "bus error". More often, this error is not detected during the program execution, but will cause unpredictable program results, since your program has modified a memory location outside of your array. It is important to be careful about exceeding the array subscripts. In our programs, we select conditions in `for` loops that specifically use the final value as a reminder to ourselves to carefully write the condition to avoid errors. Thus, in this example, we use the condition k<=20 instead of k<21, although both work properly. Also, we will generally use k as the subscript for a one-dimensional array.

Arrays are often used to store information that is read from data files. For example, suppose that we have a data file named `sensor3.txt` that contains 10 time and motion measurements collected from a seismometer. To read these values into arrays named `time` and `motion`, we could use these statements:

```
/*  Declare variables.  */
int k;
double time[10], motion[10];
FILE *sensor;
...
/*  Open file and read data into arrays.  */
sensor = fopen("sensor3.txt","r");
for (k=0; k<=9; k++)
   fscanf(sensor,"%lf %lf",&time[k],&motion[k]);
```

PRACTICE!

Show the contents of the arrays defined in each of the following sets of statements:

1. `int x[10]={-5,4,3};`
2. `char letters[]={'a','b','c'};`
3. `double z[4];`

   `. . .`
   ```
   z[1] = -5.5;
   z[2] = z[3] = fabs(z[1]);
   ```
4. `int k;`
   `double time[9];`

   `...`
   ```
   for (k=0; k<=8; k++)
      time[k] = (k-4)*0.1;
   ```

Computations and Output

Computations with array elements are specified just like computations with simple variables, but a subscript must be used to specify an individual array element. To illustrate, the program that follows reads an array y of 100 floating-point values from a data file, determines the average value of the array, and then stores it in y_ave. Then, the program will determine the number of values in the array y that are greater than the average, count them and print the count. If the purpose of this program had been to determine the average of the values in the data file, an array would not have been necessary. The loop to read values could read each value into the same variable and add its value to a sum before the next value is read. However, because we needed to compare each value to the average in order to count the number of values greater than the average, an array was needed to access each value again.

```
/*-------------------------------------------------------------*/
/*  Program chapter5_1                                         */
/*                                                             */
/*  This program reads 100 values from a data file and         */
/*  determines the number of values greater than the average.  */

#include <stdio.h>
#define N 100
#define FILENAME "lab1.txt"

int main(void)
{
    /*  Declare and initialize variables.  */
    int k, count=0;
    double y[N], y_ave, sum=0;
    FILE *lab;
```

```
/*  Open file, read data into an array,  */
/*  and compute a sum of the values.     */
lab = fopen(FILENAME,"r");
if (lab == NULL)
   printf("Error opening input file. \n");
else
{
   /*  Input and process data.  */
   for (k=0; k<=N-1; k++)
   {
      fscanf(lab,"%lf",&y[k]);
      sum += y[k];
   }

   /*  Compute average and count values that  */
   /*  are greater than the average.          */
   y_ave = sum/N;
   for (k=0; k<=N-1; k++)
      if (y[k] > y_ave)
         count++;

   /*  Print count and close file.  */
   printf("%d values greater than the average \n",count);
   fclose(lab);
}

/*  Exit program.  */
return 0;
}
/*-------------------------------------------------------------*/
```

Array values are printed using a subscript to specify the individual value desired. For example, to print the first and last values of the array y from the previous example, we would use the following statement:

```
printf("first and last array values: \n");
printf("%f %f \n",y[0],y[N-1]);
```

To print all 100 values of y, one per line, we use the following loop:

```
printf("y values: \n");
for (k=0; k<=N-1; k++)
   printf("%f \n",y[k]);
```

When printing a large array, such as this one, we probably would like to print several numbers on the same line. To invoke the modulus operator to skip to a new line before each group of five values is printed, we would use the following statements:

```
printf("y values: \n");
for (k=0; k<=N-1; k++)
   if (k%5 == 0)
      printf("\n %f ",y[k]);
   else
      printf("%f ",y[k]);
printf("\n");
```

Similar statements can also be used to write array values to a data file. For example, to print the value of y[k] on a line in a data file with a file pointer sensor, we use the following statement:

```
fprintf(sensor,"%f \n",y[k]);
```

Since the newline indicator is included, the next value written to the file will be on a new line.

The number of elements in an array is used in the array declaration and is used in loops to access the elements in the array. If the number of elements is changed, then there are several places in the program that need to be modified. Changing the size of an array is simplified if a symbolic constant is used to specify the size of the array. To change the size, only the preprocessor directive needs to be changed. This style suggestion is especially important in programs that contain many modules, or in programming environments in which several programmers are working on the same software project. Many of the following programs illustrate the use of a symbolic constant to define the size of an array.

Table 5.1 gives an updated precedence order that includes subscript brackets. Brackets and parentheses are associated before the other operators. If parentheses and brackets are in the same statement, they are associated from left to right; if they are nested, the innermost set is evaluated first.

PRACTICE!

Assume that the variable k and the array s have been defined with the following statement:

```
int k, s[]={3,8,15,21,30,41};
```

Using a hand calculation, determine the output for each of the following sets of statements:

```
1. for (k=0; k<=5; k+=2)
      printf("%d %d \n",s[k],s[k+1]);

2. for (k=0, k<=5; k++)
      if (s[k]%2 == 0)
         printf("%d ",s[k]);
   printf("\n");
```

Function Arguments

When the information in an array is passed to a function, two parameters are usually used; one parameter specifies a particular array, and the other parameter specifies the number of elements used in the array. By specifying the number of elements of the array, the function becomes more flexible. For example, if the function specifies an integer array, then the function can be used with any integer array; the parameter that specifies the number of elements assures that we use the correct size. Also, the number of elements used in an array may vary from one time to another. For example, the array may use elements read from a data file; the

Table 5.1 Operator Precedence

Precedence	Operation	Associativity
1	( ) [ ]	Innermost first
2	++ -- + - ! (type)	Right to left (unary)
3	* / %	Left to right
4	+ -	Left to right
5	< <= > >=	Left to right
6	== !=	Left to right
7	&&	Left to right
8	||	Left to right
9	?:	Right to left
10	= += -= *= /= %=	Right to left
11	,	Left to right

number of elements then depends on the specific data file used when the program is run. In all these examples, the array must be declared to be a maximum size, and then the actual number of elements used can be less than or equal to that maximum size.

The next program reads an array from a data file and then references a function to determine the maximum value in the array. The variable npts is used to specify the number of values in the array; the value of npts can be less than or equal to the defined size of the array, which is 100. The function has two arguments—the name of the array and the number of points in the array, as indicated in the function prototype statement.

This program assumes that there will not be more than 100 values in the file; otherwise, this program will not work correctly. Arrays must be specified to be as large as, or larger than, the maximum number of values to be read into them.

The purpose of program chapter5_2 is to illustrate the use of an array as a function argument. If the purpose of this program was to determine the maximum of the data values in the file, an array would not have been necessary; the maximum could have been determined as the data values were read.

```
/*-------------------------------------------------------------*/
/*  Program chapter5_2                                         */
/*                                                            */
/*  This program reads values from a data file and            */
/*  determines the maximum value with a function.             */

#include <stdio.h>
#define N 100
#define FILENAME "lab2.txt"

int main(void)
{
   /* Declare variables and function prototype.  */
   int k=0, npts;
   double y[N];
   FILE *lab;
   double max(double x[],int n);
```

```
       /*  Open file, read data into an array.  */
       lab = fopen(FILENAME,"r");
       if (lab == NULL)
          printf("Error opening input file. \n");
       else
       {
          while ((fscanf(lab,"%lf",&y[k])) == 1)
              k++;
          npts = k;

          /*  Find and print the maximum value.  */
          printf("Maximum value: %f \n",max(y,npts));

          /*  Close file and exit program.  */
          fclose(lab);
       }
       /*  Exit program.  */
       return 0;
}

/*-------------------------------------------------------------*/
/*  This function returns the maximum value in an array x       */
/*  with n elements.                                           */

double max(double x[],int n)
{
   /*  Declare variables.  */
   int k;
   double max_x;

   /*  Determine maximum value in the array.  */
   max_x = x[0];
   for (k=1; k<=n-1; k++)
      if (x[k] > max_x)
         max_x = x[k];

   /*  Return maximum value.  */
   return max_x;
}
/*-------------------------------------------------------------*/
```

There is a very significant difference between using simple values as function parameters and using arrays as parameters. When a simple variable is used as a parameter, the value is passed to the formal argument in the function, and thus the value of the original variable cannot

Call-by-value

be changed; this is a **call-by-value** reference. When an array is used as a parameter, the memory address of the array is passed to the function instead of the entire set of values in the array.

Call-by-address

Therefore, the function references values in the original array; this is a **call-by-address** reference. Because a function accesses the original array values, we must be very careful that we do

not inadvertently change values in an array within a function. Of course, there may be occasions when we wish to change the values in the array, as we will see in examples in this chapter.

he goal of this text is to teach you how to solve problems using the C computer language. To make the roblems more interesting, we are using a theme of Crime Scene Investigation, or CSI. In this nsert, we will discuss some of the technology behind crime scene investigation. In addition, t the beginning of each chapter, we present a short discussion that addresses some aspect f crime scene analysis. Then, later in the chapter, we solve a related problem using the C omputer language.

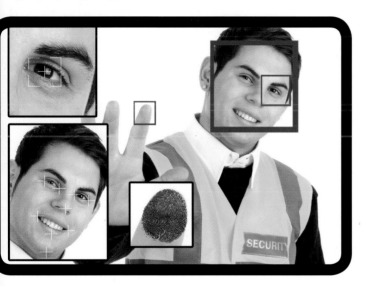

BIOMETRICS FOR IDENTIFYING PEOPLE

Identifying people involved in a crime is an important part of crime scene investigation. These people may be the victims, the suspects, or the witnesses. Biometrics is a term used to describe the physical or behavioral characteristics that can be used to identify a person. Common biometrics are fingerprints and face images. Most biometric identification can be done automatically by a computer, or it can be done by people trained in making

entifications. Iris recognition is another commonly used biometric, but it is only done automatically.

ll biometric identification systems have two processes: enrollment and identification. In the nrollment process, an individual's biometric (such as a fingerprint) is collected and stored in master database. In the identification process, an individual's fingerprint is collected, and is then compared to all the fingerprints in the master database. If there is a close enough natch, then the system will identify the individual as one that is in the database.

COMMON BIOMETRICS

ou are probably familiar with some of the most common biometrics—fingerprint, face, iris, DNA, nd speech. For these five common biometrics, we now point out some of the characteristics of he biometric that make it unique enough to identify someone, and we discuss the current state f the technology to collect the biometric and to automatically do the identification. As you would xpect, some biometrics are much more accurate than others.

Fingerprints

Fingerprints are the oldest method of identification. Today's automatic fingerprint identification system (AFIS) that is used by the FBI is based on the Henry classification system. This system classifies a fingerprint first as one of several types such as a whorl, arch, or loop. After a fingerprint has been classified with one of these categories, a number of minutiae points are then identified. Minutiae points are details within the fingerprint, such as a fingerprint ridge bifurcation (or split) or an island that is a part of a ridge that is not connected to other ridges. These minutiae points are then used to do the actual identification of an individual. The more minutiae points that match in a fingerprint from a crime scene to a suspect, the higher the likelihood that the suspect

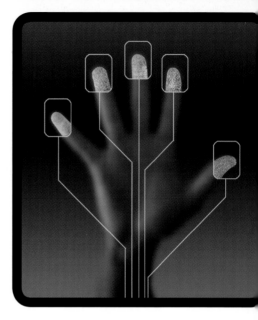

is identified. (In Chapter 7, we present more information on fingerprints, and we develop a program to solve a problem involving identification using all ten fingerprints.)

Face Recognition

Face recognition is another commonly used technique to identify an individual from an image or from a single frame (or image) taken from a surveillance video. One common technique is to use the ratio of distances between key points on a face. These ratios might include the distance between the eyes divided by the distance between the nose and the chin. Since these measurements are ratios, they can be computed from images of different sizes. The computer programs that compute these measurements must be able to locate a face in an image, and then also locate the eyes and other key points on the face. Face recognition is not a

accurate as fingerprint recognition. Many face recognition systems select the top three matches to an unknown from their master database, and then a person decides if the unknown face is one of the three matches. This process is often used when there is a security guard to review the faces selected by the automatic face recognition system. (In Chapter 3, we present more information on face recognition, and we develop a C program to solve a problem that computes and compares ratios determined from measurements of the face.)

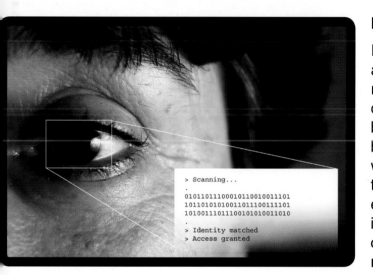

Iris Recognition

Iris recognition is one of the most accurate biometrics. Commercial iris recognition systems use an infrared camera to collect an image of the eye because an infrared image is not affected by color. (Digital camera images do not work well with iris recognition because the patterns in the iris of a dark brown eye are not clearly visible.) The infrared image is a black and white image that clearly shows the patterns in the iris. Iris recognition is often used for entry into secured areas. It is also being used in India to develop a database of all Indian citizens as part of the ongoing census collection. In the final Indian database, more than 1 billion images will be stored in a master database. (In Chapter 4, we present more information on iris recognition, and we develop a C program to solve a problem involving the identification of the boundary between the pupil and the iris.)

DNA

DNA is the most accurate biometric, but it typically requires a number of days to obtain the information regarding a potential match. DNA (deoxyribonucleic acid) is the genetic material found in cells. This genetic information is contained in genes that are double-helix strands composed of base pairs (adenine bonded with thymine or cytosine bonded with guanine) arranged in a step-like manner with phosphate groups along the side. These base pairs can occur in any order and represent the hereditary information in the gene. The number of base pairs in human DNA has been estimated to be around 3 billion. Matching long strings of these double-helix strands in two samples of DNA provides the most accurate biometric comparison. (In Chapter 6, we present more information on DNA, and we develop a C program to match segments of DNA.)

Speech Recognition

Speech is also a biometric. Your speech is affected by your vocal cords, your mouth, your tongue, your teeth, your nasal cavity, and other parts of your anatomy. Therefore, it can also be used to identify a person. However, we know that people can train themselves to sound like someone else, so a positive identity using speech recognition is a complex process. It requires collecting a number of examples of speech by a person and then developing a model of his/her speech. This can be done very accurately but requires significant amounts of effort. It is much easier to do speech recognition (determine what words are being spoken) than it is to do speaker recognition (determine who is speaking the words). For example, Apple's new iPhone 4S has a voice command feature named "Siri." Th feature allows users to make calendar appointments, ask for directions, and send message all using verbal commands. (In Chapter 5, we present more information on speech processing and we develop a C program to solve a problem involving speech analysis.)

Scanning for verification...

OTHER BIOMETRICS

There are also many other biometrics that can be used to accurately identify someone. In Chapter we present information on forensic identification with bones, and we develop a C program to solv a problem involving estimation of height from bones. In Chapter 8, we present information o hand recognition, and we develop a program on identification from a number of measurement on the hand. There are also many other biometrics. For example, gait is a biometric. How man times have you seen someone walking ahead of you and knew who it was just by the way he or she walked? Gait is both a physical and a behavioral biometric because it is based somewh on the physical structure of the body, but it is also a biometric that can be changed by a perso deliberately walking differently. Handwriting is another biometric that is primarily a behaviora characteristic. Your handwriting strokes are patterns that you have developed over years o writing, and it is very hard for someone to fool a trained expert in handwriting analysis.

PRACTICE!

Assume that we have defined the following variables:

```
int k=6;
double data[]={1.5,3.2,-6.1,9.8,8.7,5.2};
```

Using the max function presented in this section, give the value of each of the following expressions:

1. `max(data,6);`
2. `max(data,5);`
3. `max(data,k-3);`
4. `max(data,k%5);`

5.2   Problem Solving Applied: Hurricane Categories

Hurricanes

Hurricanes are tropical storms with very strong winds and heavy rains. (They are called typhoons in the western North Pacific Ocean and cyclones in the Indian Ocean.) These tropical storms, or cyclones, are low-pressure cells that typically form in the summer and early fall. The large rotating air masses are easily seen on satellite images, and the storms are carefully tracked because of the potential for damage in populated areas. If the storm's winds are between 38 and 74 miles per hour, it called a tropical storm; if the winds exceed 74 miles per hour, the storm is a tropical cyclone, or hurricane. The **Saffir–Simpson scale** defines categories of hurricane intensity based on the wind speed. In this section, we define the Saffir–Simpson scale in more detail, and we then develop a program that reads a data file containing current storms and their peak wind speeds. Finally we will analyze the wind speeds, and we will print a report with information on the storms that are strong enough to be classified as hurricanes.

Saffir–Simpson scale

The Saffir–Simpson scale of hurricane intensities is used to classify hurricanes according to the amount of damage that the storm is likely to generate if it hits a populated area. The main characteristics of the five categories are as follows:

Category 1   wind speeds of 74 to 95 mph
storm surge of 4 to 5 feet
minimal damage to property

Category 2   wind speeds of 96 to 110 mph
storm surge of 6 to 8 feet
moderate damage to property

Category 3   wind speeds of 111 to 130 mph
storm surge of 9 to 12 feet
extensive damage to property

Category 4   wind speeds of 131 to 155 mph
storm surge of 13 to 18 feet
extreme damage to property

Table 5.2 Strong Hurricanes in the U.S. during 1950–2002

Hurricane	Year	Category
Hazel	1954	4
Audrey	1957	4
Donna	1960	4
Carla	1961	4
Camille	1969	5
Celia	1970	3
Frederic	1979	3
Allen	1980	3
Gloria	1985	3
Hugo	1989	4
Andrew	1992	5
Opal	1995	3

Category 5   wind speeds over 155 mph
storm surge greater than 18 feet
catastrophic damage to property

Table 5.2 contains a list of the 12 strongest hurricanes to hit the United States from 1950 to 2002.

The most destructive hurricane ever in the United States was Hurricane Katrina in August 2005. The winds were over 125 mph, but the winds were not the most destructive part of the hurricane. The rains caused Lake Pontchartrain near New Orleans to flood, and a number of levees were breached. Over 80% of New Orleans was underwater, and over 1,800 lives were lost.

Each year, there are over 100 storms with the potential to become hurricanes. Write a program that will read a data file containing information on the current storms; assume that the data file consists of an identification number and the highest wind speed (in miles per hour) measured from the storm. The program should print a list of all storms that have wind speeds high enough to classify them as hurricanes. In addition to the identification number (an integer), print the peak wind speed and the corresponding hurricane intensity category. Also, print an asterisk after the identification number of the hurricane with the largest wind speed.

1. PROBLEM DESCRIPTION

Determine which storms are hurricanes, using a data file of current storm information.

2. INPUT/OUTPUT DESCRIPTION

The I/O diagram shows the data file as the input and the hurricane information as output.

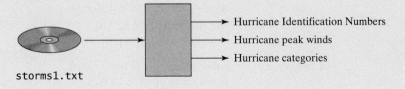

3. HAND EXAMPLE

Assume that the data file contains the following five sets of data:

Identification	Peak Wind
142	38
153	135
162	59
177	76
181	63

Our program should produce the following report:

Storms that Qualify as Hurricanes

Identification	Peak Wind (mph)	Category
153*	135	4
177	76	1

Recall that the asterisk identifies the storm with the largest wind speed.

4. ALGORITHM DEVELOPMENT

We first develop the decomposition outline because it divides the solution into a series of sequential steps. To print the information for storms that are hurricanes, we do not need an array. We could determine this information as we read through the file, since the hurricane status is dependent only on the wind speed. However, since we are required to use an asterisk to indicate the storm with the peak wind, we need to store all the information in arrays. After we have determined the maximum wind speed, we can then go back through the data, and we can print the hurricane information with the asterisk on the correct line.

Decomposition Outline

1. Read the storm data into arrays, and determine the maximum wind speed.
2. Compute the intensity categories and print information for storms that are hurricanes; place an asterisk at the maximum.

We will put the steps that determine the intensity category into a function:

Refinement in Pseudocode

main: if file cannot be opened
        print error message
    else
            read data into arrays, and determine max speed, npts
            set k to 0
            while k ≤ npts-1
                if mph[k] > 74
                    if mph[k] = max speed
                        print id[k], *, mph[k], cateory(mph[k])
                    else
                        print id[k], mph[k], category(mph[k])
            add 1 to k

```
category(mph):
      category =1;
      if mph ≥ 96
             category =2
      if mph ≥ 111
             category =3
      if mph ≥ 131
             category = 4
      if mph ≥ 155
             category = 5
```

The steps in the pseudocode are now detailed enough to convert into C:

```
/*------------------------------------------------------------------*/
/*  Program chapter5_3                                              */
/*                                                                  */
/*  This program reads storm values from a data file               */
/*  and prints a hurricane report.                                 */

#include <stdio.h>
#define N 500
#define FILENAME "storms1.txt"

int main(void)
{
   /*  Declare and initialize variables.  */
   int k=0, npts, id[N];
   double mph[N], max=0;
   FILE *storms;
   int category(double speed);

   /*  Open file, read data into an array,  */
   /*  and determine maximum wind speed.     */
   storms = fopen(FILENAME,"r");
   if (storms == NULL)
      printf("Error opening input file. \n");
   else
   {
      /*  Read data and determine maximum mph.  */
      while ((fscanf(storms,"%d %lf",&id[k],&mph[k])) == 2)
      {
         if (mph[k] > max)
            max = mph[k];
         k++;
      }
      npts = k;
```

```
            /*  Print hurricane report.  */
            if (max >= 74)
            {
               printf("Storms that Qualify as Hurricanes \n");
               printf("Identification    Peak Wind (mph)  Category \n");
            }
            else
               printf("No hurricanes in the file \n");
            for (k=0; k<=npts-1; k++)
               if (mph[k] >= 74)
                  if (mph[k] == max)
                     printf("%d*                %.0f                %d \n",
                            id[k],mph[k],category(mph[k]));
                  else
                     printf("%d                 %.0f                %d \n",
                            id[k],mph[k],category(mph[k]));

            /*  Close file.  */
            fclose(storms);
         }

      /*  Exit program.  */
      return 0;
}
/*-----------------------------------------------------------------*/
/*  This function determines the hurricane intensity               */
/*  category.                                                      */

int category(double speed)
{
   /*  Declare variables.  */
   int intensity=1;

   /*  Determine category.  */
   if (speed >= 96)
      intensity = 2;
   if (speed >= 111)
      intensity = 3;
   if (speed >= 131)
      intensity = 4;
   if (speed >= 155)
      intensity = 5;

   /*  Return intensity.  */
   return intensity;
}
/*-----------------------------------------------------------------*/
```

5. TESTING

We will test our program with a file containing the hand example. This produces the following interaction:

```
Storms that Qualify as Hurricanes
Identification    Peak Wind (mph)    Category
153*              135                4
177               76                 1
```

MODIFY!

These problems relate to the preceding program for printing a hurricane intensity report.

1. Modify the program so that it only prints the information for the hurricane with the largest wind speed.
2. Modify the program so that it also prints the number of storms from the data file.
3. Modify the program so that it also prints the number of hurricanes from the data file.
4. Modify the program so that it prints the number of hurricanes in each category.

5.3 Problem Solving Applied: Molecular Weights

Chemical reactions play an important role in many scientific and engineering systems. Understanding and controlling chemical reactions allows petroleum engineers to improve the efficiency of the refineries necessary to process oil and gas resources. Understanding the behavior and reactions of fully ionized gases at very high temperatures under the influence of strong magnetic fields will be an important step in developing controlled nuclear fusion. In genetic engineering, the identification of amino acids in DNA is a key step in developing techniques to synthesize new products.

Computing the molecular weight from a chemical formula is a common task in any application that involves chemical reactions. Write a program that reads a chemical formula from the keyboard and then computes the corresponding molecular weight. We will assume that this program will be used in a genetic engineering laboratory working with the amino acids in proteins. Amino acids contain only atoms of oxygen (O), carbon (C), nitrogen (N), sulfur (S), and hydrogen (H). For example, the chemical formula for alanine is $O_2C_3NH_7$; thus, alanine contains two atoms of oxygen, three atoms of carbon, one atom of nitrogen, and seven atoms of hydrogen. (Table 5.3 contains the atoms in amino acids.) The input to the program is a set of characters that specifies the chemical formula. The valid characters are the abbreviations O, C, N, S, and H, and we will allow these characters to be either uppercase or lowercase. One or two digits that specify the number of atoms in the element may also follow each element. Thus, the input characters for alanine could include O2C3NH7 or o2c3nh7. Errors occur if the element is not one of the five specified elements or if the formula begins with a number. Use the following molecular weights:

oxygen	15.9994	sulfur	32.066
carbon	12.011	hydrogen	1.00794.
nitrogen	14.00674		

Table 5.3 Amino Acid Molecules

Amino Acid	O	C	N	S	H
Alanine	2	3	1	0	7
Arginine	2	6	4	0	15
Asparagine	3	4	2	0	8
Aspartic	4	4	1	0	6
Cysteine	2	3	1	1	7
Glutamic	4	5	1	0	8
Glutamine	3	5	2	0	10
Glycine	2	2	1	0	5
Histidine	2	6	3	0	10
Isoleucine	2	6	1	0	13
Leucine	2	6	1	0	13
Lysine	2	6	2	0	15
Methionine	2	5	1	1	11
Phenylalanine	2	9	0	1	11
Proline	2	5	1	0	10
Serine	3	3	1	0	7
Threonine	3	4	1	0	9
Tryptophan	2	11	2	0	11
Tyrosine	3	9	1	0	11
Valine	2	5	1	0	11

I. PROBLEM STATEMENT

Compute the molecular weight of a chemical formula for an amino acid.

2. INPUT/OUTPUT DESCRIPTION

The input to the program is a chemical formula entered from the keyboard, and the output is the corresponding molecular weight displayed on the computer screen.

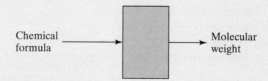

Chemical formula → Molecular weight

3. HAND EXAMPLE

If the input is the chemical formula for alanine, which is O2C3NH7, then the corresponding output should be computed in the following way:

Two atoms of oxygen:

$$2 \cdot 15.9994 = 31.9988$$

Three atoms of carbon:

$$3 \cdot 12.011 = 36.033$$

One atom of nitrogen:

$$1 \cdot 14.00674 = 14.00674$$

Seven atoms of hydrogen:

$$7 \cdot 1.00794 = 7.05558$$

The total molecular weight is 89.09412.

4. ALGORITHM DEVELOPMENT

We first develop the decomposition outline, because it breaks the solution into a series of sequential steps:

Decomposition Outline

1. Read the chemical formula.
2. Compute the molecular weight.
3. Print the molecular weight.

Parsing

Step 1 involves reading the characters from the keyboard and storing them in an integer array. Step 2 involves examining the characters (also called **parsing**) to first determine each element and then determine the number of atoms for each element. Because this comparison will require several steps, we will implement this computation in a function. In the main function, we will multiple the individual atomic weights by the appropriate number of atoms; then we will add the values to a total. Note that we will need to convert the number of atoms from character digits to a numerical value in order to perform the multiplication. Since digits are consecutive in the collating sequence, the numerical equivalent of a digit can be obtained by subtracting the value `'0'` from it.

Step 3 involves printing the final molecular weight. In this step, we will also print an error message if the input characters could not be analyzed properly. The refinement in pseudocode for the main function and the `atomic_wt` function can now be developed:

Refinement in Pseudocode

main: *print message to the user*
      set k to 0
      while more characters
         read formula[k]
         increment k by 1
      set k to 0
      while more characters
         convert current character to uppercase
         determine atomic weight
         determine if digits follow
         compute weight and add to total
      print molecular weight

The steps in the pseudocode are now detailed enough to convert into C.

```c
/*------------------------------------------------------------*/
/*  Program chapter5_4                                        */
/*                                                            */
/*  This program computes the molecular weight of an amino    */
/*  acid from its chemical formula.                           */

#include <stdio.h>
#include <ctype.h>
#define NEWLINE '\n'

int main(void)
{
   /*  Declare variables and function prototypes.  */
   int k=0, formula[20], n, current=0, done=0, d1, d2;
   double error=0, weight, total=0;
   double atomic_wt(int atom);

   /*  Read chemical formula from keyboard.  */
   printf("Enter chemical formula for amino acid: \n");
   while ((formula[k]=getchar()) != NEWLINE)
      k++;
   n = k;

   /*  Identify individual elements and add weights.  */
   while (current<=(n-1) && done==0)
   {
      if (isalpha(formula[current]))
      {
         formula[current] = toupper(formula[current]);
         weight = atomic_wt(formula[current]);
         if (weight == 0)
            done = 1;
         else
         {
            if (current < n-1)
               d1 = isdigit(formula[current+1]);
            else
               d1 = 0;
            if (d1 && current<(n-2))
               d2 = isdigit(formula[current+2]);
            else
               d2 = 0;
            if (d1 && d2)
            {
               weight *= ((formula[current+1]-'0')*10 +
                          (formula[current+2]-'0'));
               current += 3;
            }
            else
```

```
                        if (d1)
                        {
                            weight *= (formula[current+1]-'0');
                            current += 2;
                        }
                        else
                            current++;
                }
                total += weight;
            }
            else
                done = 1;
        }
        /*  Print formula and weight.  */
        printf("Formula: \n");
        for (k=0; k<=n-1; k++)
            putchar(formula[k]);
        printf("\n");
        if (done == 0)
            printf("Molecular Weight: %f \n",total);
        else
            printf("Error in formula. \n");

        /*  Exit program.  */
        return 0;
}
/*-----------------------------------------------------------*/
/*  This function returns the molecular weight of an element  */
/*  in an amino acid.                                         */

double atomic_wt(int atom)
{
    /*  Declare and initialize variables.  */
    int k=0, element[5]={'H','C','N','O','S'};
    double m_wt[5]={1.00794,12.011,14.00674,
            15.9994,32.066}, weight;

    /*  Search for element.  */
    while (k<=4 && element[k]!=atom)
        k++;

    /*  Return corresponding atomic weight.  */
    if (k <= 4)
        weight = m_wt[k];
    else
        weight = 0;
    return weight;
}
/*-----------------------------------------------------------*/
```

5. TESTING

Using the data from the hand example, the output from the program is as follows:

```
Enter chemical formula for amino acid:
O2C3NH7
Molecular Weight:  89.094116
```

MODIFY!

These problems relate to the program developed in this section for computing the molecular weight of an amino acid.

1. Test the program using an amino acid that has more than nine atoms of one of the elements.
2. Allow the user to compute the molecular weights for several amino acids. The program should stop when a period is entered. (Be sure to tell the user to enter a period when done.)

5.4 Statistical Measurements

Analyzing data collected from engineering experiments is an important part of evaluating the experiments. This analysis ranges from simple computations on the data, such as calculating the average value, to more complicated analyses. Many of the computations or measurements using data are statistical measurements because they have statistical properties that change from one set of data to another. For example, the sine of $60°$ is an exact value that is the same value every time we compute it, but the number of miles to the gallon that we get with our car is a statistical measurement, because it varies depending on parameters such as the temperature, the speed that we travel, the type of road, and whether we are in the mountains or the desert.

Simple Analysis

When evaluating a set of experimental data, we often compute the maximum value, minimum value, mean or average value, and the median. In this section, we develop functions that can be used to compute these values using an array as input. These functions (stored in a file stat_lib.c) will be useful in many of the programs that we develop later in the text and in solutions to problems at the end of the chapters. However, it is important to note that these functions assume that there is at least one value in the array.

Maximum and Minimum.   A function for determining the maximum value in an array was presented in the previous section; a similar function can be written to determine the minimum value. Both functions assume that the array contains double values; simple changes could be used to convert these functions to specify integer values.

Mean value

Average.   The Greek symbol μ (mu) is used to represent the average or **mean value**. An equation, which uses summation notation, is as follows:

$$\mu = \frac{\sum_{k=0}^{n-1} x_k}{n}, \tag{5.1}$$

where

$$\sum_{k=0}^{n-1} x_k = x_0 + x_1 + x_2 + \cdots + x_{n-1}.$$

The average of a set of values is always a floating-point value, even if all the data values are integers. To compute the mean value of a double array of n values, we will use the following function:

```
/*------------------------------------------------------------*/
/*  This function returns the average or mean value of an     */
/*  array x with n elements.                                  */

double mean(double x[],int npts)
{
   /*  Declare and initialize variables.  */
   int k;
   double sum=0;
   /*  Determine mean value.  */
   for (k=0; k<=npts-1; k++)
      sum += x[k];

   /*  Return mean value.  */
   return sum/n;
}
/*------------------------------------------------------------*/
```

Note that the variable sum was initialized to zero in the declaration statement. It could also have been initialized to zero with an assignment statement. In either case, the value of sum is initialized to zero when the function is referenced.

Median

Median.   The **median** is the value in the middle of a group of values, assuming that the values are sorted. If there is an odd number of values, the median is the value in the middle; if there is an even number of values, the median is the average of the values in the two middle positions. For example, the median of the values {1, 6, 18, 39, 86} is the middle value, or 18; the median of the values {1, 6, 18, 39, 86, 91} is the average of the two middle values, or $(18 + 39)/2$, or 28.5. Assume that a group of sorted values are stored in an array and that n contains the number of values in the array. If n is odd, then the subscript of the middle value can be represented by floor(n/2), as in floor(5/2), which is 2. If n is even, then the subscripts of the two middle values can be represented by floor(n/2)-1 and floor(n/2), as in floor(6/2)-1 and floor(6/2), which are 2 and 3, respectively.

The following function determines the median of a set of values stored in an array. We assume that the values are sorted (into either ascending or descending order). If the array is not sorted, a function developed later in this chapter can be referenced from the `median` function to sort the values.

```
/*-----------------------------------------------------------*/
/*  This function returns the median value in the sorted     */
/*  array x with npts elements.                              */

double median(double x[],int npts)
{
   /*  Declare variables.  */
   int k;
   double median_x;

   /*  Determine median value.  */
   k = floor(npts/2);
   if (n%2 != 0)
      median_x = x[k];
   else
      median_x = (x[k-1] + x[k])/2;

   /*  Return median value.  */
   return median_x;
}
/*-----------------------------------------------------------*/
```

Go through this function by hand using the two sets of data values given in this discussion.

Variance and Standard Deviation

One of the most important statistical measurements for a set of data is the variance. Before we give the mathematical definition for variance, we should develop an intuitive understanding. Consider the values of arrays `data1` and `data2`, which are plotted in Figure 5.1. If we attempted to draw a horizontal line through the middle of the values in each plot, this line would be at approximately 3.0. Thus, both arrays have approximately the same average or mean value of 3.0. However, the data in the two arrays clearly have some distinguishing characteristics. The values in `data2` vary more from the mean, or deviate more from the mean value.

Variance

Standard deviation

The **variance** of a set of values is defined as the average squared deviation from the mean; the **standard deviation** is defined as the square root of the variance. Thus, the variance and the standard deviation of the values in `data2` are greater than the variance and standard deviation for the values in `data1`. Intuitively, the larger the variance (or the standard deviation), the further the values fluctuate around the mean value.

Mathematically, the variance is represented by σ^2, where σ is the Greek symbol sigma. The variance for a set of data values (which we assume are stored in an array x) can be computed using the following equation:

$$\sigma^2 = \frac{\sum_{k=0}^{n-1}(x_k - \mu)^2}{n - 1}. \tag{5.2}$$

This equation is a bit intimidating at first, but if you look at it closely, it becomes much simpler. The term $x_k - \mu$ is the difference between x_k and the mean, or the deviation of x_k from

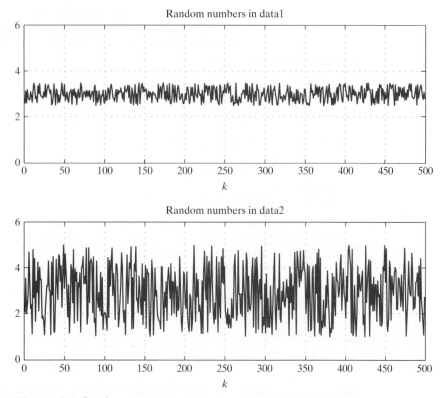

Figure 5.1 *Random sequences.*

the mean. This value is squared so that we always have a positive value. We then add the squared deviations for all data points. This sum is then divided by $n - 1$, which approximates an average. The definition of variance has two forms: The denominator of a sample variance is $n - 1$, and the denominator of a population variance is n. Most engineering applications use the sample variance, as shown in Equation (5.2). Thus, Equation (5.2) computes the average squared deviation of the data from the mean. The standard deviation is defined to be the square root of the variance:

$$\sigma = \sqrt{\sigma^2}. \qquad (5.3)$$

Both the variance and the standard deviation are commonly used in analyzing engineering data, so we give functions for computing both values. Note that the function for computing the standard deviation references the `variance` function and that the `variance` function references the `mean` function; thus, these functions must include the proper function prototype statements. Also, note that there must be at least two values in the array, or the `variance` function will attempt to divide by zero. Here are the two functions:

```
/*-----------------------------------------------------------*/
/*  This function returns the variance of an array x         */
/*  with npts elements.                                      */
```

```
double variance(double x[],int npts)
{
   /*  Declare variables and function prototypes.  */
   int k;
   double sum=0, mu;
   double mean(double x[],int npts);

   /*  Determine variance.  */
   mu = mean(x,npts);
   for (k=0; k<=n-1; k++)
      sum += (x[k] - mu)*(x[k] - mu);

   /*  Return variance.  */
   return sum/(npts-1);
}
/*------------------------------------------------------------*/
/*  This function returns the standard deviation of an array x */
/*  with npts elements.                                        */

double std_dev(double x[],int npts)
{
   /*  Declare function prototypes.  */
   double variance(double x[],int npts);

   /*  Return standard deviation.  */
   return sqrt(variance(x,npts));
/*------------------------------------------------------------*/
```

Custom Header File

The functions developed in this section are frequently used in solving engineering problems. To facilitate their use, we generate a custom header file that contains the prototype statements for these functions. Then, instead of including all the prototype statements in a main function, we can use a preprocessor directive that includes the custom header file.

The custom header file named stat_lib.h contains the following function prototype statements:

```
double max(double x[],int n);
double min(double x[],int n);
double mean(double x[],int n);
double median(double x[],int n);
double variance(double x[],int n);
double std_dev(double x[],int n);
```

The statement that includes these function prototype statements in a main function is

```
#include "stat_lib.h"
```

The use of this custom header is illustrated in the next section.

In addition to accessing the custom header file with the include statement, a program must also have access to the file stat_lib.c, which contains the statistical functions. The specific details of providing this access are system dependent and may involve adding a file name to the operating system command that performs the compilation and linking/loading operations.

PRACTICE!

Assume that the array x is defined and initialized with the following statement:

```
double x[]={2.5,5.5,6.0,6.25,9.0};
```

Using a hand calculation, compute the values returned by the following function references:

1. max(x,5)
2. median(x,5)
3. variance(x,5)
4. std_dev(x,5)
5. min(x,4)
6. median(x,4)

5.5   Problem Solving Applied: Speech Signal Analysis

A speech signal is an acoustical signal that can be converted into an electrical signal with a microphone. The electrical signal can then be converted into a series of numbers that represents the amplitudes of the electrical signal values. These numbers can be stored in data files so that the speech signal can be analyzed using computer programs. Suppose that we are interested in analyzing speech signals for the words "zero," "one," "two," ..., "nine." The goal of this analysis is to develop ways of identifying the correct digit from a data file containing

Utterance   the **utterance** of an unknown digit.

Figure 5.2 contains a plot of an utterance of the digit "zero." The analysis of this type of complicated signal often starts with computing some of the statistical measurements discussed in the last section. Other measurements used with speech signals include the average

Magnitude   **magnitude**, or average absolute value, which is computed as

$$\text{Average magnitude} = \frac{\sum\limits_{k=0}^{n-1}|x_k|}{n}, \tag{5.4}$$

where n is the number of data values.

Power   Another metric used in speech analysis is the average **power** of the signal, which is the average squared value:

$$\text{Average power} = \frac{\sum\limits_{k=0}^{n-1}x_k^2}{n}. \tag{5.5}$$

Zero crossings   The number of **zero crossings** in a speech signal is also a useful statistical measurement. This value is the number of times that the speech signal makes a transition from a negative to a positive value or from a positive to a negative value. A transition from a nonzero value to a zero value is not a zero crossing.

Speech Signal

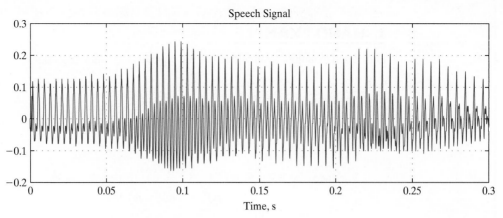

Time, s

Figure 5.2 *Utterance of the word "zero."*

Write a program to read a speech signal from a data file named `zero1.txt`. This file contains values that represent an utterance of the word "zero." Each line of the file contains a single value representing a measurement from the microphone taken in time increments of 0.000125 second, so 8000 measurements represent 1 second of data. The data file contains only valid data, with no header or trailer line; a maximum of 2500 values is contained in the file. Compute and print the following statistical measurements from the file: mean, standard deviation, variance, average power, average magnitude, and number of zero crossings.

I. PROBLEM DESCRIPTION

Compute the following statistical measurements for a speech utterance: mean, standard deviation, variance, average power, average magnitude, and number of zero crossings.

2. INPUT/OUTPUT DESCRIPTION

The I/O diagram shows the data file as the input and the statistical measurements as output.

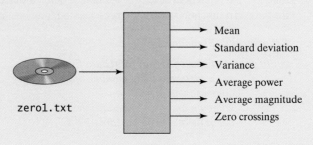

3. HAND EXAMPLE

For a hand example, assume that the file contains the following values:

```
2.5  8.2  -1.1  -0.2  1.5
```

Using a calculator, we can compute the following values:

$$\text{Mean} = \mu = \frac{(2.5 + 8.2 - 1.1 - 0.2 + 1.5)}{5}$$

$$= 2.18$$

$$\begin{aligned}\text{Variance} &= [(2.5 - \mu)^2 + (8.2 - \mu)^2 + (-1.1 - \mu)^2 \\ &\quad + (-0.2 - \mu)^2 + (1.5 - \mu)^2]/4 \\ &= 13.307\end{aligned}$$

$$\text{Standard deviation} = \sqrt{13.307}$$

$$= 3.648$$

$$\text{Average power} = \frac{[(2.5)^2 + (8.2)^2 + (-1.1)^2 + (-0.2)^2 + (1.5)^2]}{5}.$$

$$= 15.398$$

$$\text{Average magnitude} = \frac{(|2.5| + |8.2| + |-1.1| + |-0.2| + |1.5|)}{5}$$

$$= 2.7$$

Number of zero crossings $= 2$.

4. ALGORITHM DEVELOPMENT

We first develop the decomposition outline because it divides the solution into a series of sequential steps:

Decomposition Outline

1. Read the speech signal into an array.
2. Compute and print statistical measurements.

Step 1 involves reading the data file and determining the number of data points. Step 2 involves computing and printing the statistical measurements; use the functions already developed when possible. The structure chart shown in Figure 4.1 illustrates an example of a main function that references several programmer-defined functions. The refinement in pseudocode for the main function and for the necessary additional statistical functions follows:

Refinement in Pseudocode

main: *read speech signal from data file and*
                 determine the number of points, npts
            compute and print mean
            compute and print standard deviation
            compute and print variance

> *compute and print average power*
> *compute and print average magnitude*
> *compute and print zero crossings*
>
> *Additional functions*
> *ave_power(x, npts):*
> > *set sum to zero*
> > *set k to zero*
> > *while k ≤ npts – 1*
> > > *add $(x[k])^2$ to sum*
> > > *increment k by 1*
> > *return sum/npts*
>
> *ave_magn (x, npts):*
> > *set sum to zero*
> > *set k to zero*
> > *while k ≤ npts – 1*
> > > *add | x[k] | to sum*
> > > *increment k by 1*
> > *return sum/npts*
>
> *crossings(x, npts):*
> > *set count to zero*
> > *set k to zero*
> > *while k ≤ npts – 2*
> > > *if x[k] · x [k + 1] < 0*
> > > > *increment count by 1*
> > > *increment k by 1*
> > *return count*

```
/*------------------------------------------------------------*/
/*  Program chapter5_5                                        */
/*                                                            */
/*  This program computes a set of statistical measurements   */
/*  from a speech signal.                                     */

#include <stdio.h>
#include <math.h>
#include "stat_lib.h"
#define MAXIMUM 2500
#define FILENAME "zero1.txt"

int main(void)
{
   /*  Declare variables and function prototypes.  */
   int k=0, npts;
   double speech[MAXIMUM];
   FILE *file_in;
   double ave_power(double x[],int npts);
   double ave_magn(double x[],int npts);
   int crossings(double x[],int npts);
```

```
          /*  Read information from a data file.  */
          file_in = fopen(FILENAME,"r");
          if (file_in == NULL)
             printf ("Error opening input file. \n");
          else
          {
             while ((fscanf(file_in,"%lf",&speech[k])) == 1)
                k++;
             npts = k;

             /*  Compute and print statistics.  */
             printf("speech statistics \n");
             printf("    mean: %f \n",mean(speech,npts));
             printf("    standard deviation: %f \n",
                    std_dev(speech,npts));
             printf("    variance: %f \n",variance(speech,npts));
             printf("    average power: %f \n",
                    ave_power(speech,npts));
             printf("    average magnitude: %f \n",
                    ave_magn(speech,npts));
             printf("    zero crossings: %d \n",
                    crossings(speech,npts));

             /*  Close file and exit program.  */
             fclose(file_1);

          }
          /*  Exit program.  */
          return 0;
       }
/*-----------------------------------------------------------------*/
/*  This function returns the average power of an array x           */
/*  with npts elements.                                            */

double ave_power(double x[],int npts)
{
   /*  Declare and initialize variables.  */
   int k;
   double sum=0;

   /*  Determine average power.  */
   for (k=0; k<=npts-1; k++)
      sum += x[k]*x[k];
   /*  Return average power.  */
   return sum/npts;
}
/*-----------------------------------------------------------------*/
/*  This function returns the average magnitude of an array x  */
/*  with npts elements.                                        */
```

```
double ave_magn(double x[],int npts)
{
   /*  Declare and initialize variables.  */
   int k;
   double sum=0;

   /*  Determine average power.  */
   for (k=0; k<=npts-1; k++)
     sum += fabs(x[k]);

   /*  Return average magnitude.  */
   return sum/npts;
}
/*-----------------------------------------------------------*/
/*  This function returns a count of the number of zero      */
/*  crossings in an array x with npts values.                */

int crossings(double x[],int npts)
{
   /*  Declare and initialize variables.  */
   int count=0, k;

   /*  Determine number of zero crossings.  */
   for (k=0; k<=npts-2; k++)
     if (x[k]*x[k+1] < 0)
        count++;

   /*  Return number of zero crossings.  */
   return count;
}
/*-----------------------------------------------------------*/
```

Note that the number of potential zero crossings for a set of n data points is $n-1$ crossings, because each crossing is determined by a pair of values. Thus, the last pair of values tested will be at subscripts $n-2$ and $n-1$.

5. TESTING

This program requires access to the `stat_lib.h` header file and to the `stat_lib.c` file developed in the previous section. The following values were computed for the utterance "zero" using the file `zero1.txt`:

```
Speech Statistics
     mean: -0.000208
     standard deviation: 0.077035
     variance: 0.00534
     average power: 0.005932
     average magnitude: 0.060567
     zero crossings: 124
```

MODIFY

Most computers have ports for connecting a microphone. Ask your instructor or laboratory assistant to help you use tools that are widely available on the Internet for collecting speech signals, to use with the program developed in this section. Collect three separate files of your speech for the word "zero." Also collect separate files of your speech for the words for the additional digits: "one," "two," and "three."

1. Run this program on all three files of your speech for the word "zero." Note how the statistics can vary significantly from one signal to another from the same speaker.
2. Run this program on the files for the words "one," "two," and "three." The statistics for these different words should vary even more than for the ones for the same word.
3. When working with speech signals, we often compute the average value and then subtract that from each value in the file so that the resulting set of values has a zero mean. Add the feature to this program so that the rest of the statistics are computed using the zero-mean signal.
4. Modify this program to include a line in the output from this program that prints the number of data points in the file.
5. Modify this program to include a line in the output that prints the maximum value from the data file.

5.6   Sorting Algorithms

Sorting

Sorting a group of data values is another operation that is routinely used when analyzing data. Entire texts are available that present many different sorting algorithms. One of the reasons that there are so many sorting algorithms is that there is not one "best" sorting algorithm. Some algorithms are faster if the data are already close to the correct order, but these algorithms may be very inefficient if the order is random or close to the opposite order. Therefore, to choose the best sorting algorithm for a particular application, you usually need to know something about the order of the original data. Rather than try to present a complete discussion of sorting algorithms, we present two algorithms. In this section, we present a selection sort that is simple to understand and simple to code in a function. In Chapter 6, we present a quicksort function that uses a recursive algorithm to sort a set of values; this algorithm is presented in Chapter 6 because it requires material presented in that chapter.

Selection sort

The **selection sort** algorithm begins by finding the minimum value and exchanging it with the value in the first position in the array. Then the algorithm finds the minimum value beginning with the second element, and it exchanges this minimum with the second element. This process continues until reaching the next-to-last element, which is compared with the last element; the values are exchanged if they are out of order. At this point, the entire array of values will be in ascending order. This process is illustrated in the following sequences:

Original order:

5	3	12	8	1	9

Exchange the minimum with the value in the first position:

1	3	12	8	5	9

Exchange the next minimum with the value in the second position:

1	3	12	8	5	9

Exchange the next minimum with the value in the third position:

1	3	5	8	12	9

Exchange the next minimum with the value in the fourth position:

1	3	5	8	12	9

Exchange the next minimum with the value in the fifth position:

1	3	5	8	9	12

Array values are now in ascending order:

1	3	5	8	9	12

The steps in the next function are short, but it is still a good idea to go through this function using the data in this example. Follow the changes in the subscripts k, m, and j within the loops. Also, note that it takes three steps (not two) to exchange values in two variables. Because the function does not return a value, its return type is void. Here is the function:

```
/*-------------------------------------------------------------*/
/*  This function sorts an array x with npts values into        */
/*  ascending order.                                            */

void sort(int x[],int npts)
{
   /*  Declare variables.  */
   int k, j, m;
   double hold;

   /*  Implement selection sort algorithm.  */
   for (k=0; k<=npts-2; k++)
   {
      /*  Exchange minimum with next array value.  */
      m = k;
      for (j=k+1; j<=npts-1; j++)
         if (x[j] < x[m])
            m = j;
      hold = x[m];
      x[m] = x[k];
      x[k] = hold;
   }

   /*  Void return.  */
   return;
}
/*-------------------------------------------------------------*/
```

To change this function into one that sorts an array in descending values, the inner loop should search for a maximum value instead of a minimum value.

The function prototype statement that should be used to refer to this sort function is

```
void sort(int x[], int npts);
```

It is also important to note that this function modifies the original array. To keep the original order, an array should be copied into another array before this function is executed. Then the data are available in both the original order and the sorted order.

Collating sequence

The ordering of characters in a code (such as ASCII), from low to high, is called a **collating sequence**. Sorting ASCII characters into an ascending order will yield an alphabetical order.

MODIFY!

1. Write a `main` function that initializes an array, references this `sort` function, and then prints the array values in the new order.
2. Modify the `sort` function so that it sorts values in descending order instead of ascending order. Test the function with the program written in Problem 1.

5.7 Search Algorithms

Another very common operation performed with arrays is searching the array for a specific value. We may want to know if a particular value is in the array, how many times it occurs in the array, or where it first occurs in the array. Each of these searches determines a single value and thus is a good candidate for a function. In this section, we will develop several functions for searching an array; then, when we need to perform a search in a program, we can probably use one of these functions with little or no modification.

Searching algorithms fall into two groups: those for searching an unordered list and those for searching an ordered list.

Unordered List

Sequential search

We first consider searching an unordered list; thus, we assume that the elements are not necessarily sorted into an ascending numerical order (or any other order that may aid us in searching the array). The algorithm to search an unordered array is just a simple **sequential search**: it will check the first element, check the second element, and so on. There are several ways that we could implement this function. We could develop it as an integer function that either returns the position of the desired value in the array or returns a −1 if the desired value is not in the array. We could develop the function as an integer function that returns the number of times the element occurs in the array. We could also develop the function as a logical function that returns a value of true (1) if the element is in the array or false (0) if the element is not in the array. All of these ideas represent valid functions, and we could think of programs that would use each of these forms. We have developed a function that either returns the position of the desired value in an unordered array or returns a −1 if the value is not found:

```
/*-----------------------------------------------------------*/
/*  This function searches for a value in an unordered list.  */
/*  If it finds the value, it returns the index of the item   */
/*  in the list (0 for first position, 1 for second position, */
/*  and so on).  If it does not find the value, the function  */
/*  returns the value -1.                                     */

int search1(int x[],int npts,int value)
{
   /*  Declare variables.  */
   int k=0, index=-1;

   /*  Search for value.  */
   while (k<=npts-1 && x[k]!=value)
      k++;
   if (k != npts)
      index = k;

   /*  Return index.  */
   return index;
}
/*-----------------------------------------------------------*/
```

Ordered List

We now consider searching an ordered or sorted list of values. Assume that we have a list of ordered values, and we are searching for the value 25:

```
-7
2
14
38
52
77
105
```

As soon as we reach the value 38, we will know that 25 is not in the list because we know the list is ordered in ascending numerical order. Therefore, we do not have to search the entire list, as we would have to do for an unordered list; we only need to search past the point where our desired value should have been. If the list is in ascending order, we search until the current value is larger than our desired value; if the list is in descending order, we search until the current value is smaller than our desired value. The next function performs a sequential search on an ordered list. The function either returns the position of the desired value in an ordered array or returns a -1 if the value is not found:

```
/*-----------------------------------------------------------*/
/*  This function searches for a value in an ordered          */
/*  (ascending) list.  If it finds the value, it returns      */
/*  the index of the item in the list (0 for first position,  */
/*  1 for second position, and so on).  If it does not find   */
/*  the value, the function returns the value -1.             */
```

```
int search2(int x[],int npts,int value)
{
   /*  Declare variables.  */
   int k=0, index=-1;

   /*  Search for value.  */
   while (k<=npts-1 && x[k]<value)
      k++;
   if (k <= npts-1)
      if (x[k] == value)
         index = k;

   /*  Return index value.  */
   return index;
}
/*-----------------------------------------------------------*/
```

Binary search
Another popular and more efficient algorithm for searching an ordered list is a technique called a **binary search**. In this technique, we first check the middle of the array and determine whether our desired value is in the first half of the array or the second half of the array. If it is in the first half, we then check the middle of the first half and determine whether our desired value is in the first fourth of the array or the second fourth of the array. The process of dividing the array into smaller and smaller pieces continues until we find the element or find the position where it should have been. Since this technique continually divides our search area in half, it is called a binary search.

We can illustrate a binary search algorithm with an ordered list of values (−7, 2, 14, 38, 52, 77, 105) using the diagram on the next page. Assume that we are searching for the value 25. Use the variable first to store the subscript of the first value in the array, and use the variable last to store the subscript of the last value in the array. Compute the subscript of the middle position by adding first to last, and dividing by 2. (This should be done as an integer division.) Since the array contains seven values, first will contain the value 0 and last will contain the value 6; middle will be computed to be 3. Thus, compare the value with a subscript of 3 to our desired value. Since 38 is larger than 25, we can narrow our search to the top half of the array. The variable first still contains 0, and we change the value of last to the position above the middle position, or 2. We now divide that part of the array in half and compute the midpoint, which is $(0 + 2)/2$, or 1. The value with subscript 1 is 2, which is smaller than 25, so we can narrow our search to the second quarter of the array. The value of first now becomes the subscript for the first position past the middle position, or 2, and the value of last is also 2. When first and last are the same, we have determined the position where the value should be located. Thus, we have either found the value or determined that it is not in the array. In this case, the value with a subscript of 2 is 14, so the value 25 is not in the array. However, when the number of elements in the array is even, it is possible for the position of first to be greater than last if the desired value is not in the list.

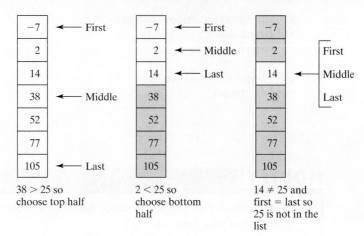

38 > 25 so
choose top half

2 < 25 so
choose bottom
half

14 ≠ 25 and
first = last so
25 is not in the
list

We now present a function that implements the binary search algorithm.

```
/*-------------------------------------------------------------*/
/*  This function searches for a value in an ordered           */
/*  (ascending) list using a binary search algorithm.  If it   */
/*  finds the value, the function returns the index of the     */
/*  item in the list (0 for first position, 1 for second       */
/*  position, and so on).  If it does not find the value,      */
/*  the function returns the value -1.                         */

int search3(double x[],int npts,int value)
{
   /*  Declare variables.  */
   int done=0, top=0, bottom, mid;
   double index=-1;

   /*  Search for value.  */
   bottom = npts-1;
   while (top<=bottom && done==0)
   {
      /*  Determine middle.  */
      mid = (top + bottom)/2;

      /*  Check value in middle.  */
      if (x[mid] == value)
         done = 1;
      else
         \*  Is value in top or bottom half?  */
         if (x[mid] > value)
            bottom = mid - 1;
         else
            top = mid + 1;
   }
```

```
    /*  Determine index value.  */
    if (done == 1)
        index = mid;

    /*  Return index value.  */
    return index;
}
/*-------------------------------------------------------------------*/
```

MODIFY!

1. Modify either of the sequential search functions to search an array of characters. Write a driver to test your function.
2. Modify the sequential search on an ordered list to return a count of the number of times a specified value occurred in an ordered list.
3. Modify the binary search function so that it correctly searches a list that is in descending order instead of ascending order. Write a driver to test your function.

5.8 Two-Dimensional Arrays

A set of data values that is visualized as a row or column is easily represented by a one-dimensional array. However, there are many examples in which the best way to visualize a set of data is with a grid or a table of data, which has both rows and columns. An array with four rows and three columns is shown in the following diagram:

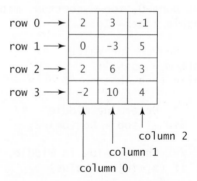

Two-dimensional array

In C, a grid or table of data is represented with a **two-dimensional array**. Each element in a two-dimensional array is referenced using an identifier followed by two subscripts—a row subscript and a column subscript. The subscript values for both rows and columns begin with 0, and each subscript has its own set of brackets. Thus, assuming that the previous array has an identifier x, the value in position x[2][1] is 6. Common errors in array references include using parentheses instead of brackets, as in x(2)(3), or using only one set of brackets or parentheses, as in x[2,3] or x(2,3).

We can also visualize this grid or table of data as a one-dimensional array, where each element is also an array. Thus, the array in the previous diagram can be interpreted as a one-dimensional array with four elements, each of which is a one-dimensional array with three elements:

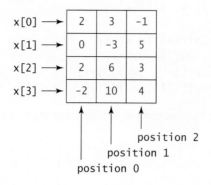

By using this representation, the notation x[2][1] can be interpreted as referring to position [1] within the one-dimensional array represented by x[2]; thus, the value of x[2][1] is 6. In general, we prefer to discuss two-dimensional arrays in terms of a grid with rows and columns, as opposed to an array of arrays.

All values in an array must have the same type. An array cannot have a column of integers followed by a column of floating-point numbers, and so on.

Definition and Initialization

To define a two-dimensional array, we specify the number of rows and the number of columns in the declaration statement. The row number is written first. Both the row number and the column number are in brackets, as shown in this statement:

```
int x[4][3];
```

A two-dimensional array can be initialized with a declaration statement. The values are specified in a sequence separated by commas, and each row is contained in braces. An additional set of braces is included around the complete set of values, as shown in the following statement:

```
int x[4][3]={{2,3,-1},{0,-3,5},{2,6,3},{-2,10,4}};
```

If the initializing sequence is shorter than the array, then the rest of the values are initialized to zero. If the array is specified with the first subscript empty, but with an initialization sequence, the size is determined by the sequence. Thus, the array x can also be defined with the following statement:

```
int x[][3]={{2,3,-1},{0,-3,5},{2,6,3},{-2,10,4}};
```

Arrays can also be initialized with program statements. For two-dimensional arrays, two nested `for` loops are usually required to initialize an array; i and j are commonly used as subscripts. To define and initialize an array such that each row contains the row number, use the following statements:

```
/*  Declare variables.  */
int i, j, t[5][4];
...
/*  Initialize array.  */
for (i=0; i<=4; i++)
   for (j=0; j<=3; j++)
      t[i][j] = i;
```

After these statements are executed, the values in the array t are as follows:

0	0	0	0
1	1	1	1
2	2	2	2
3	3	3	3
4	4	4	4

Two-dimensional arrays can also be initialized with values read from a data file. In the next set of statements, we assume that a data file contains 50 temperature values that we read and store in the array. Symbolic constants NROWS and NCOLS are used to represent the number of rows and columns. Changing the size of an array is easier to do when the numbers of rows and columns are specified as symbolic constants; otherwise, the change requires modifications to several statements. Here are the statements.

```
#define NROWS 10
#define NCOLS 5
#define FILENAME "engine1.txt"
...
/*  Declare variables.  */
int i, j;
double temps[NROWS][NCOLS};
file *sensor;
...
/*  Open file and read data into array.  */
sensor = fopen(FILENAME,"r");
for (i=0; i<=NROWS-1; i++)
   for (j=0; j<=NCOLS-1; j++)
      fscanf(sensor,"%lf",&temps[i][j]);
```

PRACTICE!

Show the contents of the arrays defined in each set of statements. Use a question mark to indicate an element that has not been initialized.

1. `int d[3][1]={{1},{4},{6}};`
2. `int g[6][2]={{5,2},{-2,3}};`
3. `float h[4][4]={{0,0}};`
4. ```
   int k, p[3][3]={{0,0,0}};
      ...
      for (k=0; k<=2; k++)
         p[k][k] = 1;
   ```
5. ```
   int i, j, g[5][5];
      ...
      for (i=0; i<=4; i++)
         for (j=0; j<=4; j++)
            g[i][j] = i + j;
   ```
6. ```
   int i, j, g[5][5];
      ...
      for (i=0; i<=4; i++)
         for (j=0; j<=4; j++)
            g[i][j] = pow(-1,j);
   ```

Computations and Output

Computations and output with two-dimensional arrays must always specify two subscripts when referencing an array element. To illustrate, consider a program that reads a data file containing power output for an electrical plant for an 8-week period. Each line of the data file contains seven values representing the daily power output for a week. The data are stored in a two-dimensional array. Then a report provides the average power for the first day of the week during the period, the average power for the second day of the week during the period, and so on. Here is the program:

```
/*-----------------------------------------------------------*/
/*  Program chapter5_6                                       */
/*                                                           */
/*  This program computes power averages over ten weeks.     */

#include <stdio.h>
#define NROWS 8
#define NCOLS 7
#define FILENAME "power1.txt"

int main(void)
{
   /* Declare variables.  */
   int i, j;
   int power[NROWS][NCOLS], col_sum;
   FILE *file_in;
```
```

```
/*  Read information from a data file.  */
file_in = fopen(FILENAME,"r");
if (file_in == NULL)
   printf("Error opening input file. \n");
else
{
   for (i=0; i<=NROWS-1; i++)
      for (j=0; j<=NCOLS-1; j++)
         fscanf(file_in,"%d",&power[i][j]);

   /*  Compute and print daily averages.  */
   for (j=0; j<=NCOLS-1; j++)
   {
      col_sum = 0;
      for (i=0; i<=NROWS-1; i++)
         col_sum += power[i][j];
      printf("Day %d: Average = %.2f \n",
              j+1,(double)col_sum/NROWS)
   }
   /*  Close file.  */
   fclose(file_in);
}
/*  Exit program.  */
return 0;
}
/*-------------------------------------------------------------*/
```

Note that the daily averages are computed by adding each column and then dividing the column sum by the number of rows (which is also the number of weeks). The column number is then used to compute the day number. A sample output from this program is as follows:

```
Day 1: Average = 253.75
Day 2: Average = 191.50
Day 3: Average = 278.38
Day 4: Average = 188.63
Day 5: Average = 273.13
Day 6: Average = 321.38
Day 7: Average = 282.50
```

Writing information from a two-dimensional array to a data file is similar to writing the information from a one-dimensional array. In both cases, a newline indicator must be used to specify when the values are to begin a new line. To write a set of distance measurements to a data file named dist1.txt with five values per line, use the following statements:

```
/*  Declare variables. */
int i, j;
double dist[20][5];
```

```
FILE *file_out;
...
/*  Write information from the array to a file.  */
file_out = fopen("dist1.txt","w")
for (i=0, i<=19; i++)
{
   for (j=0; j<=4; j++)
      fprintf(file_out,"%f ",dist[i][j]);
   fprintf(file_out,"\n");
}
```

The space after the conversion specifier in the `fprintf` statement is necessary in order to have the values separated by a space.

PRACTICE!

Assume the array g has the following declaration:

```
int i, j, g[3][3]={{0,0,0},{1,1,1},{2,2,2}};
```

Give the value of sum after each set of statements is executed.

1. ```
   sum = 0;
   for (i=0; i<=2; i++)
       for (j=0; j<=2; j++)
           sum += g[i][j];
   ```

2. ```
   sum = 1;
   for (i=1; i<=2; i++)
       for (j=0; j<=1; j++)
           sum *= g[i][j];
   ```

3. ```
   sum = 0;
   for (j=0; j<=2; j++)
       sum -= g[2][j];
   ```

4. ```
   sum = 0;
   for (i=0; i<=2; i++)
       sum += g[i][1];
   ```

Function Arguments

When arrays are used as function parameters, the references are call-by-address instead of call-by-value. In Section 5.1, which discussed one-dimensional arrays, we learned that array references in a function refer to the original array and not to a copy of the array. Thus, we must be careful so we do not unintentionally change values in the original array. Of course, an advantage of a call-by-address reference is that we can make changes in the array values, in addition to returning a value from the function call.

When using a one-dimensional array as a function argument, the function needs only the address of the array, which is specified by the array name; when using a two-dimensional array as a function argument, the function also needs information about the size of the array. In general, the function declaration and prototype statement should give complete information about the size of a two-dimensional array. To illustrate, suppose that we need to write a program that computes the sum of the elements in an array containing four rows and four columns. Computing this sum requires two nested loops; thus, the program will be more readable if we use a function to contain the steps to compute the sum. The program can then reference the function with a single statement, as in the following statements:

```
/*  Declare variables and function prototypes.   */
int a[4][4];
int sum(int x[4][4]);
...
/*  Use function to compute array sum.   */
printf("Array sum = %i \n",sum(a));
```

If we need to recompute the array sum in several places in the program, the function becomes even more effective. And, of course, if there are several different arrays of the same size, we can use the same function to compute their sums, as in the following statements:

```
/*  Declare variables and function prototypes.   */
int a[4][4], b[4][4];
int sum(int x[4][4]);
...
/*  Use function to compute array sums.   */
printf("Sum of a = %i \n",sum(a));
printf("Sum of b = %i \n",sum(b));
```

We now present the function referenced in these statements:

```
/*-----------------------------------------------------------*/
/*  This function returns the sum of the values in an array   */
/*  with four rows and four columns.                          */

int sum(int x[4][4])
{
   /*  Declare and initialize variables. */
   int i, j, total=0;

   /*  Compute a sum of the array values.   */
   for i=0; i<=3; i++)
      for (j=0; j<=3; j++)
         total += x[i][j];

   /* Return sum of array values.   */
   return total;
/*-----------------------------------------------------------*/
```

In this example, we included the number of rows and number of columns in the function definition and prototype. C allows us to omit the first subscript size, and thus we could have replaced the function definition and prototype with the following statements:

```
/*  Declare variables and function prototypes.  */
int sum(int x[][4]);
```

Style

For documentation purposes, we prefer listing both the row size and the column size of arrays in the formal argument list and in the function prototype.

In a final example, we develop a function that computes a partial sum of the elements in an array. The elements to be summed are assumed to be in a subarray in the upper-left corner of the array. The arguments of the function include the original array and the numbers of rows and columns in the subarray. The function prototype is

```
/*  function prototype  */
int partial_sum(int x[4][4], int r, int c);
```

Thus, we would use the reference partial_sum(a,2,3) if we want to sum the elements shown in the shaded area in the following array a

2	3	-1	9
0	-3	5	7
2	6	3	2
-2	10	4	6

This reference should then compute the sum of the elements in the subarray beginning in the upper-left corner and consisting of two rows and three columns. The function should return a value of 6. This function is as follows:

```
/*------------------------------------------------------------*/
/*  This function returns the sum of the values in a          */
/*  subarray of an array with four rows and four columns.     */

int partial_sum(int x[4][4],int r,int c)
{
   /*  Declare and initialize variables. */
   int i, j, total=0;

   /*  Compute a sum of the subarray values.  */
   for i=0; i<=r-1; i++)
      for (j=0; j<=c-1; j++)
         total += x[i][j];

   /* Return sum of subarray values.  */
   return total;
/*------------------------------------------------------------*/
```

When working with one-dimensional arrays, we do not have to specify the size of the array in the function; instead, we include an argument in the function definition that gives the number of values in the array. Thus, the function could be used with arrays of different sizes. For example, in Section 4.2, we developed a function to compute the average of a one-dimensional array. To use the function to compute the mean (or average) of an array a with 10 elements, the reference would be mean(a,10). If we want to compute the mean value of an array y with 50 elements, we use the reference mean(y,50). To write a function that can be used with two-dimensional arrays of various sizes, it is necessary to use pointers as function arguments; this technique is discussed in Chapter 6.

PRACTICE!

Assume that a main function contains the following statement:

```
int a[4][4]={{2,3,-1,9},{0,-3,5,7},
             {2,6,3,2},{-2,10,4,6}};
```

Using hand calculations determine the values of these references to the partial_sum function developed in this section:

1. partial_sum(a,1,4);   2. partial_sum(a,1,1);
3. partial_sum(a,4,2);   4. partial_sum(a,2,4);

The next four sections contain examples that use two-dimensional arrays. Section 5.9 contains an application related to terrain navigation; Section 5.10 uses two-dimensional arrays to represent matrices; and Sections 5.11 and 5.12 discuss and develop a solution to a system of simultaneous equations using a two-dimensional array to store the equation coefficients.

5.9 Problem Solving Applied: Terrain Navigation

Terrain navigation is a key component in the design of unmanned aerial vehicles (UAVs). Vehicles such as a robot or a car, can travel on land; and vehicles such as a drone or a plane, can fly above the land. A UAV system contains an onboard computer that has stored the terrain information for the area in which it is to be operated. Because it knows where it is at all times (often using a global positioning system [GPS] receiver), the vehicle can then select the best path to get to a designated spot. If the destination changes, the vehicle can refer to its internal maps and recompute the new path.

The computer software that guides these vehicles must be tested over a variety of land formations and topologies. Elevation information for large grids of land is available in computer databases. One way of measuring the "difficulty" of a land grid with respect to terrain navigation is to determine the number of peaks in the grid, where a peak is a point that has lower elevations all around it. For this problem, we want to determine whether the value in grid position [m][n] is a peak. Assume that the values in the four positions shown are adjacent to grid position [m][n]:

	grid[m-1][n]	
grid[m][n-1]	grid[m][n]	grid[m][n+1]
	grid[m+1][n]	

Write a program that reads elevation data from a data file named `grid1.txt`, and then prints the number of peaks and their locations. Assume that the first line of the data file contains the number of rows and the number of columns for the grid of information. These values are then followed by the elevation values, in row order. The maximum size of the grid is 25 rows by 25 columns.

1. PROBLEM STATEMENT

Determine and print the number of peaks and their locations in an elevation grid.

2. INPUT/OUTPUT DESCRIPTION

The I/O diagram shows that the input is a file containing the elevation data and that the output is a listing of the locations of the peaks.

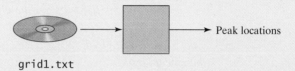

Peak locations

grid1.txt

3. HAND EXAMPLE

Assume that the following data represent elevation for a grid that has six points along the side and seven points along the top (the peaks have been underlined):

5039	5127	5238	5259	5248	5310	5299
5150	5392	5410	5401	5320	5820	5321
5290	_5560_	5490	5421	5530	_5831_	5210
5110	5429	5430	5411	5459	5630	5319
4920	5129	4921	_5821_	4722	4921	5129
5023	5129	4822	4872	4794	4862	4245

To specify the location of the peaks, we need to assign an addressing scheme to the data. Because we are going to implement this solution in C, we choose its two-dimensional array subscripting notation. Thus, we assume that the top left corner is position [0][0], the row numbers increase by 1 as we move down the page, and the column numbers increase by 1 as we move to the right. These peaks then occur at positions [2][1], [2][5], and [4][3].

To determine the peaks, we compare a potential peak with its four neighboring points. If all four neighboring points are less that the potential peak, then the potential peak is a real peak. Note that the points on the edges of the array or grid cannot be potential peaks because we do not have elevation information on all four sides of the points.

4. ALGORITHM DEVELOPMENT

First, we develop the decomposition outline because it divides the solution into a series of sequential steps.

Decomposition Outline

1. Read the terrain data into an array.
2. Determine and print the location of the peaks.

Step 1 involves reading the data file and storing the information in a two-dimensional array. Step 2 is a loop that evaluates all potential peaks, and prints their locations if they are determined to be real peaks. There are not obvious candidates for additional functions, so we develop the refinement in pseudocode using only a main function:

Refinement in Pseudocode

main: *read nrows and ncols from the data file*
            read the terrain data into an array called elevation
            set i to 1
            while i ≤ nrows − 2
                set j to 1
                while j ≤ ncols − 2
                      if elevation[i][j] > its four neighbors
                              print peak location
                    increment j by 1
                increment i by 1

The steps in the pseudocode are now detailed enough to convert to C:

```
/*-----------------------------------------------------------------*/
/*   Program chapter5_7                                            */
/*                                                                 */
/*   This program determines the locations of peaks in an          */
/*   grid of elevation data.                                       */

#include <stdio.h>
#define N 25
#define FILENAME "grid1.txt"

int main(void)
{
   /*  Declare variables.  */
   int nrows, ncols, i, j;
   double elevation[N][N];
   FILE *grid;

   /*  Read information from a data file.  */
   grid = fopen(FILENAME,"r");
   if (grid == NULL)
      printf("Error opening input file\n");
   else
   {
      fscanf(grid,"%d %d",&nrows,&ncols);
      for (i=0; i<=nrows-1; i++)
         for (j=0; j<=ncols-1; j++)
            fscanf(grid,"%lf",&elevation[i][j]);
```

```
        /* Determine and print peak locations.  */
        printf("Top left point defined as row 0, column 0 \n");
        for (i=1; i<=nrows-2; i++;)
           for (j=1; j<=ncols-2; j++)
              if ((elevation[i-1][j]<elevation[i][j]) &&
                  (elevation[i+1][j]<elevation[i][j]) &&
                  (elevation[i][j-1]<elevation[i][j]) &&
                  (elevation[i][j+1]<elevation[i][j]))
                 printf("Peak at row: %d column: %d \n",i,j)

        /* Close file.  */
        fclose(grid);
     }

     /*   Exit program.  */
     return 0;
}
/*-------------------------------------------------------------*/
```

5. TESTING

The following output was printed using a data file that corresponds to the hand example:

```
Top left point defined as row 0, column 0
Peak at row: 2 column: 1
Peak at row: 2 column: 5
Peak at row: 4 column: 3
```

Recall that this file must contain a special first line that specifies the number of rows and columns in the elevation data.

MODIFY!

Modify program chapter5_7 to determine the following information for a grid of elevation data:

1. Print a count of the number of peaks in the grid.
2. Print the location of valleys instead of peaks. Assume that a valley is a point with an elevation lower than the four surrounding elevations.
3. Find and print the location and elevation of the highest point and the lowest point in the elevation data.
4. Assuming that the distance between points in a vertical and horizontal direction is 100 feet, give the location of the peaks in feet from the lower left corner of the grid.
5. Use all eight neighboring points, instead of only four neighboring points, to determine a peak.

5.10 Matrices and Vectors*

Matrix

A **matrix** is a set of numbers arranged in a rectangular grid with rows and columns. Consider a matrix with four rows and three columns; the size of this matrix is also specified as 4×3, as shown:

$$\mathbf{A} = \begin{bmatrix} -1 & 0 & 0 \\ 1 & 1 & 0 \\ 1 & -2 & 3 \\ 0 & 2 & 1 \end{bmatrix}.$$

Row vector
Column vector
Vector

Note that the values within a matrix are written within large brackets. A matrix with one row is called a **row vector**, and a matrix with one column is called a **column vector**. The term **vector** by itself does not distinguish between a row vector and a column vector.

In mathematical notation, matrices are usually given names with uppercase boldface letters. To refer to individual elements in the matrix, the row and column number are used, with both the row and column numbers starting with the value 1. In formal mathematical notation, the uppercase name refers to the entire matrix, and the lowercase name with subscripts refers to a specific element. Thus, by using the matrix $\mathbf{A}$, the value of $a_{3,2}$ is -2. If a matrix has the same number of rows and columns, it is a **square matrix**.

Square matrix

A two-dimensional array can be used to store a matrix, but we must be careful translating equations in matrix notation into C statements because of the difference in subscripting. Matrix notation assumes that the row and column numbers begin with the value 1, whereas C statements assume that the row and column numbers of a two-dimensional array begin with the value 0. Although a vector could be stored as a two-dimensional array with either one row or one column, vectors are more commonly stored as one-dimensional arrays; thus, they do not usually keep the distinction of a row vector or a column vector.

Matrix operations are frequently used in engineering problem solutions, so we now present common operations with matrices and vectors. We will include C statements for performing some of the operations; the problems at the end of the chapter relate to developing C statements for the remaining operations.

Dot Product

Dot product

The **dot product** is a number computed from two vectors of the same size. This value is the sum of the products of the values in corresponding positions in the vectors, as shown in this summation equation, which assumes that there are n elements in the vectors $\mathbf{A}$ and $\mathbf{B}$:

$$\text{Dot product} = \mathbf{A} \cdot \mathbf{B} = \sum_{k=1}^{n} a_k b_k.$$

To illustrate, assume that $\mathbf{A}$ and $\mathbf{B}$ are the following vectors:

$$\mathbf{A} = [4 \quad -1 \quad 3] \qquad \mathbf{B} = [-2 \quad 5 \quad 2].$$

The dot product is then

$$\mathbf{A} \cdot \mathbf{B} = 4 \cdot (-2) + (-1) \cdot 5 + 3 \cdot 2$$
$$= (-8) + (-5) + 6$$
$$= -7.$$

Inner product

The dot product is also called an **inner product**.

---

*Optional section.

In C, we can compute the dot product of two one-dimensional vectors with a function:

```
/*-----------------------------------------------------------------*/
/*  This function returns the dot product of two vectors.     */

double dot_product(double a[],double b[],int n)
{
    /*  Declare and initialize variables.  */
    int k;
    double sum=0;

    /*  Compute dot product.  */
    for (k=0; k<=n-1; k++)
        sum += a[k]*b[k];

    /*  Return dot product.  */
    return sum;

}
/*-----------------------------------------------------------------*/
```

Note that the equation subscripts of 1 to n were changed to 0 to $n-1$ for the C program.

Determinant

Determinant

The **determinant** of a matrix is a value computed from the entries in the matrix. Determinants have various applications in engineering, including computing inverses and solving systems of simultaneous equations. For a 2×2 matrix $\mathbf{A}$, the determinant is defined to be the following:

$$\text{Determinant of } \mathbf{A} = |\mathbf{A}| = a_{1,1}a_{2,2} - a_{2,1}a_{1,2}.$$

Therefore, the determinant of $\mathbf{A}$ is equal to 8 for the following matrix:

$$\mathbf{A} = \begin{bmatrix} 1 & 3 \\ -1 & 5 \end{bmatrix}.$$

For a 3×3 matrix $\mathbf{A}$, the determinant is defined to be the following:

$$|\mathbf{A}| = a_{1,1}a_{2,2}a_{3,3} + a_{1,2}a_{2,3}a_{3,1} + a_{1,3}a_{2,1}a_{3,2} - a_{3,1}a_{2,2}a_{1,3}$$

$$- a_{3,2}a_{2,3}a_{1,1} - a_{3,3}a_{2,1}a_{1,2}.$$

If $\mathbf{A}$ is the matrix

$$\mathbf{A} = \begin{bmatrix} 1 & 3 & 0 \\ -1 & 5 & 2 \\ 1 & 2 & 1 \end{bmatrix},$$

then $|\mathbf{A}|$ is equal to $5 + 6 + 0 - 0 - 4 - (-3)$, or 10.

A more involved process is necessary for computing determinants of matrices with more than three rows and columns. This process is discussed in the problems at the end of this chapter.

Transpose

Transpose

The **transpose** of a matrix is a new matrix in which the rows of the original matrix are the columns of the new matrix. We use a superscript T after a matrix name to refer to the transpose. For example, consider the following matrix and its transpose:

$$\mathbf{B} = \begin{bmatrix} 2 & 5 & 1 \\ 7 & 3 & 8 \\ 4 & 5 & 21 \\ 16 & 13 & 0 \end{bmatrix}, \qquad \mathbf{B}^T = \begin{bmatrix} 2 & 7 & 4 & 16 \\ 5 & 3 & 5 & 13 \\ 1 & 8 & 21 & 0 \end{bmatrix}.$$

If we consider a couple of the elements, we see that the value in position $(3, 1)$ has now moved to position $(1, 3)$ and that the value in position $(4, 2)$ has now moved to position $(2, 4)$. In fact, we have interchanged the row and column subscript so that we are moving the value in position (i, j) to position (j, i). Also, note that the size of the transpose is different than the size of the original matrix (unless the original is a square matrix).

We now develop a function that generates the transpose of a matrix. The formal arguments of the function must include two-dimensional arrays that represent the original matrix and the matrix that is to contain the transpose of the original matrix. To allow some flexibility with this function, we assume that we have defined symbolic constants that specify the number of rows and the number of columns in the original matrix. These symbolic constants are NROWS and NCOLS. Because using a symbolic constant is equivalent to using the value it has been given, we can then use NROWS and NCOLS in the array definition and in the prototype statement. Note that the function does not return a value; hence, the return type is void. Also note that the symbolic constants NROWS and NCOLS must be defined in a program that uses this function:

```
/*------------------------------------------------------------------*/
/*  This function generates a matrix transpose.   NROWS and         */
/*  NCOLS are symbolic constants that must be declared              */
/*  in the calling program.                                        */

void transpose(int b[NROWS][NCOLS],int bt[NCOLS][NROWS])
{
   /*  Declare variables.   */
   int i, j;

   /*  Transfer values to the transpose matrix.   */
   for (i=0; i<=NROWS-1; i++)
      for (j=0; j<=NCOLS-1; j++)
         bt[j][i] = b[i][j];

   /*  Void return.   */
   return;
}
/*------------------------------------------------------------------*/
```

Matrix Addition and Subtraction

The addition (or subtraction) of two matrices is performed by adding (or subtracting) the elements in corresponding positions in the matrices. Therefore, matrices that are added (or subtracted) must be the same size; the result of the operation is another matrix of the same size. Consider the following matrices:

$$A = \begin{bmatrix} 2 & 5 & 1 \\ 0 & 3 & -1 \end{bmatrix}, \quad B = \begin{bmatrix} 1 & 0 & 2 \\ -1 & 4 & -2 \end{bmatrix}.$$

Several matrix sums and differences follow:

$$A + B = \begin{bmatrix} 3 & 5 & 3 \\ -1 & 7 & -3 \end{bmatrix}, \quad A - B = \begin{bmatrix} 1 & 5 & -1 \\ 1 & -1 & 1 \end{bmatrix},$$

$$B - A = \begin{bmatrix} -1 & -5 & 1 \\ -1 & 1 & -1 \end{bmatrix}.$$

Matrix Multiplication

Matrix
multiplication

Matrix multiplication is not computed by multiplying corresponding elements of the two matrices. The value in position $c_{i,j}$ of the product C of two matrices A and B is the dot product of row i of the first matrix and column j of the second matrix, as shown in this summation equation:

$$c_{i,j} = \sum_{k=1}^{n} a_{i,k} b_{k,j}.$$

Since the dot product requires that the vectors have the same number of elements, the first matrix (A) must have the same number of elements in each row as the second matrix (B) has in each column. Thus, if A and B both have five rows and five columns, their product has five rows and five columns. Furthermore, for these matrices, we can compute both AB and BA, but, in general, they will not be equal.

If A has two rows and three columns, and B has three rows and three columns, the product AB will have two rows and three columns. To illustrate, consider the following matrices:

$$A = \begin{bmatrix} 2 & 5 & 1 \\ 0 & 3 & -1 \end{bmatrix}, \quad B = \begin{bmatrix} 1 & 0 & 2 \\ -1 & 4 & -2 \\ 5 & 2 & 1 \end{bmatrix}.$$

The first element in the product $C = AB$ is

$$C_{1,1} = \sum_{k=1}^{3} a_{1k} b_{k1}$$

$$= a_{1,1} b_{1,1} + a_{1,2} b_{2,1} + a_{1,3} b_{3,1}$$

$$= 2 \cdot 1 + 5 \cdot (-1) + 1 \cdot 5$$

$$= 2.$$

Similarly, we can compute the rest of the elements in the product of **A** and **B**:

$$\mathbf{AB} = \mathbf{C} = \begin{bmatrix} 2 & 22 & -5 \\ -8 & 10 & -7 \end{bmatrix}.$$

In this example, we cannot compute **BA**, because **B** does not have the same number of elements in each row as **A** has in each column.

An easy way to decide if a matrix product exists is to write the sizes of the two matrices side by side. If the two inside numbers are the same, the product exists; the two outside numbers determine the size of the product. To illustrate, in the previous example, the size of **A** is 2×3, and the size of **B** is 3×3. Therefore, if we want to compute **AB**, we write the sizes side by side:

$$2 \times 3 \qquad 3 \times 3$$

The two inner numbers are both the value 3, so **AB** exists, and its size is determined by the two outer numbers, 2×3. If we want to compute **BA**, we again write the sizes side by side:

$$3 \times 3 \qquad 2 \times 3$$

The two inner numbers are not the same, so **BA** does not exist.

We now present a function to compute the product $\mathbf{C} = \mathbf{AB}$. In this function, the arrays are each of size $N \times N$, where N is a symbolic constant:

```
/*-------------------------------------------------------------------*/
/*  This function performs a matrix multiplication of two            */
/*  NxN matrices using sums of products.  N  is a symbolic           */
/*  constant that must be defined in the calling program.            */

void matrix_mult(int a[N][N],int b[N][N],int c[N][N])
{
   /*  Declare variables.  */
   int i, j, k;

   /*  Compute sums of products.  */
   for (i=0; i<=N-1; i++)
      for (j=0; j<=N-1; j++)
      {
         c[i][j] = 0;
         for (k=0; k<=N-1; k++)
            c[i][j] += a[i][k]*b[k][j];
      }

   /*  Void return.  */
   return;
}
/*-------------------------------------------------------------------*/
```

PRACTICE!

Use hand calculations to evaluate the expressions in these problems. Then write programs to test your answers using the functions developed in this section. Use the following matrices and vectors:

$$A = \begin{bmatrix} 2 & 1 \\ 0 & -1 \\ 3 & 0 \end{bmatrix}, \quad B = \begin{bmatrix} -2 & 2 \\ -1 & 5 \end{bmatrix},$$

$$C = \begin{bmatrix} 3 & 2 \\ -1 & -2 \\ 0 & 2 \end{bmatrix}, \quad D = \begin{bmatrix} 1 & 2 \end{bmatrix}$$

1. $D \cdot D$.
2. $|B|$.
3. $C^T + A^T$.
4. DB.
5. $B(C^T)$.
6. $(CB)D^T$.

The problems at the end of the chapter use the matrix operations discussed in this section; they also define additional matrix operations.

5.11   Numerical Technique: Solution to Simultaneous Equations*

The need to solve a system of simultaneous equations occurs frequently in engineering problems. A number of methods exist for solving a system of equations, and each method has its advantages and disadvantages. In this section, we present the Gauss elimination method of solving a set of **simultaneous linear equations**. The equations are called linear equations because the equations contain only linear (degree 1) terms such as x, y, and z. However, before we present the details of this technique, we first present a graphical interpretation of the solution to a set of equations.

Simultaneous linear equations

Graphical Interpretation

A linear equation with two variables, such as $2x - y = 3$, defines a straight line and is often written in the form $y = mx + b$, where m represents the slope of the line, and b represents the y-intercept. Thus, $2x - y = 3$ can also be written as $y = 2x - 3$. If we have two linear equations, they can represent two different lines that intersect in a single point, they can represent two parallel lines that never intersect, or they can represent the same line; these possibilities are shown in Figure 5.3. Equations that represent two intersecting lines can be easily identified because they will have different slopes, as in $y = 2x - 3$ and $y = -x + 3$. Equations that represent two parallel lines will have the same slope but different y-intercepts, as in $y = 2x - 3$ and $y = 2x + 1$. Equations that represent the same line have the same slope and y-intercept, as in $y = 2x - 3$ and $3y = 6x - 9$.

---

*Optional section.

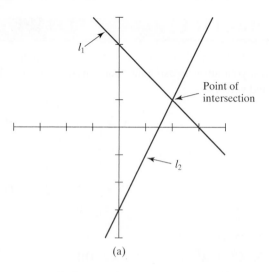

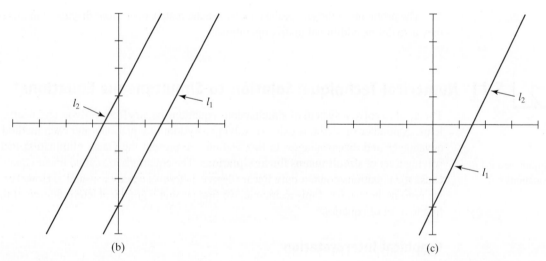

Figure 5.3 *Two lines.*

If a linear equation contains three variables, x, y, and z, then it represents a plane in three-dimensional space. If we have two equations with three variables, they can represent two planes that intersect in a straight line, they can represent two parallel planes, or they can represent the same plane; these possibilities are shown in Figure 5.4. If we have three equations with three variables, the three planes can intersect in a single point, they can intersect in a plane, they can have no common intersection point, or they can represent the same plane. Examples of the possibilities that exist if the three equations define three different planes are shown in Figure 5.5.

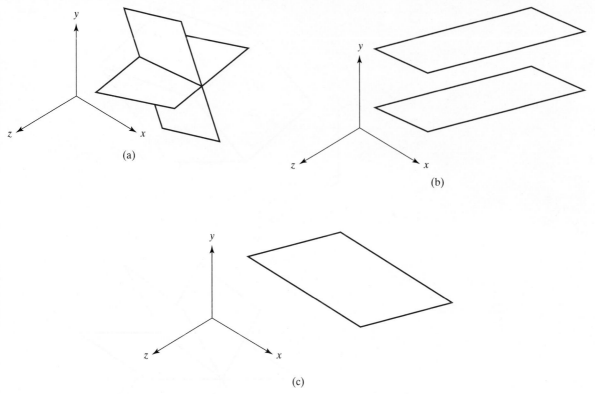

Figure 5.4 *Two planes.*

These ideas can be extended to more than three variables, although it is harder to visualize the corresponding situations. We call the set of points defined by an equation with more
Hyperplane   than three variables a **hyperplane**. In general, we consider a set of m linear equations that
contain n unknowns, where each equation defines a hyperplane that is not identical to another
hyperplane in the set of equations. If $m < n$, then the system is underspecified, and a unique
solution does not exist. If $m = n$, then a unique solution will exist if none of the equations
represents parallel hyperplanes. If $m > n$, then the system is overspecified and a unique solu-
System of equations   tion does not exist. A set of equations is also called a **system of equations**. A system with a
Nonsingular   unique solution is called a **nonsingular** system of equations, and a system with no unique solution is called a singular set of equations.

As a specific example, consider this system of equations:

$$3x + 2y - z = 10,$$
$$-x + 3y + 2z = 5,$$
$$x - y - z = -1.$$

The solution to this set of equations is the point $(-2, 5, -6)$. Substitute these values in each
of the questions to confirm that this point is a solution to the set of equations.

The material on matrices is not required for the development of the solution presented in
this section. However, if you did cover that material, it is interesting to observe that a system

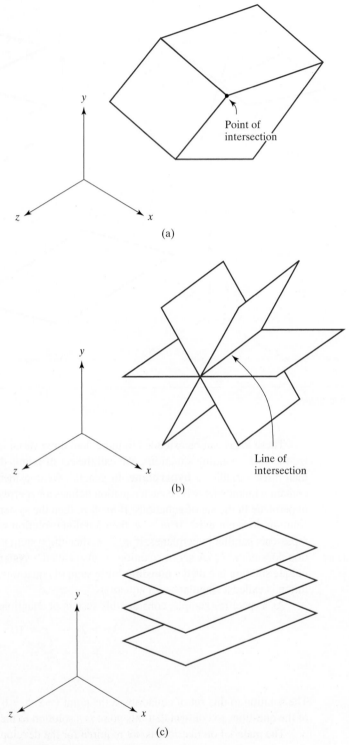

Figure 5.5 *Three distinct planes.*

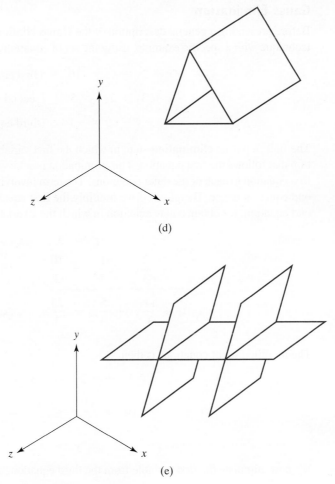

(d)

(e)

Figure 5.5 *(Continued)*

of linear equations can be expressed in terms of a matrix multiplication. To illustrate, let the information in the previous equations be expressed using the following matrices:

$$
\mathbf{A} = \begin{bmatrix} 3 & 2 & -1 \\ -1 & 3 & 2 \\ 1 & -1 & -1 \end{bmatrix}, \quad \mathbf{X} = \begin{bmatrix} x \\ y \\ z \end{bmatrix}, \quad \mathbf{B} = \begin{bmatrix} 10 \\ 5 \\ -1 \end{bmatrix}.
$$

Then, using matrix multiplication, the system of equations can be written in this form:

$$
\mathbf{AX} = \mathbf{B}.
$$

Go through the multiplication to convince yourself that this matrix equation yields the original set of equations.

In many engineering problems, we want to determine whether a common solution exists to a system of equations. If the common solution exists, then we want to find it. In the next part of this section, we present the Gauss elimination technique for solving a set of simultaneous linear equations.

Gauss Elimination

Gauss elimination

Before presenting a general description of the **Gauss elimination** technique, we illustrate the technique with a specific example, using the set of equations presented earlier:

$$3x + 2y - z = 10 \quad \text{(first equation)},$$
$$-x + 3y + 2z = 5 \quad \text{(second equation)},$$
$$x - y - z = -1 \quad \text{(third equation)}.$$

Elimination

The first step is an **elimination** step, in which the first variable is eliminated from each equation that follows the first equation. This elimination is achieved by adding a scaled form of the first equation to each of the other equations. The term involving the first variable, x, in the second equation is $-x$. Therefore, if we multiply the first equation by 1/3 and add it to the second equation, we obtain a new equation in which the x variable has been eliminated:

$$-x + 3y + 2z = 5 \quad \text{(second equation)},$$
$$x + \frac{2}{3}y - \frac{1}{3}z = \frac{10}{3} \quad \left(\text{first equation times } \frac{1}{3}\right),$$
$$\overline{}$$
$$0x + \frac{11}{3}y + \frac{5}{3}z = \frac{25}{3} \quad \text{(sum)}.$$

The modified set of equations is then

$$3x + 2y - z = 10,$$
$$0x + \frac{11}{3}y + \frac{5}{3}z = \frac{25}{3},$$
$$x - y - z = -1.$$

We now eliminate the first variable from the third equation, using a similar process:

$$x - y - z = -1 \quad \text{(third equation)},$$
$$-x - \frac{2}{3}y + \frac{1}{3}z = -\frac{10}{3} \quad \left(\text{first equation times } -\frac{1}{3}\right),$$
$$\overline{}$$
$$0x - \frac{5}{3}y - \frac{2}{3}z = -\frac{13}{3} \quad \text{(sum)}.$$

The modified set of equations is then

$$3x + 2y - z = 10,$$
$$0x + \frac{11}{3}y + \frac{5}{3}z = \frac{25}{3},$$
$$0x - \frac{5}{3}y - \frac{2}{3}z = -\frac{13}{3}.$$

We have now eliminated the first variable in all equations except for the first equation.

The next step is to eliminate the second variable in all equations except for the first and second equations. Thus, we will add the equations to a scaled form of the second equation:

$$0x - \frac{5}{3}y - \frac{2}{3}z = -\frac{13}{3} \quad \text{(third equation)},$$

$$0x + \frac{5}{3}y + \frac{25}{33}z = \frac{125}{33} \quad \left(\text{second equation times } \frac{5}{11}\right),$$

$$0x + 0y + \frac{3}{33}z = -\frac{18}{33} \quad \text{(sum)}.$$

The modified set of equations is then

$$3x + 2y - z = 10,$$

$$0x + \frac{11}{3}y + \frac{5}{3}z = \frac{25}{3},$$

$$0x + 0y + \frac{3}{33}z = -\frac{18}{33}.$$

Because there are no equations following the third equation, this part of the algorithm is completed.

Back substitution   We now perform a **back substitution** to determine the solution to the equations. The last equation has only one variable, so we can multiply the equation by a scale factor chosen to make the variable's coefficient equal to 1. Thus, we multiply the last equation by $\frac{33}{3}$, or 11, giving

$$0x + 0y + z = -6.$$

This value of z is substituted in the next-to-last equation, giving

$$0x + \frac{11}{3}y + \frac{5}{3}(-6) = \frac{25}{3}.$$

Reducing the equation so that all constant terms are on the right side, we have

$$0x + \frac{11}{3}y = \frac{55}{3}.$$

This equation has only one variable, so we now multiply it by a scale factor chosen to make the new coefficient equal to 1:

$$0x + y = 5.$$

We back up to the next equation, which is the last equation in this example:

$$3x + 2y - z = 10.$$

Substituting the values already determined, we have

$$3x + 2(5) - (-6) = 10,$$

or

$$3x = -6.$$

Thus, the value of x is -2.

The Gauss elimination technique thus has two parts—elimination and back substitution. First, the equations are modified such that the kth variable is eliminated in all equations following the kth equation. Then, starting with the last equation, we compute the value of the last variable. Using this value and the next-to-last equation, we compute the value of the next-to-last variable. This back substitution continues until we have determined the values of all the

Ill-conditioned

variables. The system is **ill-conditioned**, or does not have a unique solution, if all the coefficients for a variable are zero or are very close to zero.

Pivoting

A process called **pivoting** can be applied to improve the accuracy of Gauss elimination. Row pivoting involves reordering the rows before performing Gauss elimination, and column pivoting involves reordering the columns before performing the process. Complete pivoting involves reordering both rows and columns. These processes are discussed in the problems at the end of the chapter.

PRACTICE!

Use the Gauss elimination numerical technique to find the solution to these sets of simultaneous linear equations:

1. $-2x + y = -3$
   $x + y = \ \ \ 3.$
2. $3x + 5y + 2z = \ \ \ 8$
   $2x + 3y - \ \ \ z = \ \ \ 1$
   $x - 2y - 3z = -1.$

5.12 Problem Solving Applied: Electrical Circuit Analysis*

The analysis of an electrical circuit frequently involves finding the solution to a set of simultaneous equations. These equations are often derived using either current equations that describe the currents entering and leaving a node or using voltage equations that describe the voltages around mesh loops in the circuit. For example, consider the circuit shown in Figure 5.6. The voltages around the three loops can be described with the following equations:

$$-V_1 \quad + \quad R_1 i_1 \quad + \quad R_2(i_1 - i_2) \quad = 0,$$
$$R_2(i_2 - i_1) \quad + \quad R_3 i_2 \quad + \quad R_4(i_2 - i_3) \quad = 0,$$
$$R_4(i_3 - i_2) \quad + \quad R_5 i_3 \quad + \quad V_2 \quad = 0.$$

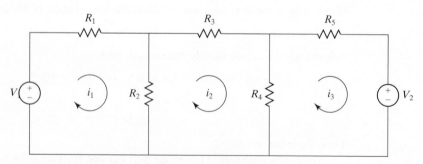

Figure 5.6 *Circuit with two voltage sources.*

---

*Optional section.

If we assume that the values of the resistors (R_1, R_2, R_3, R_4, and R_5) and the voltage sources (V_1 and V_2) are known, then the mesh currents (i_1, i_2, and i_3) are unknown. We can then rearrange the system of equations as follows:

$$
\begin{aligned}
(R_1 + R_2)i_1 &- R_2i_2 &+ \quad 0i_3 &= V_1, \\
-R_2i_1 &+ (R_2 + R_3 + R_4)i_2 &- R_4i_3 &= 0, \\
0i_1 &- R_4i_2 &+ (R_4 + R_5)i_3 &= -V_2.
\end{aligned}
$$

Write a program that allows the user to enter the values of the five resistors and the values of the two voltage sources. The program should then compute the three mesh currents.

1. PROBLEM DESCRIPTION

Compute the three mesh currents in the circuit shown in Figure 5.6.

2. INPUT/OUTPUT DESCRIPTION

The I/O diagram shows that the resistor values and the voltage values are the input values. The three mesh currents are the output values.

Resistor values ⟶ ▢ ⟶ Current values
Voltage values ⟶

3. HAND EXAMPLE

By using the resistor values and the voltage values, a system of three equations can be defined using this rearranged set of equations from the problem definition:

$$
\begin{aligned}
(R_1 + R_2)i_1 &- R_2i_2 &+ \quad 0i_3 &= V_1, \\
-R_2i_1 &+ (R_2 + R_3 + R_4)i_2 &- R_4i_3 &= 0, \\
0i_1 &- R_4i_2 &+ (R_4 + R_5)i_3 &= -V_2.
\end{aligned}
$$

For example, suppose that each of the resistor values is 1 ohm, and assume that both of the voltage sources are 5 volts. Then, the corresponding set of equations is the following:

$$
\begin{aligned}
2i_1 &- i_2 &+ 0i_3 &= 5, \\
-i_1 &+ 3i_2 &- i_3 &= 0, \\
0i_1 &- i_2 &+ 2i_3 &= -5.
\end{aligned}
$$

Once the system of equations is determined, the solution follows the steps illustrated in the hand example in the previous section. For this set of equations, the solution is $i_1 = 2.5$, $i_2 = 0$, and $i_3 = -2.5$.

4. ALGORITHM DEVELOPMENT

We first develop the decomposition outline because it breaks the solution into a series of sequential steps:

Decomposition Outline

1. Read the resistor values and the voltage values.
2. Specify the coefficients for the system of equations.
3. Perform Gauss elimination to determine currents.
4. Print currents.

Step 1 reads the information necessary to specify the circuit values. Step 2 uses this information to specify the coefficients for the system of equations. Step 3 develops the details of the elimination and back substitution steps. To keep the `main` function short and readable, functions are used for both the elimination and back substitution. The structure chart for this solution was used in Figure 4.1.

The coefficients of the simultaneous equations are stored in a two-dimensional array; the solution is stored in a one-dimensional array. The variable `index` indicates which variable is being eliminated in the `elimination` function; this variable ranges from 0 to $n - 1$ to match the subscripting in C.

The algorithm for Gauss elimination is a difficult algorithm to describe in pseudocode because of the detailed subscripting that must be specified. Go through this pseudocode with the hand example to be sure that you are comfortable with the subscript handling:

Refinement in Pseudocode

main: *read resistor values and voltage values*
specify array coefficients, a[i][j]
set index to zero
while index $\leq n - 2$
  eliminate(a,n,index)
  increment index by 1
back_substitute(a,n,soln)
print current values
eliminate(a,n,index):
  set row to index + 1
  while row $\leq n - 1$
    set scale_factor to $\dfrac{-a[row][index]}{a[index][index]}$
    set a[row][index] to zero
    set col to index + 1
    while col $\leq n$
      add a[index][col] · scale_factor
        to a[row][col]
      increment col by 1
    increment row by 1

back_substitute(a,n,soln):

> *set soln[n – 1] to* $\dfrac{-a[n-1][n]}{a[n-1][n-1]}$
>
> *set row to n – 2*
> *while row ≥ 0*
>> *set col to n – 1*
>> *while col ≥ row + 1*
>>> *subtract soln[col] · a[row][col]*
>>> *from a[row][n]*
>> *subtract 1 from col*
>> *set soln[row] to* $\dfrac{a[row][n]}{a[row][row]}$
> *subtract 1 from row*

Once we are comfortable with the pseudocode, it is relatively straightforward to convert it to C.

```
/*------------------------------------------------------------*/
/*  Program chapter5_8                                        */
/*                                                            */
/*  This program uses Gauss elimination to determine the      */
/*  mesh currents for a circuit.                              */

#include <stdio.h>
#define N 3  /*  number of unknown currents  */

int main(void)
{
   /*  Declare variables and function prototypes.  */
   int index;
   double r1, r2, r3, r4, r5, v1, v2, a[N][N+1], soln[N];
   void eliminate(double a[N][N+1],int n,int index);
   void back_substitute(double a[N][N+1],int n,
                        double soln[N]);

   /*  Get user input.  */
   printf("Enter resistor values in ohms: \n");
   printf("(R1, R2, R3, R4, R5) \n");
   scanf("%lf %lf %lf %lf %lf",&r1,&r2,&r3,&r4,&r5);
   printf("Enter voltage values in volts: \n");
   printf("(V1, V2) \n");
   scanf("%lf %lf",&v1,&v2);

   /*  Specify equation coefficients.  */
   a[0][0] = r1 + r2;
   a[0][1] = a[1][0] = -r2;
   a[0][2] = a[2][0] = a[1][3] = 0;
   a[1][1] = r2 + r3 + r4;
   a[1][2] = a[2][1] = -r4;
   a[2][2] = r4 + r5;
   a[0][3] = v1;
   a[2][3] = -v2;
```

```
            /*  Perform elimination step.  */
            for (index=0; index<=N-2; index++)
                eliminate(a,N,index);
            /*  Perform back substitution step.  */
            back_substitute(a,N,soln);
            /*  Print solution.  */
            printf("\n");
            printf("Solution: \n");
            for (index=0, index<=N-1; index++)
                printf("Mesh Current %d: %f \n",index+1,soln[index]);
            /*  Exit program.  */
            return 0;
    }
    /*------------------------------------------------------------------*/
    /*  This function performs the elimination step.               */
    void eliminate(double a[N][N+1],int n,int index)
    {
            /*  Declare variables.  */
            int row, col;
            double scale_factor;
            /*  Eliminate variable from equations.  */
            for (row=index+1; row<=n-1; row++)
            {
                scale_factor = -a[row][index]/a[index][index];
                a[row][index] = 0;
                for (col=index+1; col<=n; col++)
                    a[row][col] += a[index][col]*scale_factor;
            }
            /*  Void return.  */
            return;
    }
    /*------------------------------------------------------------------*/
    /*  This function performs the back substitution.             */
    void back_substitute(double a[N][N+1],int n,
                         double soln[N])
    {
            /*  Declare variables. */
            int row, col;
            /*  Perform back substitution in each equation.  */
            soln[n-1] = a[n-1][n]/a[n-1][n-1];
            for (row=n-2; row>=0; row--)
            {
                for (col=n-1; col>=row+1; col--)
                    a[row][n] -= soln[col]*a[row][col];
                soln[row] = a[row][n]/a[row][row];
            }
            /*  Void return.  */
            return;
    }
    /*------------------------------------------------------------------*/
```

To handle larger systems of equations, the symbolic constant N must be changed; the steps in the Gauss elimination do not need any modifications.

5. TESTING

The program interaction using the data from the hand example follows:

```
Enter resistor values in ohms:
(R1, R2, R3, R4, R5)
1 1 1 1 1
Enter voltage values in volts:
(V1, V2)
5 5

Solution:
Mesh Current 1: 2.500000
Mesh Current 2: 0.000000
Mesh Current 3: -2.500000
```

The program assumes that the system of equations has a solution, which means that none of the equations represents the same equation or parallel equations. We could modify this program to check for these conditions by adding additional statements or functions.

MODIFY!

Use the program developed in this section to answer the following questions:

1. Determine the mesh currents if all five resistors are 5 ohms and both voltage sources are 10 volts.
2. Verify your answer in Problem 1 by using matrix multiplication as discussed in this section. (This problem assumes that you covered the previous section on matrices and vectors.)
3. Determine the mesh currents if the resistors have the values of 2, 8, 6, 6, and 4 ohms, and the voltage sources have the values of 40 and 20 volts.
4. Verify your answer to Problem 3 by substituting your answer back into the original set of three equations.

5.13 Higher Dimensional Arrays*

Three-dimensional array

C allows arrays to be defined with more than two subscripts. For example, this statement defines a **three-dimensional array**:

```
int b[3][4][2];
```

The three subscripts, which are necessary to specify a specific element, will correspond to the x-, y-, and z-coordinates if you position the array at the origin of a three-dimensional space, as shown in Figure 5.7. Thus, the position that is shaded corresponds to b[2][0][1].

---

*Optional section.

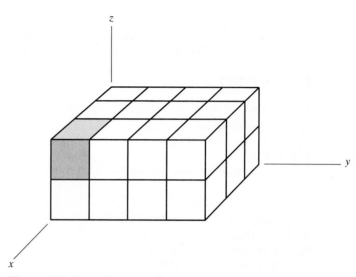

Figure 5.7 *Three-dimensional array.*

Most engineering problems that need arrays can be solved using one-dimensional or two-dimensional arrays. However, there are occasionally problems that are good candidates for using higher dimensional arrays. These problems typically involve data that are specified by several parameters; in addition, the parameters are either integers that are sequential or parameters that can easily be converted into sequential parameters. For example, suppose that a set of data representing temperature measurements is taken from the floor of a large chemical reaction chamber. Furthermore, this set of temperatures is taken at specified intervals of time during a chemical reaction. In this case, we might choose to use a three-dimensional array, using the first subscript to indicate a specific time and the other two subscripts to indicate the location within the floor. The subscripts would need to begin with zero to match the requirements of C subscripts. The subscripts [3][2][5] would then specify the value taken at the fourth time value, and at position [2][5] in the grid of temperatures.

Arrays with over three subscripts are seldom used, because it is difficult to visualize them. However, a simple way to visualize arrays with over three subscripts can be developed. First, imagine that a three-dimensional array is a building. The building has floors, and it also has a rectangular grid of rooms on each floor. Assume each room can contain a single value. The three-dimensional array representing the building uses three subscripts to specify a room; the first subscript is the floor number and the other two subscripts specify the row and column number of the room on the specified floor.

Four-dimensional array

A **four-dimensional array** is a row of buildings, as shown in Figure 5.8. The first subscript specifies the building, and the remaining three subscripts specify the room in the building.

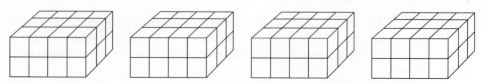

Figure 5.8 *Four-dimensional array.*

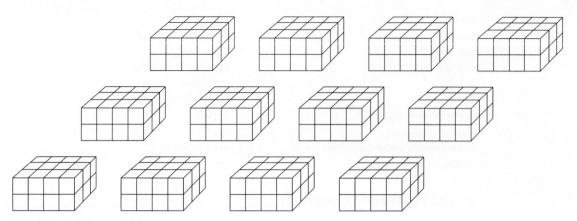

Figure 5.9 *Five-dimensional array.*

Five-dimensional array

A **five-dimensional array** is a block of buildings, as shown in Figure 5.9. The first two subscripts specify the building in the block, and the remaining three subscripts specify the room in the building.

This analogy could continue with a row of blocks, then a city of blocks, then a group of cities, then a state of cities, and so on. Although we have shown you how to visualize higher dimensional arrays, we also want to caution you about using them. Higher order arrays have a lot of overhead related to the subscripting; not only are there extra subscripts required, but extra loops are necessary each time you want to work with groups of values in the array. In general, higher order arrays also complicate the debugging and maintenance of the program. Therefore, higher order arrays should only be used when they simplify the overall visualization of the problem and the steps to solve it.

SUMMARY

An array is a data structure often used to store engineering data that are analyzed in a program. If the data are best represented by a list of information, a one-dimensional array is used; if the data are best represented by a table or grid of information, a two-dimensional array is used. Many examples were developed in this chapter to illustrate array definitions, array initializations, computations with arrays, input and output with arrays, and arrays as function parameters. A set of statistical functions was developed for computing statistical information for analyzing or sorting one-dimensional arrays. The Gauss elimination technique for solving a system of simultaneous linear equations was also presented, and a C program was developed to implement this technique.

KEY TERMS

array	Gauss elimination
binary search	hyperplane
call-by-address	ill conditioned
collating sequence	inner product
determinant	magnitude
dot product	matrix
element	matrix multiplication

mean	square matrix
median	standard deviation
nonsingular	subscripts
one-dimensional array	system of equations
parsing	transpose
power	two-dimensional array
selection sort algorithm	variance
sequential search	vector
simultaneous linear equations	zero crossing
sorting	

C STATEMENT SUMMARY

Array declaration:

```
int a[5], b[]={2,3,-1};
char vowels[]={'a','e','i','o','u'};
double x[10][5];
```

NOTES

1. The variable k is commonly used as a subscript for a one-dimensional array.
2. Use symbolic constants to declare the size of an array so that it is easy to modify.
3. In documentation, describe a two-dimensional array as a grid with rows and columns.
4. The variables i and j are commonly used as subscripts for a two-dimensional array.
5. List both the row size and the column size of arrays in the formal argument list and in the function prototype statement.

DEBUGGING NOTES

1. Only use arrays when it is necessary to keep all the data available in memory.
2. Be careful not to exceed the maximum subscript value when referencing an element in an array.
3. Select conditions in for loops to specifically use an equality with the maximum subscript value; this helps avoid errors with subscript ranges.
4. An array must be declared to be as large as, or larger than, the maximum number of values to be stored in it.
5. Because an array reference in a function is a call-by-address reference, be careful that you do not inadvertently change values in an array in the function.
6. Be sure to enclose each subscript in its own set of brackets when referencing elements of a multidimensional array.
7. When translating matrix notation to C, remember that the first row and column in a matrix is referenced with a subscript 1, not 0.
8. Multidimensional matrices complicate the logic of a program, and they should be used only when they are necessary.

PROBLEMS

SHORT ANSWER PROBLEMS

True–False Problems

Indicate whether the following statements are true (T) or false (F):

1. If the initializing sequence is shorter than an array, then the rest
    of the values are initialized to zero.   T   F
2. If the subscript of an array is a floating point number, it will cause an
    execution error.   T   F
3. When the value of a subscript or index is greater than the largest
    valid subscript of an array, it will always cause an execution error.   T   F
4. If a two dimensional array of size 5 × 4 is declared, say, arr[5][4],
    the last element is denoted as arr[4][3].   T   F

Multiple Choice

5. An array is
    (a) a group of values that all have a common variable name and all have the same data type.
    (b) a collection of elements of different data types that are stored in adjacent memory
        locations.
    (c) a variable that contains multiple values of the same data type.
    (d) a location in memory that holds multiple values of the same data type.

6. If a static array is not initialized, then the values of the elements in the array are set to
    (a) 0
    (b) '\0'
    (c) NULL
    (d) an arbitrary garbage value.

7. The subscript identifies the _____ of a particular element in the array.
    (a) location
    (b) value
    (c) range
    (d) name

8. Given that a[N][N] is a square matrix, after the execution of the following code, sum
    contains _____.

    ```
    sum = 0;
    for (row=col=0; row<N; row++)
            sum+=a[row][col++];
    ```

    (a) sum of all elements of the matrix.
    (b) trace of the matrix, i.e. sum of diagonal elements of the matrix.
    (c) sum of values of first row.
    (d) sum of values of first column.

Memory Snapshot Problems

Give the corresponding snapshots of memory after each of the following sets of statements is
executed (use a question mark to indicate an array element that is not initialized):

9.  ```
    int i=2, arr[10]={1,2};
    while(i<10)
            arr[i++]=arr[i-2]+arr[i-1];
    ```
    ```

```
10. int r,c, arr[4][2]={{4,4}, {3,2}, {2}, {1}};
    for(r=0;r<4;r++)
        for(c=0;c<2;c++)
            if(arr[r][c]==0)
                arr[r][c]= arr[r][c-1]*2;
```

Program Output

Problems 11 and 12 refer to the following statements:

```
int sum, k, i, j;
int x[4][4]={{1,2,3,4},{5,6,7,8},{9,8,7,3},{2.1,7,1}};
```

11. Give the value in sum after the following statements are executed:

```
sum = x[0][0];
for (k=1; k<=3; k++)
    sum += x[k][k];
```

12. Give the value in sum after the following statements are executed:

```
sum = 0;
for (i=1; i<=3; i++)
    for (j=0; j<=3; j++)
        if (x[i][j] > x[i-1][j])
            sum++;
```

PROGRAMMING PROBLEMS

Linear Interpolation. The following problems refer to wind-tunnel test data available on the companion website. The file contains a flight-path angle (in degrees) and its corresponding coefficient of lift on each line in the file. The flight-path angles will be in ascending order.

13. Write a program that reads the wind-tunnel test data, and then allows the user to enter a flight-path angle. If the angle is within the bounds of the data set, the program should use linear interpolation to compute the corresponding coefficient of lift. (You may need to refer to the section on linear interpolation in Section 2.5.)

14. Modify the program in Problem 13 so that it prints a message to the user. The message should give the range of angles that are covered in the data file after reading the values.

15. Write a function that could be used to verify that the flight-path angles are in ascending order. The function should return a 0 if the angles are not in order and a 1 if they are in order. Assume that the corresponding function prototype is

```
int ordered(double x[],int num_pts);
```

16. Write a function that receives two one-dimensional arrays that correspond to the flight-path angles and their coefficients of lift. The function should sort the flight-path angles into ascending order while maintaining the correspondence between the flight-path angles and their coefficients of lift. Assume that the corresponding function prototype is

```
void reorder(double x[],double y[],int num_pts);
```

17. Modify the program developed in Problem 14 such that it uses the function developed in Problem 15 to determine whether or not the data are in the desired order. If they are not in the desired order, use the function developed in Problem 16 to reorder them.

Noise Signals. In engineering simulations, we often want to generate a floating-point sequence of values with a specified mean and variance. The function developed in Chapter 4 allows us to generate numbers between limits a and b, but it does not allow us to specify the mean and variance. By using results from probability, we can derive the relationship between the limits of a uniform random sequence and its theoretical mean μ and variance σ^2:

$$\sigma^2 = \frac{(b-a)^2}{12}, \qquad \mu = \frac{a+b}{2}.$$

18. Write a program that uses the `rand_float` function developed in Chapter 4 to generate sequences of random floating-point values between 4 and 10. Then compare the computed mean and variance with the computed theoretical values. As you use more and more random numbers, the computed values and the theoretical values should become closer.

19. Write a program that uses the `rand_float` function developed in Chapter 4 to generate two sequences of 500 points. Each sequence should have a theoretical mean of 4, but one sequence should have a variance of 0.5 and the other should have a variance of 2. Check the computed means and compare them with the theoretical means. (*Hint*: Use the two previous equations to write two equations with two unknowns. Then solve for the unknowns by hand.)

20. Write a program that uses the `rand_float` function developed in Chapter 4 to generate two sequences of 500 points. Each sequence should have the same variance of 3.0, but one sequence should have a mean of 0.0 and the other should have a mean of -4.0. Compare the theoretical and computed values for mean and variance. (*Hint*: Use the two previous equations to write two equations with two unknowns. Then solve for the unknowns by hand.)

21. Write a function named `rand_mv` that generates a random floating-point value with a specified mean and variance that are input parameters to the function. Assume that the corresponding function prototype is

```
double rand_mv(double mean,double var);
```

Use the `rand_float` function developed in Chapter 4.

Power Plant Data. The data file `power1.dat` contains a power plant output in megawatts over a period of 8 weeks. Each row of data contains 7 integers that represent 1 week's data. In developing the following programs, use symbolic constants NROWS and NCOLS to represent the number of rows and the number of columns in the array used to store the data. (Generate a file to test these problems.)

22. Write a program to compute and print the average power output over this period of time. Also compute the weekly averages power output and the number of weeks the weekly average was greater than the overall average.

23. Write a program that displays the power output range of each week, i.e. the difference between the maximum output and the minimum output. Print the output in the format:

```
Week        Minimum Power        Maximum Power        Range
            Output in MW         Output in MW         in MW
```

24. Write a function to compute the average of a specified column of a two-dimensional array that has NROWS rows and NCOLS columns. The parameters should be the integer array and the desired column. Assume that the corresponding function prototype is

```
double col_ave(int x[NROWS,NCOLS],int col);
```

25. Write a program that uses the function written in Problem 24 to print a report that lists the average power output for the first day of the week, then for the second day of the week, and so on. Print the information in this format:

```
Day x: Average Power Output in Megawatts:   xxxx.xx
```

26. Write a function to compute the average of a specified row of a two-dimensional array that has NROWS rows and NCOLS columns. The parameters should be the integer array and the desired row. Assume that the corresponding function prototype is

    ```
    double row_ave(int x[NROWS,NCOLS],int row);
    ```

27. Write a program that uses the function written in Problem 26 to print a report that lists the average power output for the first week, then for the second week, and so on. Print the information in this format:

    ```
    Week x: Average Power Output in Megawatts:   xxxx.xx
    ```

28. Write a program to compute and print the mean and variance of the power plant output data.

Cryptography. The science of developing secret codes has interested many people for centuries. Some of the simplest codes involve replacing a character, or a group of characters, with another character, or group of characters. To easily decode these messages, the decoder needs the "key" that shows the replacement characters. In recent times, computers have been used very successfully to decode many codes that initially were assumed to be unbreakable. The next set of problems considers simple codes and schemes for decoding them. Generate files to test the programs.

29. A simple code can be developed by replacing each character by another character that is a fixed number of positions away in the collating sequence. For example, if each character is replaced by the character that is two characters to the right, then the letter 'a' is replaced by the letter 'c,' the letter 'b' is replaced by the letter 'd,' and so on. Write a program that reads the text in a file, and then generates a new file that contains the coded text using this scheme. Do not change the newline characters or the EOF character.

30. Write a program to decode the scheme presented in Problem 29. Test the program using files generated by Problem 29.

31. One step in decoding a simple code (such as the one described in Problem 29) without knowing the coding scheme involves counting the number of occurrences of each character. Then, knowing that the most common letter in English is 'e,' the letter that occurs most commonly in the coded message is replaced by 'e.' Similar replacements are then made based on the number of occurrences of characters in the coded message and the known occurrences of characters in the English language. This decoding often provides enough of the correct replacements that the incorrect replacements can then be determined. For this problem, write a program that reads a data file and determines the number of occurrences of each of the characters in the file. Then, print the characters and the number of times that they occurred. If a character does not occur, do not print it. (*Hint*: Use an array to store the occurrences of the characters, based on their ASCII codes.)

32. Another simple code encodes a message in text such that the true message is represented by the first letter of each word. There are no spaces between the words, but the decoded string of characters can easily be separated into words by a person. Write a program to read a data file and determine the secret message stored by the sequence of first letters of the words.

33. Assume that the true secret message in Problem 32 is stored in the second letter of each word. Write a program to read a data file and determine the secret message stored in the file.

34. Assume that the true secret message in Problem 32 is represented by the characters that are three characters to the right in the collating sequence from the first letters of the words. Write a program to read a data file and determine the secret message stored in the file using this decoding scheme.

35. Write a program that encodes the text in a data file using an integer array named key that contains 26 characters. This key is read from the keyboard; the first letter contains the

character that is to replace the letter a in the data file, the second letter contains the letter that is to replace the letter b in the data file, and so on. Assume that all punctuation is to be replaced by spaces. Check to be sure that the key does not map two different characters to the same one during the encoding.

36. Write a program that decodes the file that is the output of Problem 35. Assume that the same integer key is read from the keyboard by this program, and is used in the decoding steps. Note that you will not be able to restore the punctuation characters.

Temperature Distribution. The temperature distribution in a thin metal plate with constant (or isothermal) temperatures on each side can be modeled using a two-dimensional grid, as shown in Figure 5.10. Typically, the number of points in the grid are specified, as are the constant temperatures on the four sides. The temperatures of the interior points are usually initialized to zero, but they change according to the temperatures around them. Assume that the temperature of an interior point can be computed as the average of the four adjacent temperatures; the points shaded in Figure 5.10 represent the adjacent temperatures for the point labeled x in the grid. Each time that the temperature of an interior point changes, the temperatures of the points adjacent to it change. These changes continue until a thermal equilibrium is achieved and all temperatures become constant.

37. Write a program to model this temperature distribution for a grid with six rows and eight columns. Allow the user to enter the temperatures for the four sides. Use one array to store the temperatures. Thus, when a point is updated, its new value is used to update the next point. Continue updating the points, moving across the rows until the temperature differences for all updates are less than a user-entered tolerance value.

38. Modify the program generated in Problem 37 so that the updates are performed down the columns. Compare the equilibrium values for the two programs using different tolerance values. The equilibrium values should be very close for small tolerance values.

39. Modify the program in Problem 37 so that two arrays are used and so that the program can perform the updates as if they all happen at the same time. Thus, all temperatures are updated using one set of array values. The two arrays are needed so that all the old temperatures are available to compute each new temperature.

Gauss Elimination. The accuracy of the Gauss elimination technique can be improved using a process called pivoting. To perform row pivoting, we first reorder the equations so that the equation with the largest absolute value for the first coefficient is the first equation. We then eliminate the first variable from the equations that follow the first equation. Then, starting with the second equation, we reorder the equations such that the second equation has the largest coefficient (in absolute value) for the second variable. We then eliminate the second variable from all equations after the second equation. The process continues similarly for the rest of the variables. Assume that a symbolic constant N contains the number of equations.

40. Use the program developed in Section 5.12 as a guide to develop a function that receives a double array a of size N by N. A second parameter is a double array soln of size N. The function should solve the system of equations represented by array a, and return the solution in array soln. Assume that the corresponding function prototype is

```
void gauss(double a[N][N+1],double soln[N]);
```

41. Write a function that receives a two-dimensional array and a pivot value that specifies the coefficient of interest, j. The function should then reorder all equations starting with the

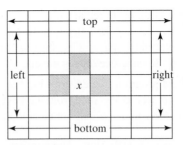

Figure 5.10 *Temperature grid in a metal plate.*

*j*th equation such that the *j*th equation will have the largest coefficient (in absolute value) in the *j*th position. Assume that the function can reference the size of the array as N by N+1 and that the corresponding function prototype is

```
void pivot_r(double a[N][N+1],int j);
```

42. Modify the function developed for Problem 40 so that the row pivoting is performed before each variable is eliminated. Use the function developed in Problem 41.

43. Column pivoting is performed in a similar fashion to row pivoting: by exchanging columns such that the largest coefficient (in absolute value) will be in the position of interest. When columns are exchanged, it is important to keep track of the changes in the order of the variables. Write a function to perform column pivoting. Include parameters to specify changes in the order of the variables. Assume that the corresponding function prototype is

```
void pivot_c(double a[N][N+1],int j,int reorder k[N]);
```

44. Modify the function developed for Problem 40 so that column pivoting is performed before each variable is eliminated. Use the function developed in Problem 43.

45. Modify the function developed for Problem 40 so that both row pivoting and column pivoting are performed before each variable is eliminated. Use the functions developed in Problems 41 and 43.

Set Operations. Problems 46–49 are based upon operations that combine sets to give new sets.

46. The intersection of the two sets A and B, is the set of elements which belong to both A and B. It is represented as $A \cap B$. For example, if A={1, 3, 5} and B={3, 4, 5, 6, 7} then $A \cap B$= { 3, 5 }. Write a function to compute the intersection of two sets. Assume that the corresponding function prototype is:

```
void intersect(int a[], int m, int b[], int n);
```

where a[] and b[] are the sets and m and n are their cardinalities.

47. The union of the two sets A and B is the set of elements which belong to A and B. It is represented as $A \cap B$. For example, if A={1, 3, 5} and B={3,4,5,6} then $A \cap B$= {1,3,4,5,6}. Write a function to compute the union of two sets. Assume that the corresponding function prototype is:

```
void setunion(int a[], int m, int b[], int n);
```

where a[] and b[] are the sets and m and n are their cardinalities.

48. A set, A, is called the proper subset of another set, B, if all the elements that belong to the set A also belong to the set B and there is at least one element in set B which does not belong to set A. It is represented as $A \subset B$. For example, if A={1, 2, 3} and B = {1, 2, 3, 5}, $A \subset B$. Write a function that tests whether a set is a subset of another set. Assume that the corresponding function prototype is:

```
void subset(int a[], int m, int b[], int n);
```

where a[] and b[] are the sets and m and n are their cardinalities.

49. The symmetric difference of two sets, A and B, is the set of elements that belong to either of the sets but not in their intersection. It is represented as $A \ominus B$. It is equivalent to

$$A \ominus B = (A-B) \cup (B-A)$$

where $A-B$ is the set of elements which belong to A but not to B. For example, if A = {1,3,5,7} and B = {4,5,6,7}, then $A-B$ = { 1, 3 } and $A \ominus B$ = { 1, 3, 4, 6 }. Write a function to compute the symmetric difference of two sets. Assume that the corresponding function prototype is:

```
void symdiff(int a[], int m, int b[], int n);
```
where a[] and b[] are the sets and m and n are their cardinalities.

Normalization Technique. There are a number of ways to normalize, or scale, a set of values. One common normalization technique scales the values such that the minimum value goes to 0, the maximum value goes to 1, and other values are scaled accordingly. For example, using this normalization, we can normalize the values in the following array:

Array values

-2	-1	2	0

Normalized array values

0.0	0.25	1.0	0.5

The equation that computes the normalized value from a value x_k in the array is

$$\text{Normalized } x_k = \frac{x_k - \min_x}{\max_x - \min_x},$$

where $\min_x$ and $\max_x$ represent the minimum and maximum values in the array x, respectively. If you substitute the minimum value for x_k in this equation, the numerator is zero, and thus the normalized value for the minimum is zero. If you substitute the maximum value for x_k in this equation, the numerator and denominator are the same value, and hence the normalized value for the maximum is 1.0.

50. Write a function that has a one-dimensional **double** array and the number of values in the array as its arguments. Normalize the values in the array using the technique presented. Assume that the corresponding function prototype is

```
void norm_1D(double x[],int num_pts);
```

51. Write a function that has a two-dimensional **double** array as an argument. Normalize the values in the array using the technique presented. Assume that symbolic contants NROWS and NCOLS specify the size of the array and that these are available to the function. Assume that the corresponding function prototype is

```
void norm_2D(double x[NROWS][NCOLS]);
```

CHAPTER SIX

6

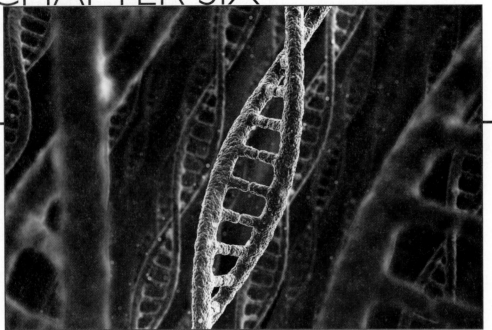

Crime Scene Investigation: DNA Analysis

A match between the DNA of a suspect and a DNA sample collected from a crime scene is considered to be a very strong piece of evidence. The FBI has a database called CODIS (Combined DNA Index System) containing over nine million DNA profiles. These profiles contain the sequence of nucleotides from a given DNA fragment. The DNA nucleotides are either adenine (A), cytosine (C), guanine (G), or thymine (T). Thus, a DNA profile is a character string that consists of letters representing the order of nucleotides, such as AAACGTACAGGT. While 99.9% of human DNA sequences are the same in every person, there are still enough differences to do DNA matching. A DNA match between an unknown DNA string and one from a database is based on a number of matches of short strings (called short tandem repeats, or STRs) between the two strings. Forensic DNA became a viable tool for law enforcement when DNA sequencing machines were developed that could automatically determine the nucleotide sequence of a given DNA fragment. These instruments use lasers to analyze light signals from fluorochromes that are attached to nucleotides. DNA matching, also called genetic fingerprinting, was developed in 1984 by Sir Alec John Jeffreys, a British geneticist who developed the techniques that are the basis of DNA matching today. In this chapter, we will develop a C program to find small strings of DNA in a larger string.

PROGRAMMING WITH POINTERS

CHAPTER OUTLINE

OBJECTIVES *In this chapter, we develop problem solutions containing:*

- pointers to variables;
- pointers to arrays;
- pointers in function references;
- pointers with character strings; and
- dynamic memory allocation.

6.1 Addresses and Pointers

Address

When a C program is executed, memory locations are assigned to the variables used in the program. Each of these memory locations has a positive integer **address** that uniquely defines the location. When a variable is assigned a value, this value is stored in the corresponding memory location. The value of a variable can be used by statements in the program, and it can be changed by statements in the program. The specific addresses used for the variables are determined each time that the program is executed and may vary from one execution to another.

It is sometimes helpful to compare memory allocation to the allocation of a group of post office boxes. If the post office has 100 mailboxes numbered from 1 to 100, then the mailbox number corresponds to the memory address. Each mailbox is assigned to an individual, using the individual's name; this name corresponds to the identifier assigned to a memory location. The contents of the mailbox correspond to the value in the memory location; this value can be examined and it can be changed.

---

*Optional section.

post office box number	individual name	contents
78	John Ruiz	catalog

memory address	identifier	contents
66572	x	105

This analogy is not completely valid, because two individuals might have the same name, but two identifiers cannot be exactly the same. Also, a mailbox might be empty, or it might contain a number of items, whereas a memory location always contains a single value.

Address Operator

In C, the address of a variable can be referenced using the **address operator** &. This operator was introduced in Chapter 2 in conjunction with the scanf statement. For example, a statement to read a floating-point value from the keyboard and to store it in the variable x is the following:

```
scanf("%f",&x);
```

This statement specifies that the value read from the keyboard is to be stored at the address specified by &x, the address of x.

To illustrate the use of the address operator to obtain the memory address of a variable, consider the following program:

```
/*--------------------------------------------------------------*/
/*  Program chapter6_1                                          */
/*                                                              */
/*  This program demonstrates the relationship between          */
/*  variables and addresses.                                    */

#include <stdio.h>

int main(void)
{
   /*  Declare and initialize variables.  */
   int a=1, b=2;

   /*  Print the contents and addresses of a and b.  */
   printf("a = %d;   address of a = %u \n",a,&a);
   printf("b = %d;   address of b = %u \n",b,&b);

   /*  Exit program.  */
   return 0;
}
/*--------------------------------------------------------------*/
```

Note that the address is printed with a %u specification that is used for printing unsigned integers. A sample output from this program is the following:

```
a = 1;   address of a = 1245052
b = 2;   address of b = 1245048
```

The following memory snapshot shows the values in the two memory locations at the time that the printf statements are executed:

a | 1 | b | 2 |

We do not usually indicate the memory addresses in these diagrams because the addresses used are system dependent.

We can modify the program so that no initial values are given to variables a and b. This modification is shown in the following program:

```
/*------------------------------------------------------------*/
/* Program chapter6_2                                         */
/*                                                            */
/* This program demonstrates the relationship between         */
/* variables and addresses.                                   */

#include <stdio.h>

int main(void)
{
   /*  Declare and initialize variables.  */
   int a, b;

   /*  Print the contents and addresses of a and b.  */
   printf("a = %d;   address of a = %u \n",a,&a);
   printf("b = %d;   address of b = %u \n",b,&b);

   /*  Exit program.  */
   return 0;
}
/*------------------------------------------------------------*/
```

A memory snapshot at the time that the printf statements are executed should show a question mark in the variable contents because the values are undefined:

a | ? | b | ? |

A sample output from this program is the following:

```
a = -858993460;   address of a = 1245052
b = -858993460;   address of b = 1245048
```

While we see that there are values in the variables (even though we have not assigned any in the program), we should not assume anything about these values. This example illustrates the importance of being sure that a program initializes a variable before using its value in other statements.

MODIFY!

1. Run program chapter6_1 two times on the computer that you are using for class assignments. Did your computer use the same addresses or different addresses? Compare the results with those of your classmates.

2. Run program chapter6_2 presented in this section. What values were in the locations assigned to a and b? If these values are zero, your compiler may automatically assign the value of zero to undefined variables. This is not an ANSI standard, and you should not assume that undefined variables will have a value of zero. Do the values change from one execution of the program to another?

Pointer Assignment

The C language allows us to store the address of a memory location in a special type of variable called a **pointer**. When a pointer is defined, the type of variable to which it will point must also be defined. Thus, a pointer defined to point to an integer variable cannot also be used to point to a floating-point variable.

Consider a statement that defines two integer variables and a pointer to an integer value. C uses an asterisk to indicate that the variable is a pointer; this asterisk is also called a **dereferencing** or **indirection** operator. This statement could be written as follows:

```
int a, b, *ptr;
```

This statement specifies that memory addresses should be assigned to three variables—two integer variables and a pointer to an integer variable. The statement does not specify the initial values for a and b, and it also does not specify an address to be stored in ptr. Thus, the memory snapshot after this declaration indicates that the initial contents of all variables are not specified; the diagram uses an arrow to indicate that ptr is a pointer variable:

To specify that ptr should point to the variable a, we could use an assignment statement that stores the address of a in ptr:

```
int a, b, *ptr;
ptr = &a;
```

This assignment could also have been made on the declaration statement:

```
int a, b, *ptr=&a;
```

In either case, the memory snapshot after the declaration is the following:

Note that it is not necessary to show the contents of ptr as long as the variable to which it points is specified.

Consider this set of statements:

```
/*  Declare and initialize variables.   */
int a=5, b=9, *ptr=&a;
...
/*  Assign the value pointed to by ptr to b.   */
b = *ptr;
```

This last statement is read as "b is assigned the value at the address contained in ptr" or "b is assigned the value pointed to by ptr." The memory snapshot after the declaration statement is executed is the following:

The memory snapshot after the assignment statement is executed is the following:

Thus, b is assigned the value pointed to by `ptr`.

Now consider this set of statements:

```
/*  Declare and initialize variables.  */
int a=5, b=9, *ptr=&a;
...
/*  Assign the value of b to the variable  */
/*  to which ptr points.                   */
*ptr = b;
```

The memory snapshot before the assignment statement is executed is the following:

Thus, the value pointed to by `ptr` is assigned the value in b.

We now extend program `chapter6_1` to demonstrate the relationship between variables, addresses, and pointers. Consider the following program:

```
/*-------------------------------------------------------------*/
/*  Program chapter6_3                                         */
/*                                                            */
/*  This program demonstrates the relationship between        */
/*  variables, addresses, and pointers.                       */

#include <stdio.h>

int main(void)
{
    /*  Declare and initialize variables.  */
    int a=1, b=2, *ptr=&a;

    /*  Print the variable and pointer contents.  */
    printf("a = %d;   address of a = %u \n",a,&a);
    printf("b = %d;   address of b = %u \n",b,&b);
```

```
    printf("ptr = %u;   address of ptr = %u \n",ptr,&ptr);
    printf("ptr points to the value %d \n",*ptr);

    /*  Exit program.   */
    return 0;
}
/*-----------------------------------------------------------------*/
```

A sample output from this program is the following:

```
a = 1;   address of a = 1245052
b = 2;   address of b = 1245048
ptr = 1245052;   address of ptr = 1245044
ptr points to the value 1
```

Note that the values of the pointer to a and the address of a are the same.

PRACTICE!

Give memory snapshots after each of these sets of statements is executed:

1. `int a=1, b=2, *ptr;`

    `. . .`

    `ptr = &b;`

2. `int a=1, b=2, *ptr=&b;`

    `. . .`

    `a = *ptr;`

3. `int a=1, b=2, c=5, *ptr=&c;`

    `. . .`

    `b = *ptr;`

    `*ptr = a;`

4. `int a=1, b=2, c=5, *ptr;`

    `. . .`

    `ptr = &c;`

    `c = b;`

    `a = *ptr;`

Pointers were used in Chapter 3 to access data files. Recall that a pointer to a file (called a file descriptor) was defined using a FILE declaration, as in

`FILE *sensor;`

where the FILE data type is defined in <stdio.h>. The pointer was associated with a specific file using the fopen statement, as shown in the following statement:

`sensor = fopen("sensor1.txt","r");`

This statement also indicates that sensor1.txt is an input file because we will be reading information from it, as specified by the parameter "r". The pointer is used again with the fscanf statement to point to the file from which we want to read data:

```
fscanf(sensor,"%f %f",&t,&motion);
```

The file pointer is necessary in the fscanf function because we may be reading information from several files in the same program. A pointer variable is used in a similar manner with an output file and the fprintf function.

Address Arithmetic

The operations that can be performed with pointers (or addresses) are limited to the following:

- A pointer can be assigned to another pointer of the same type;
- An integer value can be added to or subtracted from a pointer;
- A pointer can be assigned to or compared with the integer zero or to the symbolic constant **NULL**, which is defined in <stdio.h>; and

NULL

- Pointers to elements of the same array can be subtracted or compared as a means to accessing elements in the array.

A pointer can point to only one location, but several pointers can point to the same location. Both ptr_1 and ptr_2 point to the same variable after we execute the following statements:

```
/*  Declare and initialize variables.  */
int x=-5, y=8, *ptr_1, *ptr_2;
...
/*  Assign both pointers to x.  */
ptr_1 = &x;
ptr_2 = ptr_1;
```

The memory snapshot after these statements are executed is the following:

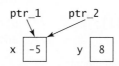

To illustrate some common errors that can be made when working with pointers, we now present several invalid statements using these variables:

```
&y = ptr_1;          /*  invalid  statement - attempts   */
                     /*  to change the address of y       */

ptr_1 = y;           /*  invalid statement - attempts     */
                     /*  to change ptr_1 to a             */
                     /*  nonaddress value                 */

*ptr_1 = ptr_2;      /*  invalid statement - attempts     */
                     /*  to move an address to an         */
                     /*  integer variable                 */
```

```
ptr_1 = *ptr_2;        /*  invalid statement - attempts    */
                       /*  to change ptr_1 to a            */
                       /*  nonaddress value                */
```

Style

It is instructive to attempt to draw memory snapshots of these invalid statements; in each statement, we are attempting to store a variable value in a pointer, or we are attempting to store a pointer in a variable. To help avoid these errors, use identifier names for pointers that clearly indicate that the identifiers are associated with pointers.

When simple variables are defined, we should not make any assumptions about the relationships of the memory locations assigned to the variables. For example, if a declaration statement defines two integers, a and b, we should not assume that the values are adjacent in memory; we also should not make assumptions about which value occurs first in memory. The memory assignment of a simple variable is system dependent. However, the memory assignment for an array is guaranteed to be a sequential group of memory locations. Thus, if array x contains five integers, then the memory location for x[1] will immediately follow the memory location for x[0], and the memory location for x[2] will immediately follow the memory location for x[1], and so on. Therefore, if ptr_x is a pointer to an integer, we can initialize it to point to the integer x[0] with this statement:

```
ptr_x = &x[0];
```

To move the pointer to x[1], we can increment ptr_x by 1, which causes it to point to the value that follows x[0], or we can assign ptr_x the address of x[1]. Thus, we could cause ptr_x, which currently points to x[0], to be changed to point to x[1] by using only the following statements:

```
++ptr_x;               /*  increment ptr_x to point to the  */
                       /*  next value in memory             */

ptr_x++;               /*  increment ptr_x to point to the  */
                       /*  next value in memory             */

ptr_x += 1;            /*  increment ptr_x to point to      */
                       /*  the next value in memory         */

ptr_x = &x[1];         /*  ptr_x is assigned the            */
                       /*  address of x[1]                  */
```

Similarly, the statement

```
ptr_x += k;
```

refers to the address of the value that is k values past the one to which ptr_x pointed before this statement was executed. These are all examples of adding integers to pointers. Similarly, integers can also be subtracted from pointers. Section 6.2 expands this discussion for one-dimensional arrays and for multidimensional arrays.

When an integer value is added to or subtracted from a pointer, we assume that the integer refers to the number of values from the one referenced by the pointer before the addition or subtraction is performed. For example, the statement

```
ptr++;
```

indicates that ptr should be modified such that it points to the next value in memory, which is the value that follows the one referenced by ptr before it is incremented. Because different

Table 6.1   Operator Precedence

Precedence	Operation	Associativity
1	( ) [ ]	innermost first
2	++ -- + - ! (type) & *	right to left (unary)
3	* / %	left to right
4	+ -	left to right
5	< <= > >=	left to right
6	== !=	left to right
7	&&	left to right
8	||	left to right
9	?:	right to left
10	= += -= *= /= %=	right to left
11	,	left to right

types of values require different amounts of memory, the actual value added to `ptr` depends on the variable type. A floating-point value requires more memory than a short integer, and thus the address increment for a floating-point value will be greater than an address increment for a short integer. For example, the memory addresses for consecutive short integers might be 45530 and 45532, and memory addresses for consecutive floating-point values might be 50200 and 50204. Fortunately, the compiler will determine the correct memory address when we add an integer to a pointer or when we subtract an integer from a pointer.

A pointer operation can be included in a statement with other operations, so it is important to be sure that the operator precedence is specified correctly. An address operator is a unary operation, and thus it is performed before a binary operation. Unary operations are also performed from right to left. These precedence rules are summarized in Table 6.1. Remember that parentheses can always be used to change the precedence of operations.

Errors with pointers can cause problems that are difficult to debug. Even worse, pointer errors can often cause a program to give incorrect results while appearing to work properly. Many pointer errors are caused by pointers that were not initialized before being used. Therefore, it is a good habit to initialize all pointers at the beginning of the program. If a pointer is not initially assigned to a memory location, give it a NULL value to indicate that the pointer has not been assigned to a memory location. To determine whether the pointer named `ptr_1` has been assigned to a memory location at some point in the program, use an `if` statement that contains a condition such as (`ptr_1` == NULL). Work the problems in the practice! exercises on the next page before continuing with Section 6.2.

6.2   Pointers to Array Elements

In Chapter 5, we covered arrays and array handling extensively, using subscripts to specify individual array elements. Pointers can also be used to specify individual array elements. Array references using pointers and addresses are almost always faster than references using subscripts; thus, pointer references for arrays are generally preferred if speed is a concern. As discussed in Section 6.1, pointer references to array values are based on the knowledge that memory assignment of array values is always sequential.

PRACTICE!

For each problem, give a memory snapshot that includes both variables and pointer references after the problem statements are executed. Include as much information as possible. Use question marks to indicate memory locations that have not been initialized.

1. `double x=15.6, y=10.2, *ptr_1=&y, *ptr_2=&x;`

   `...`

   `*ptr_1 = *ptr_2 + x;`

2. `int w=10, x=2, *ptr_2=&x;`

   `...`

   `*ptr_2 -= w;`

3. `int x[5]={2,4,6,8,3};`
   `int *ptr_1=NULL, *ptr_2=NULL, *ptr_3=NULL;`

   `...`

   `ptr_3 = &x[0];`
   `ptr_1 = ptr_2 = ptr_3 + 2;`

4. `int w[4], *first_ptr=NULL, *last_ptr=NULL;`

   `...`

   `first_ptr = &w[0];`
   `last_ptr = first_ptr + 3;`

One-Dimensional Arrays

Consider the following declaration that defines and initializes a one-dimensional array with floating-point values:

`double x[6]={1.5,2.2,4.3,7.5,9.1,10.5};`

The memory snapshot after this statement is executed is the following:

x[0]	1.5
x[1]	2.2
x[2]	4.3
x[3]	7.5
x[4]	9.1
x[5]	10.5

Reference x[0] refers to the first element in the array, reference x[1] refers to the second element in the array, and reference x[k] refers to the k + 1 element in the array. Similar references can be generated with pointers. Assume that a pointer ptr has been defined to reference double values and is then initialized with this statement:

`ptr = &x[0];`

The address of the first element in the array is then stored in the pointer. Thus, reference *ptr refers to x[0], reference *(ptr+1) refers to x[1], and reference *(ptr+k) refers to x[k].

Offset

The value of k in reference $*(ptr+k)$ is often referred to as an **offset** from the beginning element in the array. The following diagram shows the memory allocation for the array used in this example with the offsets added:

```
                              offset

               x[0]   1.5      0

               x[1]   2.2      1

               x[2]   4.3      2

               x[3]   7.5      3

               x[4]   9.1      4

               x[5]  10.5      5
```

We can compute the sum of the values in the array x using subscripts and a `for` loop:

```
/*  Declare and initialize variables.  */
int k;
double x[6], sum=0;
...
/*  Sum the values in the array x.  */
for (k=0; k<=5; k++)
   sum += x[k];
```

An equivalent set of statements that uses pointers instead of subscripts is the following:

```
/*  Declare and initialize variables.  */
int k;
double x[6], sum=0, *ptr=&x[0];
...
/*  Sum the values in the array x.  */
for (k=0; k<=5; k++)
   sum += *(ptr+k);
```

Note that reference $*(ptr+k)$ requires parentheses to perform the operations in the correct order; k is added to the address in `ptr`, and then the indirection operator refers to the value pointed to by ptr+k. Reference *ptr+k would not work correctly because it would be computed as if it were (*ptr)+k because a unary operator has precedence over a binary operator.

In the discussion on arrays in Chapter 5, we saw that using an array name as a function parameter passed the address of the first element of the array to the associated procedure. An array name can also be used to represent the address of the first element of the array in other statements. For example, the following two statements are equivalent:

```
ptr = &x[0];
ptr = x;
```

Similarly, reference $*(ptr+k)$ is also equivalent to $*(x+k)$. Thus, the statements to sum array x can be simplified to these:

```
/*  Declare and initialize variables.  */
int k;
double x[6], sum=0;
...
/*  Sum the values in the array x.  */
for (k=0; k<=5; k++)
    sum += *(x+k);
```

This example illustrates the use of an array name as an address. In most statements, the array name can replace a pointer, but it cannot be used on the left side of an assignment statement, because its value cannot be changed.

PRACTICE!

Assume that an array g is defined with the following statement:

```
int g[]={2,4,5,8,10,32,78};
int *ptr1=&g[0], *ptr2=&g[3];
```

Give a diagram of the memory allocation, including the array values. Also indicate the offset values from the initial value in the array. Using this information, give the value of the following references:

1. *g	2. *(g+1)
3. *g+1	4. *(g+5)
5. *ptr1	6. *ptr2
7. *(ptr1+1)	8. *(ptr2+2)

Two-Dimensional Arrays

A two-dimensional array is stored in sequential memory locations, in row order, as we will show in an array definition, array diagram, and corresponding memory allocation map. Note that the memory allocation map also shows the offset from the initial value in the array:

Array definition: `int s[2][3] = {{2,4,6},{1,5,3}};`

Arrary diagram:

2	4	6
1	5	3

offset

Memory allocation:

		offset
s[0][0]	2	0
s[0][1]	4	1
s[0][2]	6	2
s[1][0]	1	3
s[1][1]	5	4
s[1][2]	3	5

PRACTICE!

Draw the memory allocation for each of the following arrays, and indicate the values stored in the locations. Use a question mark to indicate positions that are not initially assigned a value. Also indicate the offset from the initial array value for each of the array elements.

1. `int d[4][2]={{1,6}};`
2. `int g[3][4]={{5,2,-2,3},{1,2,3,4}};`
3. `float h[3][3]={{0}};`

A pointer can be used to reference an element in a two-dimensional array using the offset from the initial element in the array. If pointer `ptr` has been initialized to point to `s[0][0]`, then reference `*(ptr+k)` accesses the array element with the offset of k. To illustrate, suppose that we want to compute the sum of the elements in an array `s` with two rows and three columns. The following statements compare a solution using subscripts and a solution using indirection references with pointers:

Solution 1

```
/*  Declare and initialize variables.  */
int s[2][3], srows=2, scols=3, i, j, sum=0;
...
/*  Sum the values in the array s.  */
for (i=0; i<=srows-1; i++)
   for (j=0; j<=scols-1; j++)
      sum += s[i][j];
```

Solution 2

```
/*  Declare and initialize variables.  */
int s[2][3], s_count=6, k, sum=0, *ptr=&s[0][0];
...
/*  Sum the values in the array s.  */
for (k=0; k<=s_count-1; k++)
   sum += *(ptr+k);
```

Both solutions correctly compute the sum of the array elements. Note that the first solution required nested loops, because both a row subscript and a column subscript were needed. However, the second solution required only a single loop to specify the offset from the initial array value. It is also interesting to observe that both solutions add the elements in an order that moves across the rows. Solution 1 could be modified to add the elements in an order that moves down the columns; this could be accomplished by interchanging the two `for` loops.

PRACTICE!

Assume that an integer array *x* is defined by the following statements:

`int x[2][4]={{1,8,7,6},{2,4,-1,0}}, *xptr=&x[0][0];`

Draw a memory allocation diagram, and give the value indicated by each of the following references:

1. `*xptr`
2. `*(xptr+2)`
3. `*xptr + 2`
4. `*(xptr+1) + *(xptr+3)`

To convert reference s[i][j] to an offset from pointer sptr, which has been initialized to &s[0][0], the number of columns in the array must be known. If we assume that scols contains the number of columns used in the memory allocation for s, then the offset for the value in row i and column j is equal to i*scols + j. To demonstrate the validity of this offset formula, consider the memory allocation for array s with the offset included:

		offset
Memory allocation:	s[0][0] 2	0
	s[0][1] 4	1
	s[0][2] 6	2
	s[1][0] 1	3
	s[1][1] 5	4
	s[1][2] 3	5

Suppose that we wish to convert reference s[0][1] to an offset from the initial value in the array. According to the formula, the offset should be 0*scols + 1, or 1; a corresponding reference is *(sptr+1). Similarly, because this array has three columns, a reference s[1][2] has an offset of 1*scols + 2, or 5; a corresponding reference is *(sptr+5). Similar formulas can be developed to convert higher dimensional array references to an offset from the first array value.

In C, a two-dimensional array can also be considered to be a one-dimensional array with each element being another one-dimensional array. For example, array s with two rows and three columns can be considered to be a one-dimensional array (with two elements), with each element being another one-dimensional array (with three elements). Thus, reference s[1][0] is equivalent to *s[1] because s[1] represents the address of the first element in row 1. Similarly *(s[0]+4) references the value in position s[1][1]. For readability, we usually use pointers with an offset instead of a one-dimensional reference with an offset when we want to use an indirect reference to a two-dimensional array.

PRACTICE!

Assume that array a has been defined to contain four rows and six columns. Also assume that values have been read into the array from a data file. Use pointers with an offset to perform the following operations:

1. Find the sum of the second row of values.
2. Find the sum of the third column of values.
3. Find the maximum in the first three rows of values.
4. Find the minimum in the last four columns of values.

6.3   Problem Solving Applied: El Niño–Southern Oscillation Data

El Niño

La Niña

ENSO

Along the equator, normal sea-surface temperatures are warm on the western side of the Pacific Ocean and cold on the eastern side of the Pacific Ocean. In a reoccurring phenomenon, a warm current causes the ocean temperatures on the eastern side of the Pacific (along the western shores of California, Mexico, and South America) to increase as much as 18°F. This phenomenon often occurs near Christmas; thus, it is often called **El Niño**. (In Spanish, el niño means a male child.) A reverse phenomenon also occurs in which the temperatures on the western side of the Pacific Ocean become colder; this is called **La Niña**. (In Spanish, la niña means a female child.) These conditions relate to the southern oscillation between warm currents and east–west atmospheric pressure changes. The El Niño–Southern Oscillation (**ENSO**) index is a metric that is computed from a number of variables, including atmospheric pressure, wind, and ocean temperature. When the ENSO index is positive, the ocean temperatures represent the El Niño condition; when the ENSO index is negative, the ocean temperatures represent the La Niña condition. The larger the index, the larger the variation in temperatures from normal. Write a program that reads a data file that contains the year, quarter, and ENSO index for that period of time. The program should determine and print the year and quarter with the strongest El Niño conditions.

1. PROBLEM DESCRIPTION

Determine the year and quarter with the strongest El Niño conditions.

2. INPUT/OUTPUT DESCRIPTION

The I/O diagram shows the data file as the input and the year and quarter as output.

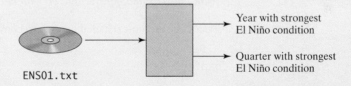

Year with strongest
El Niño condition

Quarter with strongest
El Niño condition

ENS01.txt

3. HAND EXAMPLE

Assume that the data file contains the following data:

Year	Quarter	ENSO Index
1990	1	0.6
1991	1	0.2
1992	1	1.1
1993	1	0.5
1994	1	0.1

Year	Quarter	ENSO Index
1995	1	1.2
1996	1	−0.3
1997	1	−0.1
1998	1	2.2
1999	1	−0.7
2000	1	−1.1

The corresponding output would then be the following report:

```
Maximum El Nino Conditions in Data file
Year: 1998,  Quarter: 1
```

4. ALGORITHM DEVELOPMENT

We first develop the decomposition outline because it divides the solution into a set of sequential steps.

Decomposition Outline

1. Read the ENSO data into arrays and determine the maximum positive index.
2. Print the year and quarter that match the maximum intensity.

The refinement in pseudocode follows:

Refinement in Pseudocode

main: if file cannot be opened

        print error message

   else

        read data and determine maximum intensity

        print the year and quarter that go with maximum intensity

The steps in the pseudocode are now detailed enough to convert to C.

```
/*-------------------------------------------------------------------*/
/*  Program chapter6_4                                               */
/*                                                                  */
/*  This program reads a data file of ENSO index values and         */
/*  determines the maximum El Nino condition in the file.           */

#include <stdio.h>
#define FILENAME "ENSO1.txt"
#define MAX_SIZE 1000

int main(void)
{
   /*  Declare variables and function prototypes.  */
   int k=0, year[MAX_SIZE], qtr[MAX_SIZE], max_k=0;
   double index[MAX_SIZE];
   FILE *enso;
```

```
        /*  Read sensor data file.  */
        enso = fopen(FILENAME,"r");
        if (enso == NULL)
           printf("Error opening input file. \n");
        else
        {
           while (fscanf(enso,"%d %d %lf",
                        year+k,qtr+k,index+k)==3)
           {
              if (*(index+k) > *(index+max_k))
                 max_k = k;
              k++;
           }

           /*  Print data for maximum El Nino condition.  */
           printf("Maximum El Nino Conditions in Data File \n");
           printf("Year: %d, Quarter: %d \n",
                   *(year+max_k),*(qtr+max_k));

           /*  Close file.  */
           fclose(enso);
        }

        /*  Exit program.  */
        return 0;
}
/*-----------------------------------------------------------*/
```

5. TESTING

The output from the program using the data from the hand example is as follows:

```
Maximum El Nino Conditions in Data File
Year: 1998, Quarter: 1
```

MODIFY!

1. Modify the program to find and print the maximum La Niña conditions.
2. Modify the program to find the conditions closest to zero. These would be the conditions that are closest to normal.
3. Modify the program so that it prints all years and quarters with El Niño conditions.

6.4   Pointers in Function References

In C, most function references are call-by-value references. Thus, the values of the actual parameters are copied to the formal parameters. All computations in the function use the formal parameters, and thus an actual parameter cannot be changed in a function. One exception to this rule was presented in Chapter 4; when an array name is used as an argument in a function reference, the address of the array is transferred to the function, and all references to array values use the actual array locations. Thus, values in an array can be modified by statements within a function. Other exceptions can be implemented using pointers as function parameters.

To illustrate the use of pointers as function parameters, we develop a function that exchanges the contents of two memory locations. Recall that it takes three statements to switch the values in two locations; the correct statements and corresponding memory snapshots are as follows:

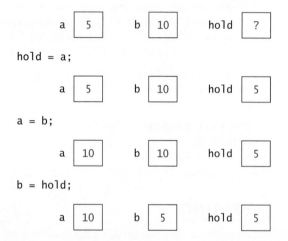

In problem solutions that switch several values, it would be convenient to access a function to perform the switch. Consider the following function, which attempts to perform the switch using simple variable parameters (call-by-value):

```
/*-------------------------------------------------------------*/
/*  Incorrect function to switch values in two variables.      */
void switch1(int a, int b)
{
    /*  Declare variables.  */
    int hold;

    /*  Switch values in a and b.  */
    hold = a;
    a = b;
    b = hold;

    /*  Void return.  */
    return;
}
/*-------------------------------------------------------------*/
```

Assume that the following statement references this function:

```
switch1(x,y);
```

If x and y contain the values 5 and −2, then the transfer of the values from the actual parameters to the formal parameters at the beginning of the function execution is the following:

After the function is executed, the values of the actual parameters and formal parameters are as follows:

Since this function uses a call-by-value reference, the value of the actual parameters (not the address), is passed to the formal parameters. The values have been switched in the formal parameters, but these values are not transferred back to the actual parameters.

After considering this incorrect solution, we are now ready to develop a function that switches the contents of two simple variables using pointers. The function has two parameters that are pointers to the two variables that we want to switch. A prototype statement for this function is the following:

```
void switch2 (int *a,int *b);
```

Thus, the function does not return a value, and its two parameters are pointers to integers. The correct function will appear as follows:

```
/*------------------------------------------------------------*/
/*  Correct function to switch values in two variables.       */
void switch2(int *a,int *b)
{
    /*  Declare variables.  */
    int hold;

    /*  Switch values in a and b.  */
    hold = *a;
    *a = *b;
    *b = hold;

    /*  Void return.  */
    return;
}
/*------------------------------------------------------------*/
```

If x and y are simple variables, then a valid call to this function is

```
switch2(&x,&y);
```

If `ptr_1` points to variable x, and `ptr_2` points to the variable y, then the values in x and y can be switched with this reference:

```
switch2(ptr_1,ptr_2);
```

Elements `x[i]` and `x[j]` could also be switched with the statements

```
switch2(&x[i],&x[j]);
switch2(x+i,x+j);
```

but not with the statement

```
switch2(x[i],x[j]);  /*  invalid statement  */
```

The actual parameter that corresponds to a pointer argument must be an address or pointer.

PRACTICE!

Consider the references to the `switch2` function. For invalid references, explain why the reference is invalid. For valid references, give a memory snapshot before and after the reference.

1. `float x=1.5, y=3.0, *ptr_x=&x, *ptr_y=&y;`

   `...`

   `switch2(ptr_x,ptr_y);`

2. `int f=2, g=7, *ptr_f=&f, *ptr_g=&g;`

   `...`

   `switch2(ptr_f,ptr_g);`

3. `int f=2, g=7, *ptr_f=&f, *ptr_g=&g;`

   `...`

   `switch2(*ptr_f,*ptr_g);`

4. `int f=2, g=7, *ptr_f=&f, *ptr_g=&g;`

   `...`

   `switch2(&ptr_f,&ptr_g);`

5. `int f=2, g=7, *ptr_f=&f, *ptr_g=&g;`

   `...`

   `switch2(&f,&g);`

6. `int f=2, g=7, *ptr_f=&f, *ptr_g=&g;`

   `...`

   `switch2(f,g);`

6.5   Problem Solving Applied: Seismic Event Detection

Seismometers

Richter scale

Seismic events

Short-time power

Long-time power

Threshold

Data window

Special sensors called **seismometers** are used to collect earth motion information. These seismometers can be used in a passive environment, in which they record the earth's motion, including earthquakes and tidal motion. By using data from several seismometers to analyze ground motion from an earthquake, it is possible to determine the epicenter of the earthquake and the intensity of the earthquake. The intensity is usually measured using the **Richter scale**, which is a scale from 1 to 10 named after U.S. seismologist C. F. Richter.

Write a program that reads a set of seismometer data from a data file named `seismic1.txt`. The first line of the file contains two values—the number of seismometer data readings that follow in the file and the time interval in seconds that occurred between consecutive measurements. This time interval is a floating-point value, and we assume that all the measurements were taken with the same time interval between them. After reading and storing the data measurements, the program should then identify possible earthquakes, which are also called **seismic events**, using a power ratio. At a specific point in time, this ratio is the quotient of a **short-time power** measurement divided by a **long-time power** measurement. If the ratio is greater than a given **threshold**, an event may have occurred at that point in time. Given a specific point in the data measurements, the short-time power is the average power, or average squared value, of the measurements using the specified point plus a small number of points that occurred just previous to the specified point. The long-time power is the average power of the measurements using the specified point plus a larger number of points that occurred just previous to the specified point. (The set of points used in a calculation is sometimes referred to as a **data window**.) The threshold is generally greater than 1 to avoid detecting events in constant data, because the short-time power is equal to the long-time power if the data values are all the same value. Assume that the numbers of measurements for the short-time power and for the long-time power are read from the keyboard. Set the threshold value to 1.5.

1. PROBLEM STATEMENT

Determine the locations of possible seismic events using a set of seismometer measurements from a data file.

2. INPUT/OUTPUT DESCRIPTION

The inputs are a data file named `seismic1.txt` and the number of measurements to use for short-time power and long-time power. The output is a report giving the times of potential seismic events.

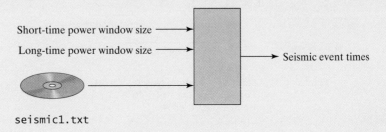

3. HAND EXAMPLE

Suppose that a data file contains the following data, which include the number of points to follow (11) and time interval between points (0.01), followed by the 11 values that correspond to a sequence of values $x_0, x_1, \ldots x_{10}$:

```
11 0.01
1    2    1    1    1    5    4    2    1    1    1
```

If the short-time power measurement is made using two samples, and the long-time power measurement is made using five measurements, then we can compute power ratios, beginning with the rightmost point in a window:

```
1    2    1    1    1    5    4    2    1    1    1
                  |___|
              short window
|_____|
      long window
```

Point x_4:   Short-time power = $(1 + 1)/2 = 1$,
Long-time power = $(1 + 1 + 1 + 4 + 1)/5 = 1.6$,
Ratio = $1/1.6 = 0.63$.

```
1    2    1    1    1    5    4    2    1    1    1
                       |___|
                  short window
     |_____|
           long window
```

Point x_5:   Short-time power = $(25 + 1)/2 = 13$,
Long-time power = $(25 + 1 + 1 + 1 + 4)/5 = 6.4$,
Ratio = $13/6.4 = 2.03$.

```
1    2    1    1    1    5    4    2    1    1    1
                            |___|
                       short window
          |_____|
                long window
```

Point x_6:   Short-time power = $(16 + 25)/2 = 20.5$,
Long-time power = $(16 + 25 + 1 + 1 + 1)/5 = 8.8$,
Ratio = $20.5/8.8 = 2.33$.

```
1    2    1    1    1    5    4    2    1    1    1
                                 |___|
                            short window
               |_____|
                     long window
```

Point x_7:   Short-time power = $(4 + 16)/2 = 10$,
Long-time power = $(4 + 16 + 25 + 1 + 1)/5 = 9.4$,
Ratio = $10/9.4 = 1.06$.

```
1    2    1    1    1    5    4    2    1    1    1
                                      └────┘
                                  short window
                   └──────────────────────┘
                         long window
```

Point x_8:   Short-time power = $(1 + 4)/2 = 2.5$,
Long-time power = $(1 + 4 + 16 + 25 + 1)/5 = 9.4$,
Ratio = $2.5/9.4 = 0.27$.

```
1    2    1    1    1    5    4    2    1    1    1
                                      └────┘
                                  short window
                        └──────────────────────┘
                              long window
```

Point x_9:   Short-time power = $(1 + 1)/2 = 1$,
Long-time power = $(1 + 1 + 4 + 16 + 25)/5 = 9.4$,
Ratio = $1/9.4 = 0.11$.

```
1    2    1    1    1    5    4    2    1    1    1
                                           └────┘
                                       short window
                             └──────────────────────┘
                                   long window
```

Point x_{10}:   Short-time power = $(1 + 1)/2 = 1$,
Long-time power = $(1 + 1 + 1 + 4 + 16)/5 = 4.6$,
Ratio = $1/4.6 = 0.22$.

By using the ratios previously computed, possible seismic events occurred at points x_5 and x_6. Because the time interval between points is 0.01 second, the times that correspond to the seismic events are 0.05 and 0.06 second. (We assume that the first point in the file occurred at 0.0 second.)

4. ALGORITHM DEVELOPMENT

We first develop the decomposition outline because it divides the solution into a series of sequential steps:

Decomposition Outline

1. Read seismic data from the data file and read numbers of measurement for power from the keyboard.
2. Compute power ratios and print possible seismic event times.

Step 1 involves reading the data file and storing the information in an array. Because we do not know the exact size of the array, we will need to specify a maximum size in the array definition. We will read the numbers of measurements for the power computations from the keyboard. Step 2 involves computing power ratios and comparing them to the threshold to determine if a possible event occurred. Because we need to compute two power measurements for each possible event location, we implement the power measurement as a function. The refinement in pseudocode for the main function and the power function can now be developed:

Refinement in Pseudocode
main: *set threshold to 1.5*
          read npts and time-interval
          read the values into sensor array
          read short-window, long-window from keyboard
          set k to long-window - 1
          while k ≤ npts-1
              set short-power to power(sensor,short-window,k)
              set long-power to power(sensor,long-window,k)
              set ratio to short-power/long-power
              if ratio > threshold
                  print k · time-interval
              increment k by 1
power(x,length,n):
          set xsquare to zero
          set k to n
          while k > n-length+1
              add x[k] · x [k] to xsquare
          return xsquare/length

We are now ready to convert the pseudocode to C.

```
/*-----------------------------------------------------------*/
/*  Program chapter6_5                                       */
/*                                                           */
/*  This program reads a seismic data file and then          */
/*  determines the times of possible seismic events.         */

#include <stdio.h>
#define FILENAME "seismic1.txt"
#define MAX_SIZE 1000
#define THRESHOLD 1.5

int main(void)
{
   /*  Declare variables and function prototypes.  */
   int k, npts, short_window, long_window;
   double sensor[MAX_SIZE], time_incr, short_power,
          long_power, ratio;
   FILE *file_ptr;
   double power_w(double *ptr,int n);

   /*  Read sensor data file.  */
   file_ptr = fopen(FILENAME,"r");
   if (file_ptr == NULL)
      printf("Error opening input file. \n");
```

```
        else
        {
            fscanf(file_ptr,"%d %lf",&npts,&time_incr);
            if (npts > MAX_SIZE)
                printf("Data file too large for array. \n");
            else
            {
                /*  Read data into an array.  */
                for (k=0; k<=npts-1; k++)
                    fscanf(file_ptr,"%lf",&sensor[k]);

                /*  Read window sizes from the keyboard.  */
                printf("Enter number of points for short window: \n");
                scanf("%d",&short_window);
                printf("Enter number of points for long window: \n");
                scanf("%d",&long_window);

                /*  Compute power ratios and search for events.  */
                for (k=long_window-1; k<=npts-1; k++)
                {
                    short_power = power_w(&sensor[k],short_window);
                    long_power = power_w(&sensor[k],long_window);
                    ratio = short_power/long_power;
                    if (ratio > THRESHOLD)
                        printf("Possible event at %f seconds \n",
                                time_incr*k);
                }

                /*  Close file.  */
                fclose(file_ptr);
            }
        }
        /*  Exit program.  */
        return 0;
}
/*-------------------------------------------------------------*/
/*  This function computes the average power in a specified    */
/*  window of a double array.                                  */

double power_w(double *ptr, int n)
{
    /*  Declare and initialize variables.  */
    int k;
    double xsquare=0;

    /*  Compute sum of values squared in the array x. */
    for (k=0; k<=n-1; k++)
        xsquare += *(ptr-k)*(*(ptr-k));

    /*  Return the average squared value.  */
    return xsquare/n;
}
/*-------------------------------------------------------------*/
```

5. TESTING

The program's output, using the data file from the hand example, is as follows:

```
Enter number of points for short-window:
2
Enter number of points for long-window:
5
Possible event at 0.050000 seconds
Possible event at 0.060000 seconds
```

MODIFY!

Modify the event-detection program to include the following new capabilities:

1. Allow the user to enter the threshold value. Check the value to be sure that it is a positive value greater than 1.
2. Print the number of events detected by the program. (Assume that events with contiguous times are all part of the same event. Thus, for the hand example, one event was detected.)

6.6 Character Strings

Character string

A character array is defined as an array in which the individual elements are stored as characters. A **character string** is a character array in which the last array element is a null character '\0', which has an ASCII integer equivalent of zero. In this section, we focus on character strings.

String Definition and I/O

Character string constants are enclosed in double quotes, as in "sensor1.txt", "r", and "15762". A character string array can be initialized using string constants, or using character constants, as shown in the following equivalent statements:

```
char filename[12] = "sensor1.txt";
char filename[] = "sensor1.txt";
char filename[] = {'s','e','n','s','o','r',
                   '1','.','t','x','t','\0'};
```

To read a line from the keyboard and store it as a character string, we use the following statements:

```
/*  Declare variables.  */
int k=0, nchars;
char line[50];
...
/*  Read characters into string until newline is read.  */
while ((line[k]=getchar()) != '\n')
    k++;
line[k] = '\0';
nchars = k + 1;
```

A similar set of statements can be used to read a character string from a line in a data file. To print the values in this character string, we can use the following statements:

```
for (k=0; k<=nchars-2; k++)
   putchar(line[k]);
putchar('\n');
```

Note that we did not print the last character, `line[nchar-1]`, since it is a null character; however, we did print a newline character so that the information printed will be on a separate line from any remaining output. We can also print a character string using the `%s` specifier; this specifier does not print the null character. Thus, the string `line` could be printed using this statement:

```
printf("String: %s \n",line);
```

If a string printed with an `%s` specifier does not end with a null character, the characters following the string will be printed until a null character is encountered.

String Functions

First, we present a programmer-defined function that uses a character string argument, and then we present a group of library functions that use character strings. Consider a programmer-defined function that determines the length of a character string argument, where the length of a character string is defined to be the number of characters up to, but not including, the null character. The prototype for this function is the following:

```
/*------------------------------------------------------------*/
/*  This function determines the length of a string.          */
int strg_len_1(char s[])
{
   /*  Declare variables.  */
   int k=0;
   /*  Count characters.  */
   while (s[k] != '\0')
      k++;
   /*  Return string length.  */
   return k;
}
/*------------------------------------------------------------*/
```

The Standard C library contains a number of functions for working with strings. As we discuss these functions and their arguments, assume that s and t are character strings. The variable n is of type `size_t`, which is an unsigned integer; c is an integer that is converted to a character. Note that some of these functions return pointers to strings. The prototype statements for these functions are included in the header file `string.h`.

strlen(s)	This function returns the length of the string s.
strcpy(s,t)	This function copies string t to string s. The function returns a pointer to s.
strncpy(s,t,n)	This function copies a maximum of n characters from string t to string s. If t has fewer characters than s, then s is padded with null characters. The function returns a pointer to s.

concatenates	strcat(s,t)	This function **concatenates** string t to the end of string s. Thus, string s will contain the characters of s followed by the characters of t; the first character of t overwrites the null character at the end of s. The function returns a pointer to s.
	strncat(s,t,n)	This function concatenates a maximum of n characters of string t to string s. If t has more than n characters, then only the first n characters of t are concatenated to s. The initial character of t overwrites the null character at the end of s; a null character is added to the new end of s. The function returns a pointer to s.
	strcmp(s,t)	This function compares string s to string t in an element-by-element comparison, starting with s[0] and t[0]. A negative value is returned if s<t, zero is returned if s is equal to t, and a positive value is returned if s>t.
	strncmp(s,t,n)	This function compares a maximum of n characters of string s to string t in an element-by-element comparison, starting with s[0] and t[0]. A negative value is returned if s<t, zero is returned if s is equal to t, and a positive value is returned if s>t.
	strchr(s,c)	This function returns a pointer to the first occurrence of the character c in the string s. If the character does not occur in s, a NULL pointer is returned.
	strrchr(s,c)	This function returns a pointer to the last occurrence of the character c in the string s. If the character does not occur in s, a NULL pointer is returned.
	strstr(s,t)	This function returns a pointer to the start of the string t within the string s. If t does not occur in s, a NULL pointer is returned.
	strspn(s,t)	This function returns the initial number of characters of string s that consists entirely of characters in string t.
	strcspn(s,t)	This function returns the initial number of characters of string s that consists entirely of characters not in string t.
	strpbrk(s,t)	This function returns a pointer to the first occurrence in string s of any character of string t. If none of the characters in t occurs in s, a NULL pointer is returned.

To illustrate the use of a few of these functions, we now present a simple example:

```
/*-----------------------------------------------------------------*/
/*  Program chapter6_6                                             */
/*                                                                 */
/*  This program illustrates the use of string functions.         */

#include <stdio.h>
#include <string.h>

int main(void)
{
   /*  Declare and initialize variables.  */
   char strg1[]="Engineering Problem Solving: ";
   char strg2[]="Fundamental Concepts", strg3[50];
```

```
    /*  print the length of strings.  */
    printf("String lengths: %d %d \n",
           strlen(strg1),strlen(strg2));

    /*  Combine two strings into one.  */
    strcpy(strg3,strg1);
    strcat(strg3,strg2);
    printf("strg3: %s \n",strg3);
    printf("strg3 length: %d \n",strlen(strg3));

    /*  Exit program.  */
    return 0;

}
/*-----------------------------------------------------------*/
```

The output from this program is the following:

```
String lengths: 29 20
strg3: Engineering Problem Solving: Fundamental Concepts
strg3 length: 49
```

Character strings are used in many engineering applications, including cryptography and pattern recognition. It is often convenient to manipulate character strings via pointers to the strings. Many of the previously discussed string functions require pointers to characters as arguments, and many return pointers to characters. Now, we will look at the syntax required to reference character strings using pointers.

Earlier in this section, we wrote a programmer-defined function using subscripts within a while loop to determine the length of a string. We will now rewrite this function using a pointer:

```
/*-----------------------------------------------------------*/
/*  This function determines the length of a string          */
/*  using a pointer with a while loop.                       */

int strg_len_2(char *s)
{
    /*  Declare variables.  */
    int count=0;

    /*  Count characters.  */
    while (*s != '\0')
    {
        count++;
        s++;
    }

    /*  Return string length.  */
    return count;
}
/*-----------------------------------------------------------*/
```

The preceding function counts every character in a character string until the null character is found. The Standard C library includes a function named `strlen` that returns the length of a string; thus it would be preferable to use the strlen function rather than write your own. Another function included in the Standard C library is the function named `strstr`. The function `strstr(s,t)` takes a pointer to character string s and a pointer to character string t as arguments and returns a pointer to the start of the string t within the string s. If t does not occur in s, a NULL pointer is returned. We will use this function in the next section.

6.7 Problem Solving Applied: DNA Sequencing

In this section, we use the new statements presented in this chapter to solve a problem related to DNA sequencing. Recall from the chapter opening discussion that a DNA profile is a character string that consists of letters representing the order of nucleotides, such as AAACG-TACAGGT. A DNA match between an unknown DNA string and one from a database is based on a number of matches of short strings (called short tandem repeats, or STRs) between the two strings.

In this section, we will count the number of occurrences of a short string in a longer string and also print the relative locations of these matches. The previous discussion with character string functions provides the basis for this solution.

1. PROBLEM STATEMENT

Given a long character string and a short character string, find the number of occurrences of the short string in the long string, and print the beginning locations of each occurrence.

2. INPUT/OUTPUT DESCRIPTION

The following diagram shows that the inputs to the program are a long character string and a short character string. The outputs are the locations of the occurrences of the short string in the first string and a count of the number of occurrences.

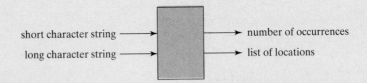

short character string ⟶ [ ] ⟶ number of occurrences
long character string ⟶ [ ] ⟶ list of locations

3. HAND EXAMPLE

Assume that the long character string is AAACTGACATTGGACCTACTTTGACT and the short character string is ACT. There are three occurrences of the shorter string, and they occur at positions 3, 18, and 24.

4. ALGORITHM DEVELOPMENT

This solution will use the character function strstr discussed in the previous section. This function will return a pointer to the start of the first occurrence of the short string in the longer string, or it will return a NULL point if the short string is not found. In order to find the next occurrence of the shorter string, we need to search the portion of the string that follows the first position of the occurrence of the shorter string. The strstr function will then return a pointer to the beginning of the first occurrence of the shorter string in the new portion of the longer string. We repeat this process until the strstr function returns a NULL value.

Decomposition Outline

1. Initialize a long character string and a short character string.
2. Compute and print the location and the number of times the shorter string occurs in the longer string.

Step 2 involves a loop in which we continue counting occurrences until we reach the end of the longer string. The refinement in pseudocode can now be developed:

> *Refinement in Pseudocode*
>
> *main:* *initialize a long character string and a short character string*
>   *set count to zero*
>   *set ptr1 to the beginning of the long string*
>   *set ptr2 to the beginning of the short string*
>   *while not at the end of the long string*
>   *if short string is in remaining long string*
>   *increment count by 1*
>   *print location*
>   *increment pointer to long string*
>   *print count*

We are now ready to convert the pseudocode to C.

```
/*-------------------------------------------------------------*/
/*  Program chapter6_7                                         */
/*                                                             */
/*  This program initializes a long character string and a short */
/*  character string.  It then prints the locations of the short */
/*  string in the long string.  It also prints the number of   */
/*  occurrences of the short string in the long string.        */

#include <stdio.h>
#include <string.h>

int main(void)
{
   /*  Declare and initialize variables.  */
   int count=0;
   char long_str[]="AAACTGACATTGGACCTACTTTGACT",
        short_str[]="ACT";
   char *ptr1=long_str, *ptr2=short_str;
```

```
                    /*  Count the number of occurrences of short_str in long_str. */
                    /*  While the function strstr does not return NULL, increment */
                    /*  count and move ptr1 to next character of the long string. */
                    while ((ptr1=strstr(ptr1,ptr2)) != NULL)
                    {
                        printf("location %i \n",ptr1-long_str+1);
                        count++;
                        ptr1++;
                    }

                    /*  Print number of occurrences.  */
                    printf("number of occurrences: %i \n",count);

                    /*  Exit program.  */
                    return 0;
                 }
                 /*--------------------------------------------------------------*/
```

5. TESTING

We first test the program with the data from the hand example. This generates the following interaction:

```
location 3
location 18
location 24
number of occurrences: 3
```

The answer matches the hand example, so we can now test the program with additional points.

MODIFY!

These problems relate to the program developed in this section for finding short strings in longer strings.

1. Modify the program so that it prints the long string and the short string.
2. Modify the program so that it allows the user to enter the long string and the short string.
3. Modify the program so that the long string is read from a data file and the short string is input from the user.
4. Modify the program so that it will work with either lowercase or uppercase characters.
5. Modify the program so that it checks to be sure that the length of the short string is less than the length of the long string.

6.8 Dynamic Memory Allocation*

Dynamic memory
allocation

Dynamic memory allocation allows us to allocate memory when a program is executed instead of allocating it when the program is compiled. This is especially important when a program uses an array whose size is not determined until the program is executed; without dynamic memory allocation, the program would have to specify the maximum size anticipated for the array. For systems with limited memory, it is possible that there would not be enough memory to run a program if all arrays had to be specified to the maximum size anticipated.

Dynamic memory allocation is specified using either the `malloc` function or the `calloc` function, which perform a "memory allocation" or a "cleared allocation." Both functions reserve a group of contiguous memory locations; in addition, `calloc` initializes the memory locations to a binary zero. The argument of the `malloc` function is the number

Byte

of bytes of memory required, where a **byte** is a unit of memory that contains 8 bits. The arguments of the `calloc` function are the number of memory locations needed and the number of bytes for each memory location. We now give the prototype statements (which are contained in the header file `<stdlib.h>`) for these functions, and then give further explanation of the parameters and the values returned by these functions:

```
void *malloc(size_t m);
void *calloc(size_t n,size_t size);
```

Because the number of bytes used for storing a value with a specific data type (such as an `int`) is system dependent, a special operator called `sizeof` is used to determine the number of bytes needed for a specific data type. The expression `sizeof(int)` represents the number of bytes used to store an integer, and `sizeof(double)` represents the number of bytes used to store a `double` value. The `sizeof` operator computes an unsigned integer that is a `size_t` type value, where the `size_t` type is system dependent, but is usually either an `unsigned int` or an `unsigned long`. Thus, to request memory for 200 integers, we could use either of these sets of statements:

```
num_pts = 200;
int *p;
p = (int *)malloc(num_pts*sizeof(int));
```

```
num_pts = 200;
int *p;
p = (int *)calloc(num_pts,sizeof(int));
```

void pointer

Both functions return a value to a pointer. If the memory is available, the pointer will contain the address of the memory; if the allocation cannot be made, the pointer will contain a NULL value. The pointer returned by these functions is called a **void pointer** because it does not specify the type of variables to be stored in the memory allocation. Therefore, a cast operator should be used with the pointer value returned by the functions in order to coerce it to the proper pointer type.

---

*Optional section.

To illustrate, assume that we want to dynamically allocate memory to store a double array named x containing npts elements. (In this example, we give a value to npts, but it could be computed by other statements or read from the keyboard or a data file.) To specify the desired allocation, we could use either of the following sets of statements:

Solution 1
```
/*  Declare variables.  */
int npts = 500;
double *x;
...
/*  Dynamically allocate memory.  */
x = (double *)malloc(npts*sizeof(double));
```

Solution 2
```
/*  Declare variables.  */
int npts = 500;
double *x;
...
/*  Dynamically allocate memory.  */
x = (double *)calloc(npts,sizeof(double));
```

If the calloc function is used, then the memory values will also be initialized to zero. With either of these solutions, we should compare the value of x to the constant NULL. This will make sure that the memory was allocated, as shown in following statement:

```
if (x == NULL)
    printf('Memory requested not available. \n");
```

After we have determined that the memory has been allocated, references to the array x, such as x[k], are valid in the program.

To release memory that has been dynamically allocated, use the free function, which has the following prototype:

```
void free(void *ptr);
```

Thus, the memory allocated to array x is released with this statement:

```
free(x);
```

In general, you should free dynamically allocated memory when it is no longer needed. Then it becomes available for possible use with another dynamic allocation.

The realloc function can be used to change the size of the memory requested by a calloc, malloc, or previously executed realloc function. Its prototype statement is

```
void *realloc(void *ptr,size_t size);
```

If ptr contains the value NULL, this function operates like the malloc function. If ptr contains a value returned earlier in a program by a calloc, malloc, or previously executed realloc function, then the size of the corresponding memory allocation is changed to the new size requested. If the new size is larger, the values of the newly allocated space are undetermined. If the new size is smaller, the values in the new size are unchanged. If the additional space cannot be allocated, the original space is unchanged and the function returns a value of NULL.

With the use of dynamic memory allocations and dynamic memory allocation releases, a program can be designed to operate with a minimal amount of memory reserved at any time during its execution. On systems that run multiple programs at the same time, the use of dynamic memory allocation and dynamic memory release may allow more programs to run simultaneously.

The next program allows you to determine the maximum amount of contiguous memory that can be dynamically allocated during its execution. This maximum amount of memory available is generally a function of the other users on the system and the other software that is stored on the system; thus, you may get different results when you run the program on different systems and at different times. Here is the program:

```c
/*-----------------------------------------------------------*/
/*  Program chapter6_8                                       */
/*                                                           */
/*  This program determines the maximum contiguous           */
/*  memory allocation that can be reserved during a          */
/*  specific program execution.                              */

#include <stdio.h>
#include <stdlib.h>
#define UNIT 1000000

int main(void)
{
   /*  Declare and initialize variables.  */
   int k=1, *ptr;

   /*  Find maximum amount of contiguous memory    */
   /*  available in units of millions of integers. */
   ptr = (int *)malloc(UNIT*sizeof(int));
   while (ptr != NULL)
   {
      free(ptr);
      k++;
      ptr = (int *)malloc(k*UNIT*sizeof(int));
   }

   /*  Print maximum amount of memory available.  */
   printf("Maximum contiguous memory available: \n");
   printf("%k integers \n",(k-1)*UNIT);

   /*  Exit program.  */
   return 0;
}
/*-----------------------------------------------------------*/
```

A typical output for this program using a personal computer is the following:

```
Maximum contiguous memory available:
31000000 integers
```

MODIFY!

Modify program `chapter6_8` such that it determines the number of thousands of contiguous values that can be stored for the data types indicated in the following:

1. long integers,   2. doubles, and   3. long doubles.

In the previous section, we developed a program that read a set of seismic data from a data file and determined possible seismic event locations within the data. The program used a symbolic constant `MAX_SIZE` to allocate an array to store the data. If the number of points in the data file exceeded `MAX_SIZE`, then the program terminated with an error message. In the following, we present statements that use dynamic memory allocation so that the maximum array size does not have to be allocated when the program is compiled (instead, the array memory is dynamically allocated based on the size of the seismic data file):

```
/*----------------------------------------------------------------*/
/*  Program chapter6_5mod                                         */
/*                                                                */
/*  This program reads a seismic data file and then              */
/*  determines the times of possible seismic events.             */
/*  Dynamic memory allocation is used.                           */

#include <stdio.h>
#define FILENAME "seismic1.txt"
#define THRESHOLD 1.5

int main(void)
{
   /* Declare variables and function prototypes. */
   int k, npts, short_window, long_window;
   double *sensor, time_incr, short_power,
          long_power, ratio;
   FILE *file_ptr;
   double power_w(double *ptr,int n);

   /* Read data header and allocate memory. */
   file_ptr = fopen(FILENAME,"r");
   if (file_ptr == NULL)
      printf("Error opening input file. \n");
   else
   {
      fscanf(file_ptr,"%d %lf",&npts,&time_incr);
      sensor = (double *)malloc(npts*sizeof(double));
      if (sensor == NULL)
         printf("Not enough memory available. \n");
      else
```
(no changes in the remainder of program chapter6_5 *on page* 312)

6.9   A Quicksort Algorithm*

Quicksort

Pivot value

In this section, a **quicksort** algorithm is implemented with a recursive function (see Section 4.8) that uses pointers as function parameters.

The quicksort algorithm selects a value, called a **pivot value**, and then separates the rest of the values into two groups—one group containing values less than the pivot value and one group containing values greater than the pivot value. For our purposes, we will select the pivot value to be the first element in the list, but the midpoint of the list is also often used as the pivot value. When this separation is done, the correct position in the list for the pivot value is determined; it goes between the two groups of values. Since the values in the two groups are not necessarily in the correct order, we take the group of smaller values and select a new pivot value. This group is separated into two new groups of values—ones smaller than the new pivot value and ones larger than the new pivot value. This process continues until we eventually have a group of smaller values that contains no values, one value, or two values. If this group contains two values, their order is switched if necessary, and then the original group of values smaller than the original pivot value are in order. We repeat this process with the original group of values that are larger than the original pivot value. When these are in order, the entire list is in order. This algorithm can be described recursively because each step is defined in terms of a similar process with a smaller group of values, and because it has a stopping point that is encountered when the group of values has two or fewer values. A hand example follows:

Original list:

4	10	3	6	−1	0	2	5

Separate into groups of values smaller and larger than the pivot value:

[3	−1	0	2]	4	[10	6	5]

Separate each remaining group into groups of values smaller and larger than the pivot value of each group:

[−1	0	2]	3	4	[6	5]	10

Separate each remaining group into groups of values smaller and larger than the pivot of each group:

−1	[0	2]	3	4	5	6	10

Separate each remaining group into groups of values smaller and larger than the pivot of each group:

−1	0	2	3	4	5	6	10

The implementation of the quicksort algorithm that we present references an additional function named `separate`. This function switches values in the array that it receives such that the pivot value is correctly positioned in the position referenced by `break_pt`. All values less than the pivot value are to the left in the array (if we visualize the array as a row), and all values greater than the pivot value are to the right. Then, the `quicksort` function is recursively called using a statement which specifies that a total of `break_pt` values, starting with `x[0]`, should be sorted:

```
quicksort(x, break_pt);
```

———————

*Optional section.

The quicksort function is also recursively called using a statement that specifies that a total of n-break_pt-1 values, starting with x[break_pt+1], should be sorted:

```
quicksort(&x[break_pt+1],n-break_pt-1];
```

Note that the actual argument uses &x[break_pt+1], a reference to the address of x[break_pt+1]. This address reference is necessary because using x[break_pt+1] generates a call-by-value, not a call-by-reference. The quicksort function and the separate function use the switch2 function developed in Section 6.2. Since the quicksort algorithm is a complicated algorithm, a good way to begin is to use the hand example and work through the statements using that data.

```
/*------------------------------------------------------------------*/
/*   Program chapter6_9                                             */
/*                                                                  */
/*   This program tests the quicksort function.                     */

#include <stdio.h>

int main(void)
{
    /*  Declare and initialize variables.  */
    int x[8]={4,10,3,6,-1,0,2,6}, npts=8,  k;
    void quicksort(int w[],int n);
    int separate(int y[],int m);
    void switch2(int *a,int *b);

    /*  Sort and print the array.  */
    printf("Before: ");
    for (k=0; k<=7; k++)
       printf("%d ",x[k]);
    printf("\n");
    quicksort(x,npts);
    printf("After: ");
    for (k=0; k<=7; k++)
        printf("%d ",x[k]);
    printf("\n");

    /*   Exit program.  */
    return 0;

}
/*------------------------------------------------------------------*/
/*   This function implements a quicksort algorithm.                */

void quicksort(int w[],int n)
{
    /*  Declare variables and function prototypes.  */
    int break_pt;
    int separate(int y[],int m);
    void switch2(int *a,int *b);
```

```
    /*  If only two elements, order them correctly.  */
    if (n == 2)
    {

        if (w[0] > w[1])
            switch2(&w[0],&w[1]);

    {
    else
    /*  If more than two elements, separate into those  */
    /*  greater than and those less than a breakpoint.  */
        if (n > 2)

        {

            break_pt = separate(w,n);
            quicksort(w,break_pt);
            quicksort(&w[break_pt+1],n-break_pt-1);

        }

    /*  Void return.  */
    return;

}

/*------------------------------------------------------------*/
/*  This function reorders the array such that y[0] is        */
/*  correctly positioned and the values less than it are      */
/*  to the right and the values greater are to the left.      */

int separate(int y[],int m)
{

    /*  Declare variables and function prototypes.  */
    int k1=1, k2=1, count=0, pivot;
    void switch2(int *a,int *b);

    /*  Separate values into two groups.  */
    pivot = y[0];
    while (k1<m && k2<m)
    {
        while ((k1<m) && (y[k1]>pivot))
            k1++;
        while ((k2<m) && (y[k2]<pivot))
            k2++;
        if ((k1<m) && (k2<m))
        {
            switch2(&y[k1],&y[k2]);
            count++;
        }
    }
```

```
           /*  Put pivot value in correct position.  */
           if (count > 0)
              switch2(&y[0],&y[count]);
           else
           {
              k1 = 0;
              while ((k1<m-1) && (y[k1]>y[k1+1]))
              {
                 switch2(&y[k1],&y[k1+1]);
                 k1++;

              }
              count = k1;
           }

           /*  Return count.  */
           return count;
}
/*-----------------------------------------------------------*/
/*  Correct function to switch values in two variables.      */

void switch2(int *a,int *b)
{
           /*  Declare variables.  */
           int hold;

           /*  Switch values in a and b.  */
           hold = *a;
           *a = *b;
           *b = hold;

           /*  void return.  */
           return;

}
/*-----------------------------------------------------------*/
```

The focus of this chapter was the relationship between a pointer and the variable to which it points. Examples were presented that demonstrated how to define pointers and how to initialize them. The valid types of operations that can be performed with pointers were listed, and the precedence relationships for operators were updated to include the address operator and the indirection operator. Examples of pointers used as function parameters and in references to arrays were given. We also presented a group of functions that work with character strings; many of these functions use pointers as function parameters. Finally, we presented the C statements that allow us to do dynamic memory allocation using pointers.

KEY TERMS

address
address operator
byte
character string
concatenate
dereference

dynamic memory allocation
indirection
NULL character
offset
pointer
void pointer

C STATEMENT SUMMARY

Pointer declaration:

```
int *ptr_1;
double a, *ptr_2=&a;
```

Dynamic memory allocation:

```
x = (double *)malloc(npts*sizeof(double));
x = (double *)calloc(npts,sizeof(double));
```

NOTES

1. Choose identifiers for pointers that clearly indicate that the identifiers are associated with pointer variables.

2. If a pointer is not initially assigned to a memory location, give it a value of NULL to indicate that it has not yet been assigned.

DEBUGGING NOTES

1. Be sure that a program initializes a variable before using its value in other statements.

2. Be sure that a pointer variable is initialized before it is used to reference a value.

3. The actual parameter that corresponds to a pointer argument in a function must be an address or a pointer.

PROBLEMS

SHORT ANSWER PROBLEMS

True–False Problems

Indicate whether the following statements are true (T) or false (F).

1. Both the address operator and the indirection operator are unary operators.   T   F
2. Printing the value of *(&x) is the same as printing the value of x.   T   F
3. A malloc function reserves contiguous memory locations dynamically when called and initializes memory locations to binary 0.   T   F
4. The memory locations given to dynamic memory space are determined when the program is compiled.   T   F

Multiple Choice Problems

Circle the letter for the best answer to complete each statement or for the correct answer of each question.

5. Which of the following operations on pointers are not valid?
    (a) Assignment of a pointer to another pointer of same type.
    (b) Addition or subtraction of a pointer with an integer.
    (c) Multiplication or division of a pointer with an integer.
    (d) Subtraction of a pointer from another pointer of same type.

6. A pointer variable
    (a) contains the data stored at a location in memory.
    (b) contains the address of a memory location.
    (c) can be used in input statements, but cannot be used in output statements.
    (d) can be changed to different values in both input and output statements.

7. How would you assign the value of a variable referenced by the pointer a to a variable name?
    (a) `a = &name;`
    (b) `name = &a;`
    (c) `a = *name;`
    (d) `name = *a;`

8. Given that x and y are integers, what will be the function prototype, if a function is called as:

    ```
    dofunc( x, &y);
    ```
    (a) `void dofunc( int a, int &b);`
    (b) `void dofunc( int a, int *b);`
    (c) `void dofunc( int a, int b);`
    (d) `void dofunc( int *a, int &b);`

Memory Snapshot Problem

9. Assuming that the address of name is 10 and the address of x is 14 (that is, name is stored in memory location 10, and x is stored in memory location 14), give the corresponding snapshots of memory after the following set of statements is executed:

    ```
    float name, x=20.5;
    float *a = &x;
    ...
    name = *a;
    ```

Program Analysis

Problems 10–13 refer to the following statements:

```
char str[]="Problem Solving";
char *i, *j;
int x;
i=str;
j=i+8;
```

10. What will be the value of *i and *j?

11. What will be the value of str[0] and str[8]?

12. What will be the output of this statement?

```
for(x=0;x<7;x++)
printf("%c", *(str+x));
```

13. What will be the output of this statement?

```
for(x=strlen(str)-1;x>=0;x--)
printf("%c", *(i+x));
```

PROGRAMMING PROBLEMS

General Functions. Pointers are often used as function arguments when we want to return more than one value from the function. In each of the following problems, write the indicated function, and then develop a main function for testing the function:

14. Write a function that converts radius, diameter, and area measurements for a circle from units of inches and square inches to units of feet and square feet. Assume that the corresponding function prototype statement is

```
void convert_ft(double *r,double *d,double *a);
```

where r, d, and a are pointers to the radius, diameter, and area variables.

15. Write a function that prints the elements of an A.P. series as well as determines the sum of the series. Assume that the corresponding function prototype is:

```
void series ( int a, int d, int n, int *sum)
```

where a is the first term, d is the common difference, n is the number of elements, and sum is a pointer to the sum of the series.

16. Write a function that determines the maximum and minimum values from a one-dimensional integer array. Assume that the corresponding function prototype statement is

```
void ranges(int x[],int npts,int *max_ptr,
            int *min_ptr);
```

where npts contains the number of values in array x, and max_ptr and min_ptr are pointers to the variables in which to store the maximum and minimum values in the array.

17. Write a function that determines the arithmetic and geometric means of a one-dimensional float array. Assume that the corresponding function prototype is:

```
void mean( int sample[], int size, float *am, float *gm);
```

where size contains the number of elements of array sample, and am and gm are pointers to the arithmetic and geometric means respectively.

18. Write a function that returns the number of positive values, zero values, and negative values in an integer array. Assume that the corresponding function prototype statement is

```
void signs(int x[],int npts,int *npos,
           int *nzero,int *nneg);
```

where npts contains the number of values in the array x and npos, nzero, and nneg are pointers to variables to store the numbers of positive values, zero values, and negative values in the array.

Vector Functions. A vector is a group of numerical values that can be represented by a one-dimensional array. These problems develop functions for manipulating values within a vector. Then a reference to the function is requested, to show how to use the function to manipulate groups of values within the vector, instead of manipulating the complete set of values in the vector. Use pointer references instead of subscripts in the functions. Assume that the first position in an array is the position referenced with a zero subscript.

19. Write a function that fills a vector with zeros. Assume that the function prototype statement is

    ```
    void zeros(int x[],int n);
    ```

    where x is a one-dimensional array with n elements. Give a reference to the function that fills positions 20 to 25 of an array a with zeros.

20. Write a function that fills a vector with the first n perfect squares. Assume that the function prototype statement is:

    ```
    void squares(int x[],int n);
    ```

    where x is a one-dimensional array with n elements.

21. Write a function that computes the sum of the odd numbers in a vector. Assume that the function prototype statement is:

    ```
    int sum_odd(int x[],int n);
    ```

    where x is a one-dimensional array with n elements.

22. Write a function that reverses the order of the values in a vector. Assume that the function prototype statement is

    ```
    void v_rev(int x[],int n);
    ```

    where x is a one-dimensional array with n elements. Give a reference to the function that reverses the values in positions 5 through 20 of array z.

23. Write a function that replaces values in an array with their absolute values. Assume that the function prototype statement is

    ```
    void v_abs(int x[],int n);
    ```

    where x is a one-dimensional array with n elements. Give a reference to the function that replaces all the values of array t (except the first five values) with their absolute values, where t has npts elements.

Character Functions. Many areas of engineering use problem solutions in which we search for a specific pattern of information in a signal. The following problems develop a set of functions for this purpose.

24. Write a function that receives a pointer to a character string and two characters. It replaces all occurrences of the first character with the second. It also counts the number of replacements. Assume that the function has the following prototype statement:

    ```
    void replace(char *ptr, char find, char replace);
    ```

25. Write a function that receives a pointer to a character string and returns the number of repeated characters that occur in the string. For example, the string "Mississippi" has

three repeated characters. Do not count repeated blanks in the string. If a character occurs more than two times, it should still only count as one repeated character; thus, "hisssss" would have only one repeated character. Assume that the function has the following prototype statement:

```
int repeat(char *ptr);
```

26. Rewrite the function from Problem 25 such that each pair of characters is counted as a repeat. Thus, the string "hisssss" would have four repeated characters. Assume that the function has the following prototype statement:

```
int repeat2(char *ptr);
```

27. Write a function that receives pointers to two character strings and returns a count of the number of times that the second character string occurs in the first character string. Do not allow overlap of the occurrences. Thus, the string "110101" contains only one occurrence of "101." Assume that the function has the following prototype statement:

```
int pattern(char *ptr1,char *ptr2);
```

28. Rewrite the function from Problem 27 such that overlap of the occurrences of the second string in the first string is allowed. Thus, the string "110101" contains two occurrences of "101." Assume that the function has the following prototype statement:

```
int overlap(char *ptr1,char *ptr2);
```

Table Function. The problems that follow develop a set of functions for computing values from a two-dimensional array or a table of data. Use pointer references instead of subscript references in the functions.

29. Write a function that computes the maximum of the last row in a table containing 4 rows and 5 columns of values. Assume that the function prototype is the following statement:

```
double max_row(double table[4][5]);
```

30. Write a function that computes the minimum of the middle column in a table containing 4 rows and 5 columns of values. Assume that the function prototype is the following statement:

```
double col_min(double table[4][5]);
```

CHAPTER SEVEN

7

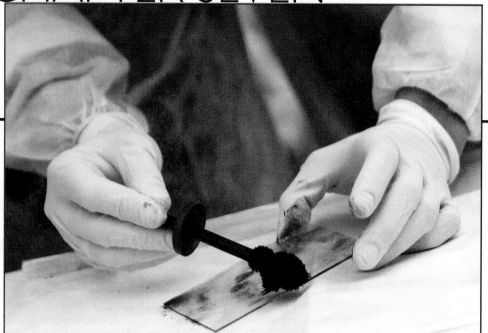

Crime Scene Investigation: Fingerprint Recognition

Fingerprints are the oldest form of biometrics. Some of the earliest references to recognizing people from fingerprints date back to the 1600s at the University of Bologna, Italy. In the late 1800s, a number of British researchers recognized the potential in the swirls and patterns on the fingertips for recognizing people, and their work lead to the worldwide systems of fingerprint databases. AFIS, the Automatic Fingerprint Identification System of the FBI, contains over 200 million digital fingerprints. Law enforcement officers around the country have access to this database. Soldiers on critical missions in Iraq were able to collect fingerprints of suspected terrorists/criminals, send the fingerprints wirelessly through satellite connections to the AFIS database in West Virginia, and get a report on potential matches in less than 15 minutes. Fingerprint information can be from a single finger (usually an index finger), or they can be "10 prints," which are prints from all 10 fingers. "Slaps" are collections of multiple fingerprints taken at the same time. Latent prints are those recovered from a crime scene that contain only part of a fingerprint; these prints often require chemical methods or alternative light sources for recovery. The matching of fingerprints is done through matching the overall structure of the ridges with categories of loops, whorls, and arches. If a match is made of the overall structure, then the matching is done on individual points in the fingerprint called minutiae points. These minutiae points include ridge bifurcations, ridge endings, and ridge islands. In Section 7.3, we will develop a C function to help analyze fingerprints.

PROGRAMMING WITH STRUCTURES

CHAPTER OUTLINE

OBJECTIVES *In this chapter, we develop problem solutions containing:*

- structures in the main function;
- structures in functions;
- arrays of structures; and
- dynamic data structures.

7.1   Structures

When solving engineering problems, it is often necessary to work with large amounts of data. In Chapter 5, we used arrays to provide a convenient way to store and manipulate large data sets. But arrays work only if all the data are of the same type. In many cases, the data that represent an object or a set of information has multiple data types. For example, recall from Chapter 5 that hurricanes are given names, and they are categorized by the intensity of their winds. Thus, to represent a hurricane, we might include the hurricane's name, the year in which it occurred, and its intensity category. A character string could represent the name, and integers could represent the year and intensity. A **structure** defines a set of data, but the individual parts of the data do not have to be the same type. Thus, a structure can be defined to represent a hurricane as follows:

Structure

```
struct hurricane
{
    char name[10];
    int year, category;
};
```

We can now define variables, and even arrays, of type `struct hurricane`. Each variable or array element would contain three data values—a character string and two integers.

Structures are often called aggregate data types, because they allow multiple data values to be collected into a single data type. Individual data values within a structure are called **data members**, and each data member is given a name. In our previous example, the names of the

Data members

355

data members are `name`, `year`, and `category`. A data member is referenced using the structure variable name followed by a period (called the **structure member operator**) and a data member name. Note the difference between referencing a value in a structure and in an array: A value in an array is referenced with the array name and a subscript.

Definition and Initialization

To use a structure in a program, you must first define the structure. The keyword `struct` is used to define the name of the structure (also called the structure **tag**) and the data members that are included in the structure. After the structure has been defined, structure variables can be defined using declaration statements. Consider the previous definition for a structure representing a hurricane. The name of the structure is `hurricane`. The three data members are `name`, `year`, and `category`. Note that a semicolon is required after the structure definition. The statements to define a structure can appear before the `main` function but they are often stored in a header file. It is important to note that these statements do not reserve any memory—they only define the structure. To define a variable of this structure type, we use the following statement in the declaration section of our program:

```
struct hurricane h1;
```

The preceding statement defines a variable named `h1` that has three data members; this is shown in the following diagram:

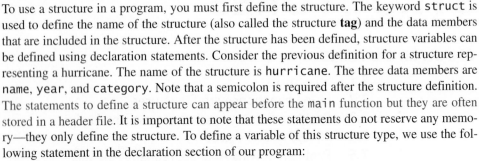

The declaration statement allocates memory for the three data members, but initial values have not been assigned; thus, their values are unknown.

The data members of a structure can be initialized in a declaration statement or with program statements. To initialize a structure with a declaration statement, the values are specified in a sequence that is separated by commas and enclosed in braces. The following declaration statement defines and initializes the example structure `h1`:

```
struct hurricane h1={"Camille",1969,5};
```

The data members of `h1` are now initialized:

h1	"Camille"	name
	1969	year
	5	category

To initialize the structure `h1` using program statements, we can use statements such as the following:

```
h1.name = "Camille";
h1.year = 1969;
h1.category = 5;
```

Thus, to reference an individual data member, we use the structure variable name and the data member name separated by a period.

PRACTICE!

Consider the following structure:

```
struct hurricane
{
    char name[10];
    int year, category;
};
```

Show the contents of the data members of the structures defined in each set of statements.

1. `struct hurricane h1={"Andrew",1969,5};`

2. `struct hurricane h2;`

3. `struct hurricane h3;`
    ```
    ...
    h3.name = "Hugo";
    ```

Input and Output

We can use `scanf` or `fscanf` statements to read values into the data members of a structure, and `printf` or `fprintf` to print the values of the data members. The structure member operator must be used to specify an individual data member. In the next program, the information for a group of hurricanes is read from a file named `storms2.txt`, and the information is printed to the screen.

```
/*--------------------------------------------------------------*/
/*  Program chapter7_1                                          */
/*                                                              */
/*  This program reads the information for a hurricane          */
/*  from a data file and then prints it.                        */

#include <stdio.h>
#define FILENAME "storms2.txt"

/*  Define structure to represent a hurricane.  */
struct hurricane
{
    char name[10];
    int year, category;
};

int main(void)
{
    /*  Declare variables.  */
    struct hurricane h1;
    FILE *storms;
```

```
    /*  Read and print information from the file.  */
    storms = fopen(FILENAME,"r");
    if (storms == NULL)
        printf("Error opening data file. \n");
    else
    {
        while (fscanf(storms,"%s %d %d",h1.name,&h1.year,
                &h1.category) == 3)
        {
            printf("Hurricane: %s \n",h1.name);
            printf("Year: %d, Category: %d \n",h1.year,
                    h1.category);
        }
        fclose(storms);
    }
    /*  Exit program.  */
    return 0;
}
/*------------------------------------------------------------*/
```

Note that the reference in the fscanf statement for the hurricane name is h1.name instead of &h1.name. Since name is a character string, the variable name represents an address or pointer.

Assume that a sample run of this program contains the information in Table 5.2, page 222. The first few lines would appear as follows:

```
Hurricane: Hazel
Year: 1954, Category: 4
Hurricane: Audrey
Year: 1957, Category: 4
```

PRACTICE!

Assume that the structure variables h1 and h2 have been defined with the following statements:

```
struct hurricane
{
    char name[10];
    int year, category;
};

int main(void)
{
    /*  Declare variables.  */
    struct hurricane h1={"Audrey",1957,4};
    struct hurricane h2={"Frederic",1979,3};
```

Show the output for each of the following sets of statements.

1. `printf("%s \n%s \n",h2.name,h1.name);`
2. `printf("Category %d hurricane: %s \n",h1.category,h1.name);`

Computations

We have seen that the structure member operator (.) is used with the name of the structure variable to access individual data members of the structure. When the name of the structure variable is used without the structure member operator, it refers to the entire structure. The assignment operator can be used with the structure variables of the same type to assign an entire structure to another structure, as shown in these statements:

```
struct hurricane
{
   char name[10];
   int year, category;
};

int main(void)
{
   /*  Declare variables.  */
   struct hurricane h1={"Audrey",1957,4}, h2;
...
   h2 = h1;
```

However, relational operators cannot be applied to an entire structure. To compare one structure with another, you must compare the individual data members.

To illustrate computations with structures, consider a program that reads storms2.txt, a file containing hurricane information, and then prints the names of all category 5 hurricanes.

```
/*-------------------------------------------------------------*/
/*  Program chapter7_2                                    */
/*                                                        */
/*  This program reads hurricane information from a data  */
/*  file and then prints the names of category 5 hurricanes.  */

#include <stdio.h>
#define FILENAME "storms2.txt"

/*  Define structure to represent a hurricane.  */
struct hurricane
{
   char name[10];
   int year, category;
};

int main(void)
{
   /*  Declare variables.  */
   struct hurricane h1;
   FILE *storms;
```

```
      /*  Read and print information from the file.  */
      storms = fopen(FILENAME,"r");
      if (storms == NULL)
         printf("Error opening data file. \n");
      else
      {
         printf("Category 5 Hurricanes: \n");
         while (fscanf(storms,"%s %d %d",h1.name,&h1.year,
               &h1.category) == 3)
            if (h1.category == 5)
               printf("%s \n",h1.name);
         fclose(storms);
      }

      /*  Exit program.  */
      return 0;
}
/*-------------------------------------------------------------*/
```

7.2   Using Functions with Structures

Structures can be used as arguments to functions, and functions can return structures. We will consider each of these cases separately.

Structures as Function Arguments

Entire structures can be passed as arguments to functions. When a structure is used as a function argument, it is a call-by-value reference. When a function reference is made, the value of each data member of the actual parameter is passed to the function and is used as the value of the corresponding data member of the formal parameter. Thus, changing the value of a formal parameter does not change the corresponding actual parameter. We have modified the program in the previous section. In this program, the information for a hurricane is printed from a function.

```
/*-------------------------------------------------------------*/
/*  Program chapter7_3                                         */
/*                                                             */
/*  This program reads the information for a hurricane         */
/*  from a data file and then prints it, using a function.     */

#include <stdio.h>
#define FILENAME "storms2.txt"

/*  Define structure to represent a hurricane.  */
struct hurricane
{
   char name[10];
   int year, category;
};

int main(void)
{
   /*  Declare variables and function prototype.  */
   struct hurricane h1;
   FILE *storms;
   void print_hurricane(struct hurricane h);
```

```
/*  Read and print information from the file.  */
storms = fopen(FILENAME,"r");
if (storms == NULL)
   printf("Error opening data file. \n");
else
{
   while (fscanf(storms,"%s %d %d",h1.name,&h1.year,
            &h1.category) == 3)
        print_hurricane(h1);
      fclose (storms);
}

/*  Exit program.  */
   return 0;
}
/*-----------------------------------------------------------*/
/*  This function prints the hurricane information.         */

void print_hurricane(struct hurricane h)
{
   printf("Hurricane: %s \n",h.name);
   printf("Year: %d, Category: %d \n",h.year,h.category);
   return;
}
/*-----------------------------------------------------------*/
```

When functions are written to modify the value of a data member within a structure, we must use a pointer to the structure as the function argument. This allows the function direct access to the data members of the structure. When data members are accessed via a pointer to the structure, the **pointer operator** (->) is used instead of the structure member operator. This technique is demonstrated in Section 7.6, where we present dynamic data structures.

Pointer operator

Functions that Return Structures

A function can be defined to return a value of type `struct`. After the function is called, an entire structure is returned to the calling function. To illustrate, we present a new structure that relates to tsunamis. Assume that we are working with information related to a tsunami. The information includes the date (month, day, year) of the tsunami, its maximum height in feet, the number of fatalities, and the location. The month, day, year, and number of fatalities are represented by integers, a floating-point number represents the maximum height, and the location is represented by a character string. Thus, a structure to represent the information could be defined as follows:

```
struct tsunami
{
   int mo, da, yr, fatalities;
   double max_height;
   char location[20];
);
```

We now present a function that reads information entered from the keyboard, saves it in the structure, and then returns the information to the `main` function. Assume that the preceding structure is defined in the main function and that a function prototype is defined with the following statement:

```
struct tsunami get_info(void);
```

We will use a function that interacts with the user to get the information and then returns it in a structure. The code is as follows:

```
/*----------------------------------------------------------------*/
/*  This function gets information from the user to enter          */
/*  into a tsunami structure.                                      */

struct tsunami get_info(void)
{
   /*  Declare variables.  */
   struct tsunami t1;

   printf("Enter information for tsunami in following order: \n");
   printf("Enter month, day, year, number of deaths: \n");
   scanf("%d %d %d %d",&t1.mo,&t1.da,&t1.yr,&t1.fatalities);
   printf("Enter location (<20 characters, no spaces): \n");
   scanf("%s",t1.location);

   return(t1);
}
/*----------------------------------------------------------------*/
```

MODIFY!

1. Write a function that reads the user's information for a variable of the structure type `hurricane`.
2. Write a function that prints the tsunami information from a variable of the structure type `tsunami`.

7.3   Problem Solving Applied: Fingerprint Analysis

In this section, we use the new statements presented in this chapter to solve a problem related to fingerprint analysis. Fingerprints can be divided into three categories: loops, whorls, and arches. Approximately 60% to 65% of all fingerprints are loops; approximately 30% to 35% are whorls, and approximately 5% are arches. In order to reduce the number of fingerprints in a database that need to be matched to an unknown fingerprint, we can eliminate matches to fingers that do not have the same category. For example, there is no need to match a fingerprint that is a whorl to a fingerprint that is an arch. In fact, there have been some techniques developed that can look at the categories of all 10 fingers and immediately eliminate much of a database from comparison. These techniques are roughly based on the Henry technique, one of the early methods for doing fingerprint recognition.

Assume that the technique that we are going to implement assigns a number to a fingerprint record with 10 fingers. We are going to assign a number to each fingertip based on whether or not it is a whorl. The fingertips will be identified as *R* (for right) or *L* (for left). In addition, a second letter will identify the finger as *t* (for thumb), *i* for (index), *m* (for middle),

r (for ring), and p (for pinky, or little finger). The value for each fingertip is assigned as below:

> If Rt or Ri is a whorl, its value is 16.
>
> If Rm or Rr is a whorl, its value is 8.
>
> If Rp or Lt is a whorl, its value is 4.
>
> If Li or Lm is a whorl, its value is 2.
>
> All other values are zero.

We now compute the following fraction:

$$\text{overall category} = \frac{(Ri + Rr + Lt + Lm + Lp + 1)}{(Rt + Rm + Rp + Li + Lr + 1)}$$

Note that this category cannot be 0, and the denominator cannot be 0. When this category is computed, it is then stored in the record with the fingerprints. Once the category of the unknown fingerprint is computed, then there is a small range of categories that could match from the database. Thus, a search for a match may only need to compare a few thousands fingerprints instead of many millions. This search would then compare the details of the minutiae points of the unknown to the minutiae points of the fingerprints with an overall category value close to the unknown's overall category value.

In this section, we will develop a C function that receives the set of information on a fingerprint that is stored in a structure. We will compute the overall category of the fingerprint and store that in the structure. Thus, this function could be used to assign the overall category number to an unknown, or it could be used in a loop to compute the overall category numbers for the fingerprints in a database.

1. PROBLEM STATEMENT

Write a function that will compute the overall category number for a fingerprint.

2. INPUT/OUTPUT DESCRIPTION

The following diagram shows that the inputs to the program are the categories for the 10 fingerprints. The output is the overall category computed from this information.

categories of 10 fingerprints ⟶ [ ] ⟶ overall category

3. HAND EXAMPLE

Assume that the whorls from a 10-print occur in the right thumb, right ring finger, and left middle finger. The values of the individual fingers are all 0, with these exceptions:

$$Rt = 16$$
$$Rr = 8$$
$$Lm = 2$$

If we then compute the overall category, we have:

$$\text{overall category} = (8 + 2 + 1)/(16 + 1) = 11/17 = 0.65$$

4. ALGORITHM DEVELOPMENT

We first develop the decomposition outline because it divides the solution into a set of sequential steps.

Decomposition Outline for function

1. Get the fingerprint data from the function inputs.
2. Compute the overall category from the fingerprint information.
3. Store the overall category in the fingerprint structure.

We will also need to develop a program that can be used to test this function. The decomposition outline for the test program is the following:

Decomposition Outline for the test program

1. Read fingerprint information from the user.
2. Use function to compute the overall category.
3. Print the overall category.

These are straightforward algorithms, so we can go directly to the C code from the decomposition outlines.

```
/*------------------------------------------------------------------*/
/*  Program chapter7_4                                              */
/*                                                                  */
/*  This program stores fingerprint information in a structure. */
/*  It then references a function to compute the overall category.*/

#include <stdio.h>

/*  Define a structure for the fingerprint information.       */
/*  The order for fingertips is right hand, thumb to pinky,   */
/*  and left hand, thumb to pinky. The codes are L for loops, */
/*  W for whorls, and A for arches.                           */

struct fingerprint
{
  int ID_number;
  double overall_category;
  char fingertip[10];
};

int main(void)
{
    /*  Declare and initialize variables.                     */
    struct fingerprint new_print;
    double compute_category(struct fingerprint f);
```

```
        /*  Specify information for the new fingerprint.         */
        new_print.ID_number = 2491009;
        new_print.overall_category = 0;
        new_print.fingertip[0] = 'W';
        new_print.fingertip[1] = 'L';
        new_print.fingertip[2] = 'L';
        new_print.fingertip[3] = 'W';
        new_print.fingertip[4] = 'A';
        new_print.fingertip[5] = 'L';
        new_print.fingertip[6] = 'L';
        new_print.fingertip[7] = 'W';
        new_print.fingertip[8] = 'A';
        new_print.fingertip[9] = 'L';

        /*  Reference function to compute overall category.       */
        new_print.overall_category = compute_category(new_print);

        /*  Print overall category computed by the function.      */
        printf("Fingerprint Analysis for ID: %i \n",
               new_print.ID_number);
        printf("Overall Category: %.2f \n",new_print.overall_category);

        /*  Exit program.                                          */
        return 0;
}

/*-------------------------------------------------------------*/
/*  This function computes the overall category                */
/*  for a fingerprint.                                         */

double compute_category(struct fingerprint f)
{
        /* Declare and initialize variables.                     */
        double Rt=0, Ri=0, Rm=0, Rr=0, Rp=0, Lt=0, Li=0, Lm=0, Lr=0,
               Lp=0, num, den;

        /*  Set values based on whorls.                          */
        if (f_fingertip[0] == 'W')
           Rt = 16;
        if (f_fingertip[1] == 'W')
           Ri = 16;
        if (f_fingertip[2] == 'W')
           Rm = 8;
        if (f_fingertip[3] == 'W')
           Rr = 8;
        if (f_fingertip[4] == 'W')
           Rp = 4;
        if (f_fingertip[5] == 'W')
           Lt = 4;
        if (f_fingertip[6] == 'W')
           Li = 2;
        if (f_fingertip[7] == 'W')
           Lm = 2;
```

```
        /*  Compute the numerator and denominator for overall category. */
        num = Ri + Rr + Lt + Lm + Lp + 1;
        den = Rt + Rm + Rp + Li + Lr + 1;

        return num/den;
}
/*----------------------------------------------------------------*/
```

Note that we printed the overall category from the structure just to be sure that the value was stored back in the structure.

5. TESTING

The test program has initialized the fingertip categories to match those from the hand example. The output from the program is the following:

```
Fingerprint Analysis for ID Number 24910049
Overall Category:  0.65
```

The answer matches the hand example, so we can now test the program with additional data by changing the values in the test program.

MODIFY!

These problems relate to the program developed in this section.

1. Modify the program so that it also counts and prints the number of whorls for the hand.
2. Modify the program so that it also counts and prints the number of arches for the hand.
3. Modify the program so that it also counts and prints the number of loops for the hand.
4. Modify the program so that it combines problems 1, 2, and 3 and thus prints the numbers of the three different categories for the fingertips.
5. Modify the program in problem 4 so that it prints the percentages of the three different categories for the fingertips.

7.4   Arrays of Structures

In engineering applications, it is often convenient to store information in arrays for analysis. However, an array can contain only a single type of information, such as an array of integers or an array of character strings. If we need to use an array to store information that contains different types of values, then we can use an array of structures. For example, if we want to store an array of information for hurricanes, we can use an array of the structure type hurricane; if we need to store an array of information for tsunamis, we can use an array of the structure type tsunami. We can write a statement to define an array of 25 elements, each of the type hurricane (we will repeat the definition of the structure type hurricane here, too):

```
struct hurricane
{
  char name[10];
  int year, category;
};
...
struct hurricane h[25];
```

Each element in the array is a structure containing the three variables, as illustrated in the following diagram:

To access an individual data member of a structure in the array, we must specify the array name, a subscript, and the data member name. As an example, we will assign values to the first hurricane in the array h:

```
h[0].name = "Camille";
h[0].year = 1969;
h[0].category = 5;
```

To access an entire structure within the array, we must specify the name of the array and a subscript. As an example, we can call the output function defined in Section 7.1. We will print the first hurricane in the array with this statement:

```
print_hurricane(h[0]);
```

The output would be

```
Hurricane: Camille
Year: 1969, Category: 5
```

We will now present a program that reads the information for a group of hurricanes from a data file into an array. The program determines the maximum category in the array and then prints the names of all hurricanes with the maximum category. It should be clear that we cannot print this information as we read the data file. We will not know the maximum category until we have reviewed all the information in the file. At that point, we must go back through the file to print the information for hurricanes with the maximum category.

```
/*-----------------------------------------------------------*/
/*  Program chapter7_5                                       */
/*                                                           */
/*  This program reads hurricane information from a data     */
/*  file and then prints the names of all hurricanes that    */
/*  have the maximum category in the file.                   */

#include <stdio.h>
#define FILENAME "storms2.txt"

/*  Define structure to represent a hurricane.  */
struct hurricane
{
   char name[10];
   int year, category;
};

int main(void)
{
   /*  Declare variables and function prototype.  */
   int max_category=0, k=0, npts;
   struct hurricane h[100];
   FILE *storms;
   void print_hurricane(struct hurricane h);

   /*  Read and print information from the file.  */
   storms = fopen(FILENAME,"r");
   if (storms == NULL)
      printf("Error opening data file. \n");
   else
   {
      printf("Hurricanes with Maximum Category \n");
      while (fscanf(storms, "%s %d %d",h[k].name,&h[k].year,
             &h[k].category) == 3)
      {
         if (h[k].category > max_category)
            max_category = h[k].category;
         k++;
      }
      npts = k;

      for (k=0; k<=npts-1; k++)
         if (h[k].category == max_category)
            print_hurricane(h[k]);

      fclose(storms);
   }

   /*  Exit program   */
   return 0;
}
```

```
/*-------------------------------------------------------------*/
/*  This function prints the hurricane information.            */
void print_hurricane(struct hurricane h)
{
   printf("Hurricane: %s \n",h.name);
   printf("Year: %d, Category: %d \n",h.year,h.category);
   return;
}
/*-------------------------------------------------------------*/
```

7.5   Problem Solving Applied: Tsunami Analysis

A tsunami is a large destructive wave. These large waves typically are generated by sudden changes in the seafloor—changes caused by earthquakes, underwater volcanic eruptions, and underwater landslides. In shallow water, these waves can travel more than 125 miles per hour; in deep water, they can travel over 400 miles per hour. Along the coast of Chile and Peru, records of tsunamis go back hundreds of years. For example, on October 28, 1562, an earthquake in Chile generated a wave with a height of 52 feet.

Some of the larger tsunamis on record include one on September 10, 1899, in the Gulf of Alaska. This tsunami was the result of an earthquake and landslide that generated a wave with a height of 197 feet. On March 28, 1964, again in the Gulf of Alaska, an earthquake caused a tsunami with a height of 230 feet. More recently, on June 3, 1994, an earthquake in eastern Java, Indonesia, generated a tsunami with a height of 197 feet. On July 17, 1998, an earthquake in Papua, New Guinea, brought with it a tsunami with a height of 49 feet; this was clearly not among the largest tsunamis, but it caused over 2,200 fatalities. On March 11, 2011, a magnitude 9.0 earthquake off the coast of Japan caused a tsunami with a height over 130 feet. The tsunami also caused over 13,000 deaths.

We now develop a program that will read a file containing information on large tsunamis from the 1990s (see Table 7.1) and print a report identifying the tsunami with the largest wave height (in feet), the average wave height, and the locations of all tsunamis with waves above the average.

Table 7.1 Large Tsunamis from the 1990s

Date	Location	Maximum Wave (m)	Fatalities
September 2, 1992	Nicaragua	10	170
December 2, 1992	Flores Island	26	1,000*
July 12, 1993	Okushiri, Japan	31	239
June 3, 1994	Eastern Java	14	238
November 14, 1994	Mindoro Island	7	49
October 9, 1995	Jalisco, Mexico	11	1
January 1, 1996	Sulawesi Island	3.4	9
February 17, 1996	Irian Jaya	7.7	161
February 21, 1996	Peru	5	12
July 17, 1998	Papua, New Guinea	15	2,200*

*This is an estimate of the number of fatalities.

Source: Harold V. Thurman and Alan P. Trujillo. *Essentials of Oceanography,* 7th ed. Prentie-Hall, Upper Saddle River, NJ, 2002.

I. PROBLEM STATEMENT

Print a report giving the maximum wave height for the tsunamis in a data file named `waves2.txt`. Include the average wave height (in feet) and the location of all tsunamis with a wave height higher than the average.

2. INPUT/OUTPUT DESCRIPTION

The I/O diagram shows the data file as the input and the report information as the output.

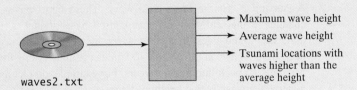

waves2.txt

→ Maximum wave height
→ Average wave height
→ Tsunami locations with waves higher than the average height

3. HAND EXAMPLE

If we assume that the data file contained the information in Table 7.1, then the maximum wave height is 31 meters. Add all of the wave heights, and divide by 10, to compute the average wave height; this computation gives a value that needs to be converted to feet. (Recall that one meter is equal to 3.28 feet.) We can print this information, along with the locations of tsunamis with wave heights larger than the average, in the following report:

```
Summary Information for Tsunamis
Maximum Wave Height (in feet):  101.68
Average Wave Height (in feet):  42.67
Tsunamis with greater than the average height:
Flores_Island
Okushiri,_Japan
Eastern_Java
Papua,_New_Guinea
```

In the data file, we have replaced spaces with underscores in the character strings so that they will be read as a single character string.

4. ALGORITHM DEVELOPMENT

We first develop the decomposition outline because it divides the solution into a set of sequential steps.

Decomposition Outline

1. Read the tsunami data into arrays and determine the maximum and average wave height.
2. Print the maximum and average wave heights.
3. Print the locations with a wave height greater than the average.

Refinement in Pseudocode

main: if file cannot be opened
            print error message
      else
            read data and determine maximum height and average height
            print the maximum height and average height
            print all locations with heights greater than the average

The steps in the pseudocode are now detailed enough to convert to C.

```
/*-----------------------------------------------------------------*/
/*  Program chapter7_6                                             */
/*                                                                */
/*  This program reads tsunami information from a data            */
/*  file and then prints the maximum wave height, the average     */
/*  wave height, and the location of all tsunamis with            */
/*  greater-than-average wave heights.                            */

#include <stdio.h>
#define FILENAME "waves2.txt"

/* Define structure to represent a tsunami.  */
struct tsunami
{
  int mo, da, yr, fatalities;
  double max_height;
  char location[20];
};

int main(void)
{
   /*  Declare variables.  */
   int k=0, npts;
   double max=0, sum=0, ave;
   struct tsunami t[100];
   FILE *waves;

   /*  Read and print information from the file.  */
   waves = fopen(FILENAME,"r");
   if (waves == NULL)
      printf("Error opening data file. \n");
   else
   {
      while (fscanf(waves,"%d %d %d %d %lf %s",&t[k].mo,&t[k].da,
                    &t[k].yr,&t[k].fatalities,&t[k].max_height,
                    t[k].location) == 6)
      {
         sum = sum + t[k].max_height;
         if (t[k].max_height > max)
            max = t[k].max_height;
         k++;
      }
      npts = k;
```

```
            ave = sum/npts;
            printf("Summary Information for Tsunamis \n");
            printf("Maximum Wave Height (in feet): %.2f \n",max*3.28);
            printf("Average Wave Height (in feet): %.2f \n",ave*3.28);
            printf("Tsunamis with greater than average heights: \n");
            for (k=0; k<=npts-1; k++)
                if (t[k].max_height > ave)
                    printf("%s \n",t[k].location);

            fclose(waves);
        }

        /*  Exit program.  */
        return 0;
}
/*------------------------------------------------------------*/
```

5. TESTING

The output from the program using the data from the hand example follows:

```
Summary Information for Tsunamis
Maximum Wave Height (in feet): 101.68
Average Wave Height (in feet): 42.67
Tsunamis with greater than the average heights:
Flores_Island
Okushiri,_Japan
Eastern_Java
Papua,_New_Guinea
```

MODIFY!

These problems relate to the program developed in this section to analyze tsunami data.

1. Modify the program to find and print the number of tsunamis in the file during the same year as the maximum wave height.

2. Modify the program to find and print the date of the tsunami with the largest number of fatalities.

3. Modify the program to find and print the locations for all tsunamis with over 100 fatalities.

4. Modify the program so that it counts the number of tsunamis that occurred during the month of July.

5. Modify the program so that it prints the number of tsunamis in Peru. Assume that the location might also include the city in Peru, so you will need to search the character string for the substring `"Peru"`.

7.6   Dynamic Data Structures*

Static data structures, such as arrays, have a fixed size during the execution of a program. Recall from Chapter 5 that the size of an array must be declared using a constant. When an array is defined in a declaration statement, a contiguous block of memory is allocated and the elements of the array are referenced using the name of the array (address of first element) and a subscript (offset from first element). Static data structures require that the programmer has information about the size of the data set being used so that sufficient memory is allocated but not wasted, and care must be taken not to exceed the maximum size of the data structure. **Dynamic data structures** are data structures that can grow and shrink during the execution of a program. Memory is allocated and freed as needed, and pointers are used to connect the data since the data is not necessarily stored in a contiguous memory space.

Dynamic data structures

We will use a **linked list** to illustrate the use of dynamic data structures. A linked list is organized as a group of **nodes** that are connected by pointers. A node consists of data (which may be a collection of one or more variables) and a pointer to the next node. We generally assume that there is some order to the data stored in the nodes, such as an ascending order, and that we may want to insert new nodes or delete old nodes from this ordered list. The simplest way to think of a linked list is to visualize a small group of nodes (each containing some information) that are linked together by pointers. Figure 7.1 shows a linked list with four nodes, containing the ordered information 10, 14, 21, and 35. A separate pointer, which we will call **head**, points to the first node in the linked list.

Linked list
Nodes

Head

To access this linked list, we use the pointer head to reference the information in the first node (the one containing the value 10). Then, since the first node contains a pointer to the next node (the one containing the value 14), we can move to the second node. Similarly, we use the pointer in the second node to move to the third node (the one containing the value 21). The last node in a linked list will contain a pointer value of NULL to indicate that we are at the last node in the list. We will use the symbol Ω to indicate the end of the list in our diagrams. (Omega, Ω, is the last letter in the Greek alphabet.)

To implement a linked list in C, you can begin to see the steps that are required. Each node of the linked list is a structure. For our example, the structure would contain an integer value and a pointer to the next node. The memory for each node is allocated, as it is needed, using one of the dynamic memory allocation statements. The pointer in the last node is a NULL pointer.

To **insert a value in a linked list**, we need to find the location for the insertion. We use the pointer to the first node to access the first data value in the list. If the data value is less than the value to be inserted, we move to the next data value in the list using the pointer in the

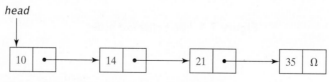

Figure 7.1 *Linked list.*

———————

*Optional section.

current node. We will assume that if the data value to be inserted is already in the list, we should print a message, but not insert a duplicate data value. (Depending on the application, it may be valid to insert duplicate data values.) The desired insertion location can be one of four places—before the first node, between two nodes, after the last node, or in an empty list. We must handle each of these cases carefully, so we can make sure that the pointers are updated properly. Figures 7.2 through 7.5 outline each of these cases assuming that we started with the linked list given in Figure 7.1 for each of the first three cases.

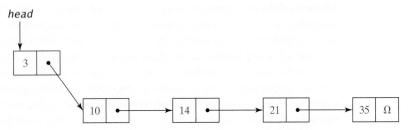

Figure 7.2 *Insert before the first node (requires updating the head pointer).*

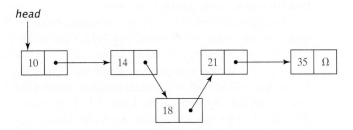

Figure 7.3 *Insert between two nodes.*

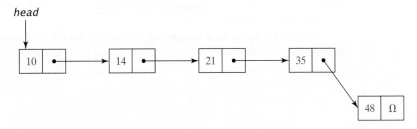

Figure 7.4 *Insert after last node.*

Figure 7.5 *Insert in an empty list (require updating the head pointer).*

To **delete a value from a linked list**, we need to find the node to be deleted. We again use the pointer to the head of the list to access the first data value in the list. We then move from one node to another using the pointer in the current node. If we reach a data value greater than the one to be deleted, we print a message that the value to be deleted is not in the list. The node to be deleted can be the first node, a node between two other nodes, or the last node. We must handle each of these cases carefully, so we can make sure that the pointers are updated properly. Figures 7.6 through 7.8 outline each of these three cases. For each case, we start with the linked list given in Figure 7.2. Note that there are nodes (and data values) that cannot be accessed after these deletions. For example, in Figure 7.6, we can no longer access the value 10; in Figure 7.7, we can no longer access the value 21; and in Figure 7.8, we can no longer access the value 35. Also, note that deleting the first node may result in an empty list if the list only contains one node.

Empty list

We will now develop a set of four functions to create and maintain a linked list. The first function determines if a linked list is an **empty list**, the second function prints the contents of a linked list, the third function inserts a node in its correct ordered position, and the fourth function deletes a node. Each node in the linked list will be represented by the following structure:

```
struct node
{
    int data;
    struct node *link;
};
```

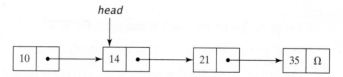

Figure 7.6 Delete first node (requires updating the head pointer).

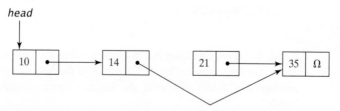

Figure 7.7 Delete a node between two nodes.

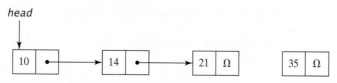

Figure 7.8 Delete the last node.

The structure node has two data members: an integer to store the data value and a pointer to the next node in the linked list. The function to determine if a list is empty has the following prototype:

```
int empty(struct node *head);
```

The function empty examines the head of a linked list and returns a value of 1 if the linked list is empty and a value of zero otherwise. This function will be used by the remaining three functions.

```
/*-----------------------------------------------------------*/
/*  This function returns a value of one if the linked list  */
/*  is empty.                                                */
int empty(struct node *head)
{
   /*  Declare variables.  */
   int k=0;

   /*  Determine if the list is empty.  */
   if (head == NULL)
      k = 1;

   /*  Return integer.  */
   return k;
}
/*-----------------------------------------------------------*/
```

The prototype for the function that prints the list is

```
void print_list(struct node *head);
```

The print_list function begins at the start of the list and prints the data value of each node until the end of the list is reached. The condition in the if statement in this function is the reference to the empty function:

```
   if (empty(head))
   ...
```

Recall that if the condition is a numerical value, then the value zero is considered to be "false" and a non-zero value is "true." Therefore, if the list is empty, the function returns a value of 1, and hence is "true"; otherwise, the function returns a value of zero, and the condition is then "false."

```
/*-----------------------------------------------------------*/
/*  This function prints a linked list.                      */
void print_list(struct node *head)
{
   /*  Declare variables.  */
   struct node *next;
```

```
   /*  Print linked list.  */
   if (empty(head))
      printf("Empty list \n")
   else
   {
      printf("List Values: \n");
      next = head;
      while (next->link != NULL)
      {
         printf("%d \n",next->data);
         next = next->link;
      }
      printf("%d \n",next->data);
   }

   /*  Void return.  */
   return;
}
/*-------------------------------------------------------------*/
```

The `insert` function begins at the head of the list and inserts the new node in the correct location in the list. If the new data value is already in the list, a message will be printed and the new node will not be inserted. Recall that the head of a linked list is a pointer to the first node in the list. When a pointer is used as a function argument, it is a call-by-value reference. Since it is necessary to update the head pointer when a node is inserted before the first node or a node is inserted in an empty list, we must pass a pointer to the head pointer so that it will be a call-by-address reference. This allows us to update the head pointer when necessary inside the function. Look carefully at the syntax.

```
/*-------------------------------------------------------------*/
/*  This function inserts a new node in a linked list.         */

void insert(struct node **ptr_to_head, struct node *nw)
{
   /*  Declare variables and function prototypes.  */
   struct node **next;

   /*  Check for insert to empty list.  */
   if (empty(*ptr_to_head))
      *ptr_to_head = nw;
   else
   /*  Traverse list to find location for insert.  */
   {
      next = ptr_to_head;
      while ( ((*next)->data < nw->data) &&
              ((*next)->link != NULL) )
         next = &(*next)->link;
      /*  Check for duplicate data.  */
      if ((*next)->data == nw->data)
         printf("Node already in list. \n");
```

```
            else
            /*  Check for insert after last node.  */
                if ((*next)->data < nw->data)
                    (*next)->link = nw;
                else
                {
                    nw->link = *next;
                    *next = nw;
                }
        }
    /*  Void return.  */
    return;
}
/*--------------------------------------------------------------------*/
```

The prototype for the function to delete nodes is:

```
void remove(struct node **ptr_to_head, int old);
```

The function begins at the head of the list and searches for the node with a data value of old. If old is not found in the list, a message is printed. If old is found, the node is deleted and memory is freed. Since the head pointer must be updated whenever the first node in the list is deleted, we must pass a pointer to the head of the list. (We did not use the name delete because many compilers use this as a reserved word.)

```
/*--------------------------------------------------------------------*/
/*  This function deletes a node from a linked list.               */
void remove(struct node **ptr_to_head, int old)
{
    /*  Declare variables and function prototypes.  */
    struct node *next, *last, *hold, *head;

    /*  Check for delete to empty list.  */
    head = *ptr_to_head
    if (empty(head))
        printf("Empty list. \n");
    else
    /*  Check for deletion of first node.  */
    {
        if (head->data == old)
        {
            /*  Delete first node.  */
            hold = head;
            *ptr_to_head = head->link;
            free(hold);
        }
        else
        /*  Traverse list to find old node.  */
        {
            next = head->link;
            last = head;
```

```
        while ((next->data < old)) &&
               (next->link != NULL))
        {
           last = next;
           next = next->link;
        }
        /*  Delete node if found.  */
        if (next->data == old)
        {
           hold = last;
           last->link = next->link;
           free(hold);
        }
        else
           printf("Value %d not in list. \n",old);
     }
  }

  /*  Void return.  */
  return;
}
/*-------------------------------------------------------------*/
```

We will now present a main function to test these functions. The print_list function is called after every call to insert and remove to verify that the functions are working correctly.

```
/*-------------------------------------------------------------*/
/*  Program chapter7_7                                        */
/*                                                            */
/*  This program tests the functions to insert and delete     */
/*  items in a linked list.                                   */
#include <stdio.h>
#include <stdlib.h>

/*  Define structure to represent a node in a linked list.  */
struct node
{
   int data;
   struct node *link;
};
int main(void)
{
   /*  Declare variables and function prototypes.  */
   int k=0, old, value;
   struct node *head, *next, *previous, *nw, **ptr_to_head=&head;
   void insert(struct node **ptr_to_head, struct node *nw);
   void remove(struct node **ptr_to_head, int n);
   int empty(struct node *head);
   void print_list(struct node *head);

   /*  Generate and print a linked list with five nodes.  */
   head = (struct node *)malloc(sizeof(struct node));
   next = head;
```

```
        for (k=1: k<=5; k++)
        {
            next->data = k*5;
            next->link = (struct node *)malloc(sizeof(struct node));
            previous = next;
            next = next->link;
        }
        previous->link = NULL;
        print_list(head);

        /*  Allow user to insert or delete nodes in the list.  */
        while (k != 2)
        {
            printf("Enter 0 to delete node, 1 to add node, 2 to quit. \n");
            scanf("%d",&k);
            if (k == 0)
            {
                printf("Enter data value to delete: \n");
                scanf("%d,&old);
                remove(ptr_to_head,old);
                print_list(head);
            }
            else
                if (k == 1)
                {
                    prinf("Enter data value to add: \n");
                    scanf("%d,&value);
                    nw = (struct node *)malloc(sizeof(struct node));
                    nw->data = value;
                    nw->link = NULL;
                    insert(ptr_to_head,nw);
                    print_list(head);
                }
        }

    /*  Exit program.  */
    return 0;
}
/*-------------------------------------------------------------*/
```

A sample run of this program generates the following output:

```
List Values:
5
10
15
20
25
Enter 0 to delete node, 1 to add node, 2 to quit.
1
```

```
Enter data value to delete:
16
List Values:
5
10
15
16
20
25
Enter 0 to delete node, 1 to add node, 2 to quit.
0
Enter data value to delete:
10
List Values:
5
15
16
20
25
```

Additional Dynamic Data Structures

We will present five additional linked data structures that are very powerful. Although each of these data structures could be the topic of a separate section, we will introduce them in this section to illustrate some of the special structures that can be developed with linked lists.

Circularly linked list

Circularly Linked List. A **circularly linked list** is generated when the last node in a linked list points to the first node in the list as shown in Figure 7.9. Note that the NULL constant is not used with the circularly linked list because there really is not an end to the list. However, in an empty circularly linked list, the pointer first will contain the NULL constant.

Inserting and deleting in a circularly linked list is very similar to inserting and deleting in a regular linked list. We do need to be careful that we properly handle the pointer first, because it points to the beginning of the list and also determines when we have followed the links back around to the beginning.

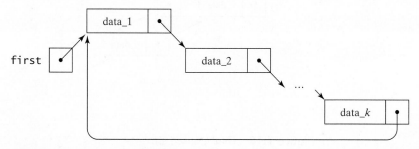

Figure 7.9 *Circularly linked list.*

Operating systems, which are really very sophisticated programs, have a number of applications suitable to a circularly linked list. For example, suppose that a particular program in an operating system is keeping track of the interactive users on the system. Each time a new user logs into the system, the user is added to the list; each time a user logs off the system, the user is deleted from the list. This list has an order because new users are added at the end. When running interactive programs, the computer executes a number of steps in the first user's program, then a number of steps in the next user's program, and so on, until it is back to the first program. It then continues again and again, in a circle. Thus, the information about the number of users on a system and the order in which the system will step through them is an ideal application for a circularly linked list. This type of circularly linked list is sometimes called a round-robin data structure.

Doubly Linked List.   In a linked list, the pointer in each node is used to point ahead to the next node. There are applications in which we would like to have the data linked together so that we can move forward or backward in the list; this type of list is called a **doubly linked list**. Figure 7.10 contains a linked list with both forward links and backward links.

Doubly linked list

Although there is clearly an advantage to being able to move in either direction in a linked list, the routines for manipulating such a list become longer. For example, each insertion requires changing two forward links and two backward links. We must also be sure that the backward link of the first node in the list contains the NULL constant, as well as the forward link of the last node in the list.

Doubly linked lists are useful when we want to be able to insert or delete items without returning to the beginning of the list for each insertion or deletion. For example, suppose we have just inserted a new data value at the tenth item of the list. If we now wish to insert another item, then, instead of starting at the beginning of the list, we can compare the new item to the one that we are currently accessing in the list. If the new item should come before the one that we are accessing, we can use the backward links to back through the list until we find the proper spot for the new insertion. If the new item should follow the one that we are accessing, we can use the forward links to continue through the list until we find the proper spot for the new insertion. This type of insertion can be very efficient in certain situations.

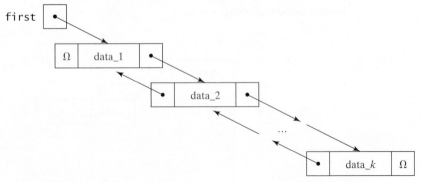

Figure 7.10   *Doubly linked list.*

Stack

Stack. A **stack** is one of the most frequently–used dynamic data structures. It is often described in terms of a bucket, as shown in Figure 7.11. Adding an item to a stack is analogous to dropping it in the bucket. The top item on the stack is always the last item added. Thus, when we remove an item from a stack, it is the top item, or the last item that we added.

LIFO

This data structure is called a **LIFO**, or last in–first out structure. The function for adding items to a stack is called a push function, and the function for removing items from a stack is called a pop function. Figure 7.12 illustrates the contents of a stack as items are added (push functions) and deleted (pop functions). From the diagram, it is clear that a stack is a dynamic data structure. Thus, it can be implemented with a linked structure as shown in Figure 7.13.

The routines for handling the push and pop functions must track both the links between nodes in the stack and the top and bottom of the stack. The top of the stack is the position for the next insertion. Thus, if the top of the stack and the bottom of the stack point to the same position, the stack must be empty.

Stacks are used in a wide variety of applications. For example, if we want to print a set of values in reverse order, we can access the values in their regular order and place each item in the stack. When we reach the end of the values, we begin removing them from the stack. Since the last item added to the stack is the first one removed, we remove the items from the stack in the reverse order in which we added them. There are also a number of applications in which we need to hold values temporarily and then retrieve the most recently stored values first. Compilers frequently need this type of storage as they analyze the syntax in a program statement and convert it into machine language or assembly language.

Queue

Queue. The data structure called a **queue** (pronounced "cue") should be very familiar, although you may not have realized that this structure had a name. Every time you stand in line, whether it is at the grocery store, the video store, or the fast-food restaurant, you are in a queue. A queue is a data structure in which items are added at one end and removed from the

FIFO

other, as shown in Figure 7.14. The queue is also called a **FIFO**, or first in–first out, structure.

Queues are commonly used in operating systems to track users waiting for some computer resource. For example, suppose a computer network has one color printer. If several users attempt to print reports at the same time, the operating system will generally "queue" the users so that the reports are printed one at a time, in the order in which the requests are made.

The functions for handling the queue data structure must be able to handle the steps of inserting at one end of the queue and deleting from the other end. Thus, we need pointers for the front of the queue and for the back of the queue (sometimes called the head and tail, respectively). Obviously, we also need to be able to detect an empty queue. The links within the queue are similar to the regular linked list, but we can only insert at one end and delete from the other. Figure 7.15 shows how the links and pointers are used to implement the queue.

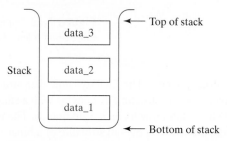

Figure 7.11 *Stack structure.*

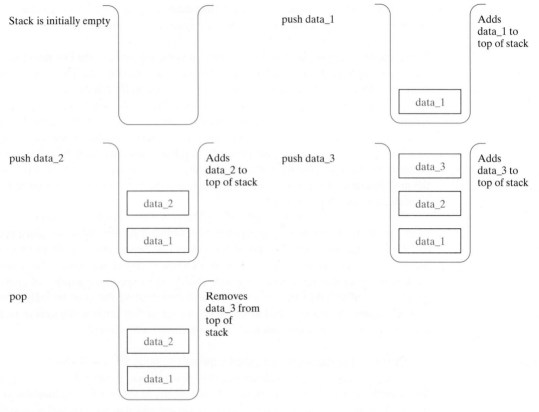

Figure 7.12 *Push and pop operations.*

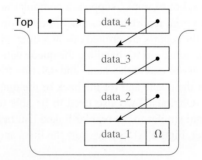

Figure 7.13 *Implementation of a stack using a linked list.*

Binary tree

Binary Tree. The final dynamic data structure that we present is a binary tree. A **binary tree** is a data structure that begins with a single node, often called the root of the tree. This node has a left branch and a right branch. The node at each branch also has a left branch and a right branch. The overall structure of a binary tree is shown in Figure 7.16.

Binary trees are especially useful in certain types of searches. For example, assume that the data stored in a tree are ordered, and the smaller values are always on the left, with the

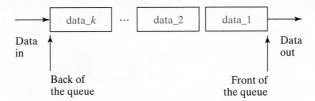

Figure 7.14 Queue structure.

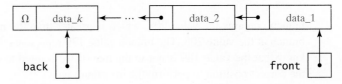

Figure 7.15 Implementation of a queue using a linked list.

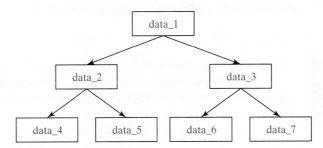

Figure 7.16 Tree structure.

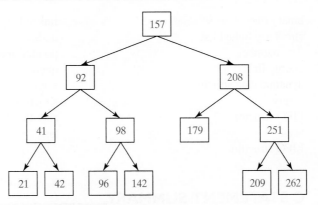

Figure 7.17 An ordered list stored in a tree.

larger values always on the right. Thus, the algorithm to determine whether a particular value is in the list is very efficient. We first compare the root to the value for which we are searching. The results of the comparison immediately determine which half of the tree must be searched. The comparison in the first branch of the correct half of the tree reduces the search to one-fourth of the tree and so on. This process can be illustrated using Figure 7.17. Suppose we wish to determine if the number 189 is contained in the tree. We begin with the root node. Since 189 is greater than 157, we know we must search the right branch. The first right

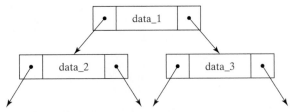

Figure 7.18 *Implementation of a tree using a linked list.*

branch contains the value 208, which is greater than the number 189. We thus take the left branch at the value 208. The branch value 179 represents the end of the branch, and we now know that the value 189 is not in the tree. If we wanted to insert it in the tree, we are now at the correct position to perform the insertion.

The routines to search a tree and to add and delete nodes in the tree require the use of the root position and the left and right links. The implementation of the tree using a linked data structure can be performed as shown in Figure 7.18.

Structures provide a convenient way to define a collection of data values that may or may not be of the same data type. Several examples illustrated how structures are used in C programs. Dynamic data structures were introduced and an algorithm for implementing a linked list was presented. Other dynamic structures were discussed, including circularly linked lists, doubly linked lists, queues, stacks, and trees.

KEY TERMS

binary tree	linked list
circularly linked list	node
data member	pointer operator
doubly linked list	queue
dynamic data structures	stack
empty list	structure
FIFO structure	structure member operator
head	tag
LIFO structure	

C STATEMENT SUMMARY

Structure definition:

```
struct tsunami
{
    int mo, da, yr, fatalities;
    double max_height;
    char location[20];
};
```

```
struct node
{
    int data;
    struct node *link;
};
```

Structure declaration:

```
struct tsunami t1;
struct node n1, n2=(2,NULL);
```

Accessing data members:

```
t1.mo = 10;
n1.link = &n2;
```

Arrays of structures:

```
struct tsunami t2[100];
```

NOTES_____

1. Structure definitions are often included in `.h` files outside of `main`.

DEBUGGING NOTES_____

1. A structure definition does not allocate memory. To allocate memory, a variable must be declared after the structure has been defined.

2. The structure member operator must be used with the name of the structure variable to reference individual data members.

3. The name of the structure variable without the structure member operator refers to the entire structure.

4. The relational operators cannot be used on an entire structure.

PROBLEMS_____

SHORT-ANSWER PROBLEMS

True–False Problems

Indicate if the following statements are true (T) or false (F).

1. Memory is allocated for a structure during structure definition and is equal to the total memory occupied by the individual data members.   T   F

2. To access individual members of a structure, the syntax *variable_name.data_element* is used.   T   F

3. A structure may have a data element of the same name as the structure itself.   T   F

4. The structure member operator is used with the name of a structure
   to reference a data member of a structure.   T   F

5. The pointer operator is used with a pointer to a structure to access
   the data members of a structure.   T   F

Multiple Choice Problems

Use the following structure definition to answer Problems 6 to 9:

```
struct cable
{
    char name[30];
    int diameter;
    float datarate;
    float range;
}c1,c2,*c3;
```

6. What will be the correct statement to input the name of cable c1, considering that the name
   of a cable may consist of more than one word?
   (a) `scanf("%c",&name);`
   (b) `scanf("%s",&c1.name);`
   (c) `gets(c1.name);`
   (d) `gets(name);`

7. To compare whether c1 and c2 are identical cables,
   (a) it is necessary to compare all the individual elements in c1 and c2.
   (b) it is necessary to compare the first and the last elements of c1 and c2.
   (c) c1 and c2 may be directly compared by the statement `if(c1==c2)`
   (d) c1 and c2 may be directly compared by the statement `if(c1=c2)`

8. What will be the correct set of statements to assign a diameter to the cable pointed by c3?
   (a) `c3->diameter=30;`
       `c3=&c1;`
   (b) `c3=&c1;`
       `c3->diameter=30;`
   (c) `c1=*c3;`
       `c1.diameter=30;`
   (d) `c1.diameter=30;`
       `c1=*c3;`

9. Give the correct function prototype of a function `cable_f` that accepts a cable as an
   argument and returns another cable that has the same datarate but different name,
   diameter, and range.
   (a) `void cable_f ( struct cable c1);`
   (b) `void cable_f ( cable );`
   (c) `cable cable_f ( struct cable c1);`
   (d) `struct cable cable_f ( struct cable c1);`

Memory Snapshot Problem

Use the following structure definition to answer Problems 10 through 13:

```
struct date
{
   int month, day, year;
};
```

Give the corresponding snapshot of memory for the structures start_date and end_date; assume that each statement builds on the previous statements.

10. ```struct date start_date, end_date;```

11. ```start_date.month = 9;```

12. ```start_date.year = 2005;```
    ```end_date.year = start_date.year + 3;```

13. ```
if (end_date.month < 7)
       end_date.day = 1;
    else
       end.date.day = 30;
```

PROGRAMMING PROBLEMS

Hurricanes. In this chapter, we defined a structure that can be used to represent information for hurricanes:

```
struct hurricane
{
   char name[20];
   int year, category;
};
```

These problems refer to the file storms2.txt, which contains information for the strongest hurricanes in the United States during 1950 to 2002. Note that the information in the file is ordered by year.

14. Write a program to read the information in storms2.txt. Use the preceding structure and print a report with the average category of these hurricanes. Use an output format similar to the following:

    ```
    Hurricane Summary for Strongest Hurricanes in 1950-2002
    Average Category:  xx.x
    ```

    Note: You do not need an array to solve this problem.

15. Write a program to read the information in storms2.txt. Use the preceding structure and print the number of hurricanes in each category. Use an output format similar to the following:

    ```
    Hurricane Summary for Strongest Hurricanes in 1950-2002
    Category      Hurricanes
    1             x
    2             x
    3             x
    ```

    Note: You do not need an array to solve this problem.

16. Write a program to read the information in `storms2.txt`. Use the preceding structure and print the information for the hurricanes that occurred between 1960 and 1969. Use an output format similar to the following:

```
Strongest Hurricanes between 1960 and 1969
Name   Year   Category
```

Note: You do not need an array to solve this problem.

17. Write a program to read the information in `storms2.txt`. Use the preceding structure and print the information for the hurricanes that occurred between two years entered by the user. Use an output format similar to the following:

```
Hurricane Summary for Strongest Hurricanes between x and x
Name   Year   Category
```

Note: You do not need an array to solve this problem.

18. Write a program to read the information in `storms2.txt`. Use the preceding structure and print the information for the hurricanes such that you first print the information for category 5 hurricanes, then the information for category 4 hurricanes, and so on.

```
Strongest Hurricanes between 1950 and 2002
Category   Number of Hurricanes
```

Instead of sorting the information in the file, read it into an array and then make multiple passes through the array.

19. Write a program to read the information in `storms2.txt`. Use the preceding structure and print the hurricane names in alphabetical order:

```
Strongest Hurricanes between 1950 and 2002
Hurricane Name
```

You may want to review the sections on sorting data in arrays (Chapter 5).

20. Write a program to read the information in `storms2.txt`. Use the preceding structure and print the hurricane names in alphabetical order, along with the other information:

```
Strongest Hurricanes between 1950 and 2002
Hurricane   Year   Category
```

Hint: First solve the problem presented in Problem 19. Then modify the program so that each time you switch values in the hurricane name array, you do the same switches in the year array and the category array.

Earthquakes. The following set of problems refers to the file `earthquake.txt` that contains information about the deadliest earthquakes of the 1990s.

Use the following structure for your reference:

```
struct quakes
{
    int serial, year;
    char name[20], country[20];
```

```
        unsigned long int fatalities;
        float magnitude;
};
```

21. Write a program that reads from `earthquake.txt` and prepares a report of the major earthquakes in China using the above structure.

22. Write a program that reads from `earthquake.txt` and prepares a report of the earthquakes in chronological manner using the above structure.

23. Write a program that reads data from `earthquake.txt` and then inputs a range of years from the keyboard. Then print information about the earthquakes using the following format:

```
Deadliest Earthquakes in between _____ and _____

Name          Year        Fatalities        Magnitude
```

24. Write a program that reads from `earthquake.txt` and print a report of the earthquakes that have caused fatalities of over 100,000 in the same format as question 23. Also compute the total numbers of such earthquakes and their average magnitude.

25. Write a program that reads from `earthquake.txt` and print a report of the earthquakes that have

    (a) caused the maximum fatalities.
    (b) had the maximum magnitude.

CHAPTER EIGHT

8

Crime Scene Investigation: Hand Recognition

Hand recognition systems are typically used for access control. For example, Disney World uses hand recognition for entrance by people with season passes. Many athletic gyms use hand recognition for entrance instead of a swipe card. These systems allow high-volume traffic use in applications that do not need high security. Hand recognition is a good biometric for these applications because the measurements can be taken very quickly and the computations are not complicated. However, hand recognition is not typically used by itself in applications that need very accurate identification, because the accuracy is not as good as for fingerprint or iris recognition. To use hand recognition, you slide your hand into a structure that has pins that fit between your fingers and that hold your hand in place. The system measures the length, width, and height of your fingers, as well as the distances between finger joints. In fact, most hand recognition systems make a total of nearly 100 measurements. Using these measurements, the system can compare the measurements of your hand to information stored in a master database. If the measurements of your hand are close enough to those of one of the hands in the database, you will be granted access to the system. If there is not a close match, you may be required to show identification in order to gain access. In Section 8.5 we develop a C function to help analyze measurements taken from a hand in a hand recognition system.

AN INTRODUCTION TO C++

CHAPTER OUTLINE

OBJECTIVES *In this chapter, we develop problem solutions containing:*

- C++ language objects for standard input and output;
- C++ language objects for file input and output; and
- C++ user-defined classes.

8.1   Object-Oriented Programming

Object-oriented programming

Bjarne Stroustrup of AT&T Bell Laboratories developed C++ in the early 1980s. C++ is essentially C with additional features to support **object-oriented programming**. C++ supports all C operators and control structures, as well as the definition and use of functions. The C++ language is a superset of C, so most C programs can be compiled using a C++ compiler. The reverse, however, is not true: C compilers will not recognize the additional object-oriented features of C++.

Object-oriented programming requires a change in the way we think about problems that we want to solve. When approaching the task of designing an object-oriented program to solve a problem, we begin by identifying the key elements in the problem. We identify what defines these elements and how these elements will be created, modified, and used within the program. Since the elements we identify often do not match an existing data type, we choose to define our own data type. In C++, new data types are defined using classes.

Class
Object

Object-oriented programming is characterized by the use of classes, objects, polymorphism, and inheritance. A **class** is a programmer-defined data type that includes both data and functions that operate on the data. An **object** is a variable of a defined class type, often referred to as an instance of a class. Objects call functions that are defined for their class. These functions are called

Member functions
Polymorphism

Inheritance

member functions, and they operate on the data members of the calling object. **Polymorphism** is the ability to assign many meanings to the same name. Overloading of functions is one example of polymorphism in C++. The same function name can be given multiple definitions. At run time, the system can determine which function definition to invoke. **Inheritance** allows a class to inherit attributes from an existing class. The new class (referred to as the child class or derived class) inherits all of the data and functions from the existing class (referred to as the parent class or the base class) and may have additional data members or member functions of its own. Inheritance is a key concept to object-oriented design, but the discussion of inheritance is beyond the scope of this chapter.

8.2   C++ Program Structure

In this section, we begin our discussion of C++ by illustrating some of the differences between the basic structure of a C++ program and a C program. We will use the program introduced in Chapter 1 on page 18 as our example; it computes and prints the distance between two points. That program, converted to C++, is shown below:

```
//-------------------------------------------------------------
//   Program chapter8_1
//
//   This program computes the distance between two points.

#include <iostream>
#include <cmath>
using namespace std;

int main(void)
{
   // Declare and initialize variables.
   double x1=1, y1=5, x2=4, y2=7,
          side_1, side_2, distance;

   // Compute sides of a right triangle.
   side_1 = x2 - x1;
   side_2 = y2 - y1;
   distance = sqrt(side_1*side_1 + side_2*side_2);

   // Print distance.
   cout.setf(ios::fixed);
   cout.precision(2);
   cout << "The distance between the two points is "
        << distance << endl;

   // Exit program.
   return 0;
}
//-------------------------------------------------------------
```

Many new `include` files exist for C++. The file `iostream` is included for standard input and output and will be discussed in the next section.

Multiline comments can be enclosed within /* and */. C++ also recognizes the double slash (//) as the delimiter for a one-line comment. The // delimiter may be used anywhere on the line. On a program line, everything to the right of the // is treated as a comment.

8.3   Input and Output

C++ uses the predefined object cin (pronounced see-in) to perform standard input and the predefined object cout (pronounced see-out) to perform standard output. These objects are defined in the header file iostream. To use either of these objects in a program, we must include the following preprocessor directive:

```
#include <iostream>
```

This directive includes the ostream and istream class definitions, as well as other information necessary to use the cin and cout objects. Note that cout is an object of the ostream class, and cin is an object of the istream class.

The preprocessor directive for the math library is:

```
#include <cmath>
```

Another directive tells the compiler to use the library filenames declared in namespace std, and it should be used along with the other preprocessor directives:

```
using namespace std;
```

The cout Object

Stream

Insertion operator

The cout object is defined to stream output to the standard output device. The word **stream** suggests a continual stream of characters that is generated by the program and sent to the output device. The stream **insertion operator** (<<) is used with cout (or any ostream object). Any combination of values may be sent to the output device. For our examples, we will assume that the standard output device is the screen. The following statement outputs four values to the screen:

```
cout << "The radius of the circle is " << radius << " centimeters"
    << endl;
```

Each value to be output must be preceded by the << operator. In the preceding example, the first value to be output is the string "The radius of the circle is", the second value to be output is the value of the variable radius, the third value to be output is the string "centimeters", and the fourth value is the predefined iostream manipulator endl. Since endl inserts a newline character into the output stream, it causes the line to be printed.

Another use of cout can be illustrated by the following example:

```
double radius=10, area;
const double PI=3.141579;
...
cout << "The radius of the circle is: " << radius << " centimeters"
    << endl
    << "The area is " << PI*radius*radius << " square centimeters"
    << endl;
```

In this example, we use the modifier const to declare a named constant for PI. The output from these statements is

```
The radius of the circle is: 10 centimeters
The area is 314.158 square centimeters
```

Notice that no decimal point is displayed in the value of radius, even though radius is declared to be of type double. You can control the format of your output in C++ using stream functions and manipulators.

Stream Functions

Dot operator

Format flags

Recall that cout is an object of the ostream class. Stream functions are member functions of the ostream class and can be called by ostream objects. A special operator called the **dot operator** (.) is used when objects call member functions. In this example, we will show you a few of the stream functions and **format flags** that can be used to format output. The setf function is used to set the format flags. The precision function, when used with ios::fixed, specifies how many places to print to the right of the decimal point. When the precision function is used without setting ios::fixed, it specifies the number of significant digits to be displayed. Table 8.1 lists several of the more commonly used format flags.

```
double radius = 10, area;
const double PI=3.141579;

//set format flags
cout.setf(ios::fixed);  // setf function is called by cout
cout.setf(ios::showpoint);
cout.precision(2);      //set precision
...
cout << "The radius of the circle is: " << radius << " centimeters"
     << endl
     << "The area is " << PI*radius*radius << " square centimeters"
     << endl;
```

The output from these statements is as follows:

```
The radius of the circle is: 10.00 centimeters
The area is 314.16 square centimeters
```

PRACTICE!

Assume that the integer variable sum contains the value 150 and that the double variable average contains the value 12.368. Show the output generated by the following code segments:

1. `cout << sum << average;`

2. `cout << sum;`
   `cout << average;`

3. `cout << sum << endl << average;`

4. `cout.precision(2);`
   `cout << sum << endl << average;`

5. `cout.setf(ios::showpoint);`
   `cout.precision(3);`
   `cout << sum << ',' << average;`

6. `cout.setf(ios::fixed);`
   `cout.setf(ios::showpoint);`
   `cout.precision(3);`
   `cout << sum << ',' << average;`

Table 8.1 Common Format Flags

Flag	Meaning
ios::showpoint	display the decimal point
ios::fixed	decimal notation
ios::scientific	scientific notation
ios::right	print right justified
ios::left	print left justified

The cin Object

Extraction operator

White space

The cin object is defined to stream input from the standard input device. For our examples, we will assume that the standard input device is the keyboard. The stream **extraction operator** (>>) is used with cin (or any istream object) to input values and assign a value to a variable. The >> operator discards all **white space** (i.e., blanks, tabs, and newlines). The following statement inputs three values from the keyboard:

```
cin >> var1 >> var2 >> var3;
```

The cin statement waits for input. In the foregoing example, the first value typed into the keyboard will be assigned to the variable var1, the second value to var2, and the third value to var3. The input is not read until the Enter key is pressed. This allows for backspacing and correcting mistakes. The values entered from the keyboard must be separated by white space, but it does not matter how much white space is present. The cin statement will continue to discard white space until it receives values for each of the variables in the statement. The values entered should be compatible with the data type of the variables in the cin statement.

The use of cin is illustrated by the following example:

```
int id;
double rate, hours;
char code;
...
cin >> rate >> hours >> id >> code;
cout << rate << endl << hours << endl << id << endl
     << code << endl;
```

Assume that the input stream from the keyboard contains the following two lines of input:

```
10.5    40
556     r
```

Then the variables in the input statement would be assigned the following values:

rate	10.5
hours	40
id	556
code	'r'

The cout statement would print the following output to the screen:

```
10.5
40
556
r
```

No specifiers are required with cin. The >> operator interprets the input value according to the data type of the variable that follows it. The >> operator also discards all white space. For some applications requiring character data, it may not be desirable to discard white space. The member function get can be used with istream objects to get a single character from the input stream. The statements

```
char ch;
cin.get(ch);
```

will read the next character from the keyboard and assign the character to the variable ch. The get function does not discard white space; it treats white space as valid character data.

Defining File Streams

File stream

We have been using cin to read input from the keyboard and cout to print output to the screen. If we want to read values from a file or print information to a file, we must define a **file stream** object and associate this object with a file.

C++ provides two file stream classes: one for file input and one for file output. The ifstream class is used to define objects that stream input from a file, and the ofstream class is used to define objects that stream output to a file. The ifstream class and the ofstream class are defined in the header file fstream. Declaration statements are used to define ifstream and ofstream objects, as shown in the following statements:

```
ifstream indata;    //defines indata as an input file stream object
ofstream outdata;   //defines outdata as an output file stream object
```

After a file stream-object has been defined, it must then be associated with a specific file. The member function open is called by the file stream object, and a file name is passed as an argument. Thus, the following statements associate a data file with the objects indata and outdata, respectively:

```
indata.open("sensor1.txt");    //opens the file sensor1 for input
outdata.open("plot1.txt");     //opens the file plot1 for output
```

Now that the objects indata and outdata have been defined and each has been associated with a data file, they can be used for input and output in the same way that cin and cout are used. If we assume that the data file sensor1 holds some experimental data values, we can read one value from the file with the following statement:

```
indata >> x;
```

We can write the value of x, along with the natural log of x and e^x to the output file plot1.txt, with the following statement:

```
outdata << x << " " << log(x) << " " << exp(x) << endl;
```

White space is printed to separate the values.

The function close is used to close a file after we are finished with it. The file stream object calls the function. To close the two files used in this example, we use the following statements:

```
indata.close();
outdata.close();
```

When opening data files for input, it is always a good idea to verify that the open function has been successful. If the open function fails to open the file, no error message will be generated, but all attempts to read from the file will fail. The member function fail can be used to determine if the open function was successful. The member function eof can be used to determine when the end of the file has been encountered. The use of these functions is illustrated in the next section.

8.4   C++ Program Examples

One of the best ways to understand the differences between C and C++ is to compare two programs that solve the same problem—but one is written in C and the other is written in C++. In this section, we present several C++ programs that solve the same problems that were solved earlier in this text. We will include the page number for the C program so that the two programs can easily be compared. In these programs, most of the differences will be in the input and output statements.

Simple Computations

In Section 2.10, we developed a program to compute new velocity and acceleration for an aircraft after a change in the power level. Compare the C program developed on page 69 with the following C++ program:

```
//-------------------------------------------------------------
//   Program chapter8_2
//
//   This program estimates new velocity and acceleration
//   values for a specified time.

#include <iostream>
#include <cmath>
using namespace std;

int main(void)
{
   // Declare variables.
   double time, velocity, acceleration;

   // Get time value from the keyboard.
   cout << "Enter new time value in seconds:" << endl;
   cin >> time;

   // Compute velocity and acceleration.
   velocity = 0.00001*pow(time,3) - 0.00488*pow(time,2)
            + 0.75795*time + 181.3566;
   acceleration = 3 - 0.000062*velocity*velocity;
```

```
    //  Print velocity and acceleration.
    cout.setf(ios::fixed);
    cout.precision(3);
    cout << "Velocity = " << velocity << " m/s" << endl;
    cout << "Acceleration = " << acceleration << " m/s^2" << endl;

    //  Exit program.
    return 0;
}
//-------------------------------------------------------------
```

Loops

In Section 3.5, we developed a program to convert degrees to radians. Compare the C program developed on page 106 with the following C++ program:

```
//-------------------------------------------------------------
//  Program chapter8_3
//
//  This program prints a degree-to-radian table using a
//  for loop structure.

#include <iostream>
using namespace std;

int main(void)
{
    //  Declare constants and variables.
    const double PI=3.141593;
    int degrees;
    double radians;

    //  Print radians and degrees in a loop.
    cout.setf(ios::fixed);
    cout.precision(6);
    cout << "Degrees to Radians" << endl;
    for (degrees=0; degrees<=360; degrees+=10)
    {
        radians = degrees*PI/180;
        cout << degrees << " "
             << radians << endl;
    }

    //  Exit program.
    return 0;
}
//-------------------------------------------------------------
```

Functions, One-Dimensional Arrays, and Data Files

In Section 5.1, we developed a program to read up to 100 values from a data file and then determine the maximum value. The steps to determine the maximum value in the array were performed using a function. Compare the C program developed on page 219 with the following C++ program:

```cpp
//---------------------------------------------------------
//   Program chapter8_4
//
//   This program reads values from a data file and determines
//   the maximum value with a function.

#include <iostream>
#include <fstream>
using namespace std;
#define FILENAME "lab2.txt"

int main(void)
{
    //  Declare variables and function prototype.
    const int N=100;
    int k=0, npts=N;
    double y[N];
    double max(double x[], int n);
    ifstream lab;

    //  Open file, read data into an array.
    lab.open(FILENAME);
    if (lab.fail())
        cout << "Error opening input file." << endl;
    else
    {
        while (!lab.eof())
        {
            lab >> y[k];
            k++;
        }
        npts = k;

        //  Find and print the maximum value.
        cout << "Maximum value: "
             << max(y,npts)) << endl;

        //  Close file and exit program.
        lab.close();
    }

    //  Exit program.
    return 0;
}
//---------------------------------------------------------
//   This function returns the maximum value in an array x
//   with n elements.

double max(double x[],int n)
{
    //  Declare variables.
    int k;
    double max_x;
```

```
              //  Determine maximum value in the array.
              max_x = x[0];
              for (k=1; k<=n-1; k++)
                 if (x[k] > max_x)
                    max_x = x[k];

              //  Return maximum value.
              return max_x;
           }
           //------------------------------------------------------------
```

8.5   Problem Solving Applied: Hand Recognition

In this section, we use the new C++ statements presented in this chapter to solve a problem related to hand recognition. For simplicity, we are going to use just five measurements: the length of the five fingers on the right hand. Write a function that has the inputs from an unknown hand and one from an entry in the database. The function should compute the sum of the absolute value of the differences between corresponding fingers. This sum represents the distance between the two hands. The recognition program would use this function to compare the unknown to each hand in the database and then choose the hand from the database that is closest to the unknown. If this difference is less than a specified threshold, then the unknown identify is identified as a match to the database and access is allowed. A program will also be developed to allow us to test this function.

1. PROBLEM STATEMENT

Write a function that will compute the overall distance between measurements of two hands.

2. INPUT/OUTPUT DESCRIPTION

The following diagram shows that the inputs to the function are the 5 finger lengths for the unknown and the 5 finger lengths from a record in the database. The output is the overall distance measurement.

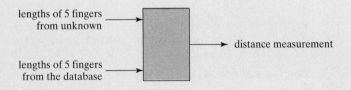

3. HAND EXAMPLE

Assume that these are the distances, in cm:

	unknown	database entry	|difference|
thumb	5.4	6.2	0.8
index finger	7.2	7.0	0.2
middle finger	7.9	8.0	0.1
ring finger	7.4	7.4	0.0
little finger	5.1	5.8	0.7

The sum of the differences is 1.8.

4. ALGORITHM DEVELOPMENT

We first develop the decomposition outline because it divides the solution into a set of sequential steps.

Decomposition Outline for function

1. Get the finger data from the function inputs.

2. Compute the overall distance value.

3. Return the overall distance value.

We will also need to develop a program that can be used to test this function. The decomposition outline for the test program is the following:

Decomposition Outline for the test program

1. Specify finger lengths for unknown and for database entry.

2. Use function to compute the distance measurement.

3. Print the distance measurement.

These are straightforward algorithms, so we can go directly to C++ code from the decomposition outlines.

```
//------------------------------------------------------------
//  Program chapter8_5
//
//  This program computes and prints the distance between two hand
//  measurements.

#include <iostream>
#include <cmath>
using namespace std;
```

```
int main(void)
{
    //Declare and initialize variables.
    double unknown[5]={5.4,7.2,7.9,7.4,5.1},
           known[5]={6.2,7.0,8.0,7.4,5.8};
    double distance(double hand_1[5],double hand_2[5]);

    // Compute and print distance.
    cout << "Distance: " << distance(unknown,known) << endl;

    // Exit program.
    return 0;
}
//-----------------------------------------------------------------
//  This function computes the distance between two hand measurements.
double distance(double hand_1[5],double hand_2[5])
{
    //  Declare variables.
    int k;
    double sum=0;

    //  Compute sum of absolute value differences.
    for (k=0; k<=4; k++)
       sum = sum + fabs(hand_1[k]-hand_2[k]);

    //  Return distance value.
    return sum;
}
//-----------------------------------------------------------------
```

5. TESTING

The test program has initialized the unknown and the known hand measurements to match those from the hand example. The output from the program is the following:

Distance: 1.8

The answer matches the hand example, so we can now test the program with additional data by changing the values in the test program. This would be a good problem to solve using hand measurements from the people in your class. Then take a new measurement from one of the students and compare the distance measurements with the entire class to see if the smallest distance is the one that it should match. Also, does it matter which order you use for the hand measurements in the function? It doesn't, but try switching the order to convince yourself. What if the distance measurement was just a sum, instead of an absolute value?

MODIFY!

These problems relate to the program developed in this section for comparing hand measurements.

1. Modify the program so that it reads the unknown measurement from the keyboard.
2. Modify the program so that it reads the known measurement from a data file and contains a loop that compares the unknown to all measurements in the file.
3. Modify the program in problem 2 so that it also prints the minimum distance measurement.
4. Modify the program in problem 3 so that it prints the entry number with the minimum distance, as in "Known 4 has best match."
5. Modify the program in problem 4 so that it prints all entries with the minimum distance.

8.6   Problem Solving Applied: Surface Wind Directions

Surface winds drive the surface currents in the oceans, while water density drives deep currents. The surface winds can be measured by satellite, and they tend to follow the major wind belts around the globe. In the Northern Hemisphere, the trade winds typically blow from the northeast to the southwest; in the Southern Hemisphere, the trade winds typically blow from the southeast to the northwest. (These winds are called trade winds because the ocean's shipping paths were determined by them.) The boundary between the trade winds and the equator is often referred to as the "doldrums," because there are periods of little wind. For sailing ships, this area could be particularly frustrating because of the slow passage through it. In this section, we develop a program that will read a file containing surface wind directions for a region of the ocean. Our program will determine the main direction that the wind is blowing, and the program will also compute the percentage of points in the file blowing in that direction.

Figure 8.1 contains a diagram showing the eight directions on a compass. Assume that a data file contains a set of directions that corresponds to a set of points on a grid in the ocean. For this example, assume that the grid is 5 points by 5 points, and each row of information is stored on a separate line in a data file named `wind1.txt`. The wind directions have been coded as shown in Table 8.2.

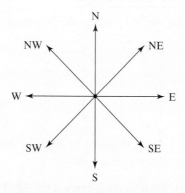

Figure 8.1 *Compass directions.*

Table 8.2 Wind Direction Codes

Direction	Code
N	1
NE	2
E	3
SE	4
S	5
SW	6
W	7
NW	8

I. PROBLEM STATEMENT

Read a set of wind directions for an ocean grid. Determine and print the main direction of the wind and the corresponding number of points blowing from that direction.

2. INPUT/OUTPUT DESCRIPTION

The I/O diagram shows the data file as the input and the report information as output.

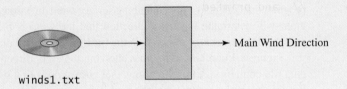

winds1.txt

Main Wind Direction

3. HAND EXAMPLE

Assume that the grid contains the following information:

```
4     4     5     5     4
3     4     4     4     5
4     5     4     4     4
4     4     4     4     4
4     4     4     4     5
```

The value 4 occurs 19 out of 25 times, or 76% of the time. Therefore, the output from the program should appear as follows:

```
The wind is blowing from the SE 76% of the time.
```

4. ALGORITHM DEVELOPMENT

First, we develop the decomposition outline because it divides the solution into a set of sequential steps.

Decomposition Outline

1. Read the wind direction data into an array and determine the direction with the maximum values.

2. Compute and print the percentage of time the wind blows from that direction.

> *Refinement in Pseudocode*
>
> *main: if file cannot be opened*
> > *print error message*
> > *else*
> > > *read data and determine grid points for each direction*
> > > *determine the direction with maximum points*
> > > *compute and print the direction with maximum points*

The steps in the pseudocode are now detailed enough to convert to C++:

```
//---------------------------------------------------------------
//  Program chapter8_6
//
//  This program reads wind direction information from a data
//  file and then determines the direction with the most points.
//  The percentage of points with the maximum direction is computed
//  and printed.

#include <iostream>
#include <fstream>
using namespace std;

int main(void)
{
   /*  Declare variables.  */
   int r, c, k, maxk=0;
   int grid[5][5], category[8]={0,0,0,0,0,0,0,0};
   double perc;
   char* direction[8]={"N  ","NE ","E  ","SE ","S  ",
                       "SW ","W  ","NW "};
   ifstream winds;

   /*  Read and print information from the file.  */
   winds.open("winds1.txt");
   if (winds.fail())
      cout << "Error opening input file." << endl;
   else
   {
      for (r=0; r<=4; r++)
         winds >> grid[r][0] >> grid[r][1] >> grid[r][2]
                >> grid[r][3] >> grid[r][4];
```

```
                // Determine sums for the direction categories.
                for (r=0; r<=4; r++)
                    for (c=0; c<=4; c++)
                    {
                        k = grid[r][c];
                        category[k]++;
                    }
                // Determine category with maximum sum.
                for (k=0; k<=7; k++)
                    if (category[k] > category[maxk])
                        maxk = k;

                // Print report.
                cout.setf(ios::fixed);
                cout.precision(1);
                perc = (double)category[maxk]/25*100;
                cout << "The wind is blowing from the "
                     << direction[maxk-1]
                     << perc << "% of the time." << endl;

                // Close file.
                winds.close();
            }
            // Exit program.
            return 0;
        }
//-------------------------------------------------------------
```

5. TESTING

Using the data from the hand example, the output from the program should appear as follows:

```
The wind is blowing from the SE 76 % of the time.
```

MODIFY!

These problems relate to the program developed in this section for determining primary wind direction.

1. Modify the program to print the number of values in each category, using direction notation (such as SE).
2. Modify the program to print more than one output line if there is more than one direction with the same number of maximum values.
3. Modify the program to print the message "Possible cyclone or hurricane" if there are winds in all four quadrants of the compass.
4. Modify the program so that it prints a grid to demonstrate the direction of the winds. For example, it could print the character '>' if the wind is from the W, and '<' if the wind is from the E. Choose characters for all eight wind directions.
5. Modify the program so that it reads the grid size (number of rows and number of columns) from the first row of the data file. Assume a maximum of 100 rows by 100 columns.

8.7   Classes

In C++, classes are the building blocks of object-oriented programming. A class is similar to a structure, except that a class is designed to include function members, as well as data members. A well-designed class can be used as easily as a predefined data type. Instances of a class are called objects. We have used predefined classes and objects in the previous sections of this chapter. For example, we used the `cin` object and the `cout` object to perform standard input and output operations. These objects are also called functions, including `precision` and `setf`. In this section, we will discuss programmer-defined classes.

Defining a Class Data Type

Class declaration
Class
implementation
Data members

A class definition consists of two parts: a **class declaration** and a **class implementation**. We will discuss each part separately.

In the class declaration, the name of the class is specified using the keyword `class`. The body of the class declaration consists of type declaration statements for the **data members** and function prototypes for the function members. Suppose we want to define a data type to represent a data point that is represented in rectangular coordinates; thus, the data type is represented by an x-coordinate and a y-coordinate, as shown in Figure 8.2. When designing a class type, we need to consider the data members that are required to represent the new data type, as well as the operations that we would like to define for the data type.

The class design is also described in the class declaration. The class design includes function members to implement the desired operations. Consider the class declaration that we present for the `xy_coordinate` class, which consists of two data members to define the point and two member functions that operate on the coordinates to determine the radius r and the angle θ, as shown in Figure 8.2. We will add additional member functions to the design of

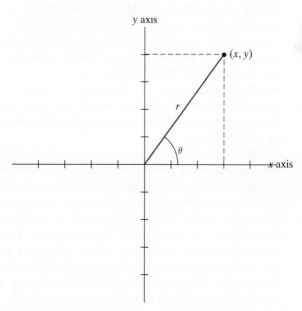

Figure 8.2 *Rectangular coordinates for a data point.*

the `xy_coordinate` class later in this section. The class declaration, is typically stored in a header file, as shown in the following code:

```
//------------------------------------------------------------
//  Class declaration (version 1)
//
//  These statements define a class for xy-coordinates.
//  This declaration is stored in xy_coordinate.h.

#include <iostream>
#include <cmath>

class xy_coordinate
{
    //  Declare function prototypes for public members.
    public:
        void input()
        void print()
        double radius();
        double angle();

    private:
    //  Declare private data members.
        double x, y;
};
//------------------------------------------------------------
```

The `xy_coordinate` class has two data members and four function members. The keywords `public` and `private` are used to control access to the members of the class. Members that are specified as **public members** may be referenced anywhere in the user program. Members that are specified as **private members** may only be referenced by member functions of the `xy_coordinate` class. We recommend that all data members be specified as private to adhere to object-oriented design principles. This restricted access is known as information hiding. If modifications are made to the data representation of a class, only the member functions need to be modified; no change to the user program is required. Any private member functions can only be called by other member functions. These private member functions are often referred to as helper functions because they are designed to help the other member functions.

Public members
Private members

The class implementation consists of all the member function definitions. When defining a member function, the **scope resolution operator** (`::`) is used in the function definition. This operator is placed between the class name and the function name to specify that the function is a member of the class. When writing a member function definition, first specify the type of value being returned by the function. The class name, the scope resolution operator, the name of the function, and the parameter list will follow. Recall that all member functions have direct access to the data members, so the data members do not appear in the parameter list. Helper functions can also be called directly by a member function. A class implementation can also be stored in a separate file. The `xy_coordinate` class implementation is now presented. Note that the function `angle` is written to compute a four-quadrant angle.

Scope resolution
operator

```
//------------------------------------------------------------
//  Class implementation (version 1)
//
//  These statements define implementation of an
//  xy_coordinate class. They are stored in xy-coordinate.h.

//  This function reads the xy coordinates from the keyboard.
void xy_coordinate::input()
{
   cin >> x >> y;
}
//  This function prints the xy coordinates to the screen.
void xy_coordinate::print()
{
   cout << "(" << x << "," << y << ")" << "\n";
}


//  This function computes the radius.
double xy_coordinate::radius()
{
   return sqrt(x*x + y*y);
}


//  This function computes the angle in radians.
double xy_coordinate::angle()
{
   // Compute angle of polar form.
     double z, pi=3.141593;
     if (x >= 0)
        z = atan(y/x);
     if (x<0 && y>0)
        z = atan(y/x) + pi;
     if (x<0 && y<=0)
        z = atan(y/x) - pi;
     if (x==0 && y==0)
        z = 0;
     return z;
}
//------------------------------------------------------------
```

We now present a program to test this new data type.

```
//------------------------------------------------------------
//  Program chapter8_7
//
//  This program demonstrates the use of
//  the xy_coordinate class and its functions.
#include <iostream>
#include <cmath>
#include "xy_coordinate.h"
using namespace std;
```

```
int main(void)
{
    //  Declare and initialize variables.
    xy_coordinate pt1;

    //  Read input point.
    cout <<  "Enter x and y coordinates:"  << endl;
    pt1.input();

    //  Print coordinate in xy form and polar form.
    cout.setf(ios::fixed);
    cout.precision(2);
    cout << "Coordinate in xy form:" << endl;
    pt1.print()
    cout << "Coordinate in polar form:" << endl;
    cout << "magnitude: " << pt1.radius() << endl;
    cout << "phase (in degrees): << pt1.angle()*180/3.141593 << endl;

    //  Exit program.
    return 0;
}
//------------------------------------------------------------------
```

To execute this program, you need to be sure that the `xy_coordinate` header file and the `xy_coordinate` implementation file are accessible to the compiler, along with the program file. This separation of the components allows us to build our own libraries for programmer-defined classes. These libraries can be used by many application programs in the same way that the standard libraries, such as `iostream` are used. The application program must include the declarations file and be linked to the implementation file. Using these files, we can now test the program. Here is a sample output from the program:

```
Enter x and y coordinates:
4 4
Coordinate in xy form:
(4.00,4.00)
Coordinate in polar form:
magnitude: 5.66
phase (in degrees): 45.00
```

Constructor Functions

When we define a variable, we often want to assign an initial value to the variable, as in

```
double sum=0;
```

Constructor
functions

If we do not initialize a variable at the time we define it, the variable holds an unknown value until valid data is assigned. **Constructor functions** are special member functions that are called automatically when an object of that class is declared. The constructor function is used to initialize the data members of the object being defined. When designing a class, a complete set of constructor functions should be provided. Constructor functions have three unique properties:

- a constructor is called automatically when an object of that class is declared;
- the name of a constructor function is the name of the class;
- no return value is associated with a constructor function, and it is not a void function.

To illustrate, we define two constructor functions to initialize the class members. One function initializes x and y to zero; the other initializes x and y to values in the declaration statement. The two declarations are the first ones in the public declaration list in the following code:

```
//------------------------------------------------------------
//  Class declaration (version 2)
//
//  These statements define a class for xy-coordinates.
//  Assume that this declaration is stored in xy_coordinate.h.
//  The update is the addition of two constructor functions.

#include <iostream>
#include <cmath>
using namespace std;

class xy_coordinate
{
    //  Declare two constructor functions and six function
    //   prototypes for public members.
    public:
        xy_coordinate();
        xy_coordinate(double a, double b);
        void input();
        void print();
        double radius();
        double angle();

    //  Declare private data members.
    private:
        double x, y;
};
//------------------------------------------------------------
```

Default constructor

The **default constructor** is called automatically whenever an object is defined in a declaration as follows:

```
xy_coordinate pt1;
```

Thus pt1 is an object of the class xy-coordinate. The data members of the object pt1 are initialized to the values assigned by the default constructor function xy-coordinate(). The constructor function with parameters is called automatically whenever an object is defined in a declaration statement as follows:

```
xy_coordinate pt2(3,4);
```

The data members of the object pt2 will be initialized to the values passed through the parameter list using the constructor function xy-coordinate(double a, double b). Thus, x is given the value 3 and y is given the value 4.

Constructor functions are progammer-defined functions. Here are the constructor functions that should be added to the class implementation file:

```
//------------------------------------------------------------
//  This constructor function initializes x and y to zero.
xy_coordinate::xy_coordinate()
{
    x = 0;
    y = 0;
}

//  This constructor function initializes x and y to parameters.
xy_coordinate::xy_coordinate(double a, double b)
{
    x = a;
    y = b;
}
//------------------------------------------------------------
```

Class Operators

The assignment operator is defined for objects of the same class type. If `pt1` and `pt2` are both `xy_coordinate` objects, then this statement is a valid statement:

```
pt1 = pt2;
```

Each data member of the object `pt1` is assigned the value of the corresponding data member of the object `pt2`. However, the C++ arithmetic operators and relational operators cannot automatically be used with a programmer-defined class type. This comparison is not valid:

```
if (pt1 == pt2)          (invalid comparison)
```

A class definition can include a set of operators to be used with objects of the class.

Overload

The ability to **overload** operators is a powerful feature in C++. Overloading operators allows a programmer-defined data type to be used as easily as any predefined data type. As an example, consider the arithmetic operators defined in C++. These operators are defined to operate on all predefined data types. However, they are not defined for objects from programmer-defined data types. When designing a class data type, a set of arithmetic operators that work on objects of the class may be included. Operators are included in a class definition in the same way as member functions, except that the keyword operator is used. It is followed by the name of the function, where the name of the function is one of the predefined C++ operators. Only predefined operators may be overloaded. You may not, for example, define a new operator ** to perform exponentiation since this operator is not one of the predefined operators in C++. In the next section, we illustrate the use of overloaded operators in the definition of a complex number class that includes the arithmetic operators (i.e., + , - , *, /).

8.8 Numerical Technique: Complex Roots

Complex numbers are needed to solve many problems in science and engineering, particularly in physics and electrical engineering. Therefore, it is often very useful to define a complex class in C++. (Some compilers have added a complex class, but since it is not part of the standards, you will not find it consistently available.) Recall that a complex number has the form $a + bi$, where i is $\sqrt{-1}$ and a and b are real numbers. Thus, the real part of the number is represented by a, and the imaginary part of the number is represented by b. An obvious way to represent a complex number is with an ordered pair of real values. Thus, complex number representations are very similar to the xy-coordinate representation in the previous section. In fact, if we just relabel the x-axis to the real axis, and relabel the y-axis to the imaginary axis, we will have a representation for complex values, as shown in Figure 8.3.

To input a complex number from the keyboard, we enter two values; hence, the input function is essentially the same as for xy-coordinates. The output of a complex number is different, because we want to include a reference to i. For example, the complex value represented the pair of numbers (3, 5) could be printed as $3+5i$. If the imaginary part is negative, then we would prefer the output to be $3-2i$ instead of $3+-2i$. This preference is handled in the member function definition for printing a complex number.

When an arithmetic operation is performed between two complex values, the result is also a complex value. The rules for complex arithmetic are not as familiar as those for integers or real values. Table 8.3 lists the results of the basic operations with two complex numbers.

We often refer to the magnitude and phase of a complex number. These are the r and θ values, respectively, shown in Figure 8.3. These are exactly the same as the magnitude and phase of an xy-coordinate, so we will be able to use similar definitions.

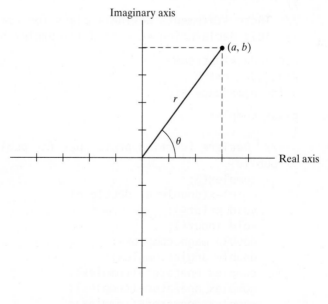

Figure 8.3 *Graphical representation of a complex number.*

Table 8.3 Arithmetic Operations with Complex Numbers

Operation	Result
$c_1 + c_2$	$(a_1 + a_2) + (b_1 + b_2)i$
$c_1 - c_2$	$(a_1 - a_2) + (b_1 - b_2)i$
$c_1 \cdot c_2$	$(a_1 a_2 - b_1 b_2) + (a_1 b_2 + a_2 b_1)i$
c_1 / c_2	$\dfrac{a_1 a_2 + b_1 b_2}{a_2^2 + b_2^2} + \dfrac{a_2 b_1 - b_2 a_1}{a_2^2 + b_2^2} i$
$(c_1 = a_1 + b_1 i,\quad c_2 = a_2 + b_2 i)$	

Complex Class Definition

We now develop the class declaration and class implementation for complex numbers. Then, in the final part of this section, we develop a program that computes and prints the complex roots for a quadratic equation.

In the class declaration, we need to include the private members for the real and imaginary components of a complex number. In the public members, we include a function to input a complex number from the keyboard and a function to print a complex number on the screen. In addition, we need to include member functions to define the arithmetic operators. The resulting class declaration and class implementation are shown in the following code.

Review these statements carefully to observe the steps. However, this is only an introduction to C++, so we have presented only the details needed for these examples. We have not tried to give you a complete understanding of class declarations and implementations.

Class Declaration and Implementation:

```
//------------------------------------------------------------
//  Class declaration
//
//  These statements define a class for complex numbers.
//  This declaration is stored in complex.h.

#include <iostream>
#include <cmath>
using namespace std;

class complex
{
   //  Declare function prototypes for public members.
   public:
      complex();
      complex(double a, double b);
      void print();
      void input();
      double magn(complex);
      double angle(complex);
      complex operator+(complex);
      complex operator-(complex);
      complex operator*(complex);
      complex operator/(complex);
```

```cpp
//  Declare private members.
   private:
      double real, imag;
};
//------------------------------------------------------------
//  Class implementation
//
//  These statements define implementation of a complex class.

//  This function is the default constructor to initialize a
//  complex number that is not given a value.
complex::complex()
{
   real = 0;
   imag = 0;
}

//  This function is the constructor to initialize a complex
//  number to a specified value.
complex::complex(double a, double b)
{
   real = a;
   imag = b;
}

//  This function prints a complex number.
void complex::print()
{
   if (imag > 0)
      cout << real << "+" << imag << "i" << endl;
   else
      if (imag == 0)
         cout << real << endl;
      else
         cout << real << imag << "i" << endl;
}

//  This function reads two values for a complex number.
void complex::input()
{
   cin >> real >> imag;
}

//  This function defines the sum of complex numbers.
complex complex::operator+(complex c)
{
   // Definition of complex addition.
   complex temp;
   temp.real = c.real + real;
   temp.imag = c.imag + imag;
   return temp;
}
```

```
// This function defines the difference of complex numbers.
complex complex::operator-(complex c)
{
   // Definition of complex subtraction.
   complex temp;
   temp.real = real - c.real;
   temp.imag = imag - c.imag;
   return temp;
}

// This function defines the product of complex numbers.
complex complex::operator*(complex c)
{
   // Definition of complex multiplication.
   complex temp;
   temp.real = (real*c.real - imag*c.imag);
   temp.imag = (imag*c.real + real*c.imag);
   return temp;
}

// This function defines the quotient of complex numbers.
complex complex::operator/(complex c)
{
   // Definition of complex division.
   complex temp;
   temp.real = (real*c.real + imag*c.imag)/
               (pow(c.real,2) + pow(c.imag,2));
   temp.imag = (imag*c.real - real*c.imag)/
               (pow(c.real,2) + pow(c.imag,2));
   return temp;
}
//------------------------------------------------------------------
```

We now present a simple example to test some of the characteristics of the complex class that we have developed: When building your own classes, test each member function as it is written. Debugging the complete class definition at once can be difficult.

```
//------------------------------------------------------------------
// Program chapter8_8
//
// This program demonstrates the use of
// the complex number class and its operations.

#include <iostream>
#include "ccomplex"
including namespace std;

int main(void)
{
   // Declare and initialize variables.
   complex c1(4,1), c2(-3,-2), c3;
```

```
    //  Print initial values.
    cout << "c1:"
    c1.print();
    cout << "c2:"
    c2.print();
    cout << "c3:";
    c3.print();

    //  Compute and print new values.
    c3 = c1 + c2;
    cout << "c1+c2 = "
    c3.print();
    c3 = c1 - c2;
    cout << "c1-c2 = "
    c3.print();
    c3 = c1*c2;
    cout << "c1*c2 = "
    c3.print();
    c3 = c1/c2;
    cout << "c1/c2 = "
    c3.print();

    //  Exit program.
    return 0;
}
//----------------------------------------------------------
```

The output from this program is as follows:

```
c1: 4+1i
c2: -3-2i
c3: 0
c1+c2 = 1-1i
c1-c2 = 7+3i
c1*c2 = -10-11i
c1/c2 = -1.07692+0.384615i
```

Complex Roots for Quadratic Equations

A general equation for a quadratic equation with real coefficients is

$$ax^2 + bx + c = 0.$$

There are two roots to this equation, and they can be computed using the following equations:

$$\text{root}_1 = \frac{-b + \sqrt{b^2 - 4ac}}{2a}$$

$$\text{root}_2 = \frac{-b - \sqrt{b^2 - 4ac}}{2a}$$

The term under the square root symbol, $b^2 - 4ac$, is called the discriminant. If the discriminant is greater than or equal to zero, then the two roots are real values, and they can be expressed as follows:

$$\text{root}_1 = -\frac{b}{2a} + \frac{\sqrt{\text{discriminant}}}{2a}$$

$$\text{root}_2 = -\frac{b}{2a} - \frac{\sqrt{\text{discriminant}}}{2a}$$

If the discriminant is less than zero, then the two roots are complex values, and they can be expressed as follows:

$$\text{root}_1 = -\frac{b}{2a} + \frac{\sqrt{-\text{discriminant}}}{2a}i$$

$$\text{root}_2 = -\frac{b}{2a} - \frac{\sqrt{-\text{discriminant}}}{2a}i$$

We now present a program that reads the values of a, b, and c for a quadratic equation from the keyboard. The program then determines and prints the two roots. If the roots are real, then they appear as real values. If the roots are complex, then they are printed as complex values. The code is as follows:

```
//----------------------------------------------------------------
//  Program chapter8_9
//
//  This program computes and prints the roots of a quadratic
//  equation.  It uses the user-defined complex class.

#include <iostream>
#include <cmath>
#include "ccomplex"
using namespace std;

int main(void)
{
    // Declare and initialize variables.
    double a, b, c, term1, disc;
    complex root1, root2;

    // Read values for a, b, c from the keyboard.
    cout << "Enter real values a, b, c:" << endl;
    cin >> a >> b >> c;
```

```
// Compute roots of quadratic equation.
term1 = -b/(2*a);
disc = b*b - 4*a*c;
if (disc >= 0)
{
    root1.real = term1 + sqrt(disc)/(2*a);
    root2.real = term1 - sqrt(disc)/(2*a);
}
else
{
    root1.real = term1;
    root1.imag = sqrt(-disc)/(2*a);
    root2.real = term1;
    root2.imag = -sqrt(-disc)/(2*a);
}

// Print roots.
cout << "Roots:" << endl;
root1.print();
root2.print();

// Exit program.
return 0;
}
//------------------------------------------------------------
```

Here are two example interactions with this program:

```
Enter real values a, b, c:
1 -3 -4
Roots:
4
-1

Enter real values for a, b, c:
1 0 4
Roots:
0+2i
0-2i
```

MODIFY

These problems refer to the `complex` class developed in this section and the program that computes the roots of a quadratic equation.

1. Test the program with an equation that will give a double root. Does it handle that situation correctly?
2. Modify the program so that it also prints one of these phrases: `"real roots"`, `"double roots"`, or `"complex roots"`.
3. Modify the program so that it prints two decimal positions for the roots.

SUMMARY

The C++ language is a superset of C. We introduced the structure of a C++ program and presented the C++ statements necessary to perform standard I/O and file I/O. Object-oriented programming is characterized by the use of classes and objects. A class is a programmer-defined data type that combines data members and function members. An object is an instance of a class. Objects call member functions to operate on the data members. Examples were developed in this chapter to illustrate class definitions and computation with classes using xy-coordinates and complex numbers.

KEY TERMS

class
class declaration
class implementation
constructor function
data member
default constructor
dot operator
extraction operator
format flag
inheritance
insertion operator

member function
object
object-oriented programming
overload
polymorphism
private member
public member
scope resolution operator
stream
white space

C++ STATEMENT SUMMARY

Preprocessor directives to include information for standard I/O and file I/O:

```
#include <iostream>
#include <fstream>
using namespace std;
```

One-line comments:
```
//program chapter8-1
```

Keyboard input statement:
```
cin >> data;
```

Screen output statement:
```
cout << data;
```

Class declaration:
```
class class_name
{
public:
    function prototypes;
private:
    declarations;
    function prototypes;
};
```

Member function definition:
```
return_type class_name::function_name(parameter list)
{
    declarations;
    statements;
}
```

Declaring an object (instance of a class):
```
xy_coordinate pt1;
```

Initializing an object:
```
xy_coordinate pt2(3,4);
```

NOTES

1. All data members of a class should be specified as private.

2. A well-designed class should have one or more constructor functions.

DEBUGGING NOTES

1. Test each member function definition as it is written. Coding the complete class implementation before testing makes debugging very difficult.

PROBLEMS

SHORT ANSWER PROBLEMS

True–False Problems

Indicate whether the following statements are true (T) or false (F).

1. All operators except ".","::","?:", and " - > " can be overloaded.　　T　　F

2. The statement `cout.precision(3)` formats the output such that the total number of digits is 3.　　T　　F

3. In a class, there can be more than one constructor of the same name, provided that the list of arguments is different in different constructors.　　T　　F

4. The characteristic of object-oriented programming by which data and information are grouped together is called data hiding.   T   F

5. Member functions are called using the dot operator.   T   F

6. The dot operator is required as part of a member function definition.   T   F

7. An object is a member function of a class.   T   F

8. Member functions have access to all private data members.   T   F

9. Constructor functions are member functions.   T   F

10. A constructor function may not be overloaded.   T   F

Multiple Choice Problems

Circle the letter for the best answer to complete each statement or for the correct answer of each question.

11. The output of the following statements will be:

```
double y = 0.000055;
cout.setf(ios::scientific);
cout<<y;
```

(a) 5.500000e-05
(b) 0.000055
(c) 5.5e-05
(e) 0.550000e-04

12. Private data or member functions in a class?
(a) cannot be accessed at all.
(b) can be accessed from only private sections of that class.
(c) can be accessed only from within the class.
(d) can be accessed from within the class and from derived classes.

13. Functions designed to initialize data members of a class are called
(a) accessor functions.
(b) constructor functions.
(c) object functions.
(d) class functions.
(e) none of the above.

PROGRAMMING PROBLEMS

14. Define a class to represent a date. A date is defined using three integer variables: month, day, and year. Include member function definitions to

- input a date;
- print a date as month/day/year (10/1/1999);
- print a date as month day, year (October 1, 1999);
- initialize a date object.

15. Define a class `point` that defines the co-ordinates of a point in a 2-D graph. It has two attributes that depict the x-coordinate and the y-coordinate. Include member functions that will perform the following operations:

    • Compare whether two points are identical by overloading operator ==.
    • Compute the distance between two points.
    • Convert to polar co-ordinates.

16. Define a class `HugeInteger` that can store positive integers having as many as 40 digits. Include member functions to perform addition of two long positive integers using overloaded + operator. Your class should allow only 39 digit numbers to be input, so as to contain carry over digits from the addition of two huge positive integers.

APPENDIX A

ANSI C Standard Library

While entire texts have been written to discuss the ANSI C Standard Library, the intent of this appendix is to present only a short discussion on the information defined in each of the header files in the ANSI C Standard Library. This discussion is not intended to provide all the details necessary to use the functions, but to give you enough information so that you can determine whether the functions may be of use in a particular application; you can then obtain more details from other references. What follows assumes that you are familiar with the various data types, including pointers and character strings.

`<assert.h>`

The header file `<assert.h>` gives a definition of the assert function that provides diagnostic information in testing a program. The system-dependent diagnostic information is stored in the standard error file, which can be accessed after the program is completed.

`<ctype.h>`

The header file `<ctype.h>` defines several functions for testing and converting characters. (See also Chapter 2.) The function prototype statements and corresponding discussions use the following definitions:

digit one of the characters 0123456789
hexadecimal digit a digit or one of the characters ABCDEFabcdef
uppercase letter one of the characters ABCDEFGHIJKLMNOPQRSTUVWXYZ
lowercase letter one of the characters abcdefghijklmnopqrstuvwxyz
alphabetic character an uppercase or a lowercase letter
alphanumeric character a digit or an alphabetic character
punctuation character one of the characters !"#%&'();<=>?[\]*+,−./:^
graph character an alphanumeric character or a punctuation character
print character a graph character or the space character
motion control character one of the control characters FF (form feed), NL (new line), CR (carriage return), HT (horizontal tab), and VT (vertical tab)
white space the space character or one of the motion control characters
control character one of the motion control characters or BEL (bell) or BS (backspace)

We now list each function prototype and give a brief definition of the corresponding function:

```
int isalnum(int c);
```
returns a nonzero (true) value if and only if the input character is a digit or an uppercase or a lowercase letter
```
int isalpha(int c);
```
returns a nonzero (true) value if and only if the input character is an uppercase or a lowercase letter
```
int iscntrl(int c);
```
returns a nonzero (true) value if and only if the input character is one of the control characters
```
int isdigit(int c);
```
returns a nonzero (true) value if and only if the input character is a digit

```
int isgraph(int c);
```
returns a nonzero (true) value if and only if the input character is a graph character
```
int islower(int c);
```
returns a nonzero (true) value if and only if the input character is a lowercase letter
```
int isprint(int c);
```
returns a nonzero (true) value if and only if the input character is a printing character
```
int ispunct(int c);
```
returns a nonzero (true) value if and only if the input character is a punctuation character
```
int isspace(int c);
```
returns a nonzero (true) value if and only if the input character is a white-space character
```
int isupper(int c);
```
returns a nonzero (true) value if and only if the input character is an uppercase character
```
int isxdigit(int c);
```
returns a nonzero (true) value if and only if the input character is a hexadecimal character
```
int tolower(int c);
```
converts an uppercase letter to a lowercase letter
```
int toupper(int c);
```
converts a lowercase letter to an uppercase letter

<errno.h>

The header file <errno.h> provides a definition of the macros EDOM and ERANGE and an external function errno. EDOM and ERANGE are integer constants with nonzero values that are system dependent. The purpose of the errno function is to report error conditions, and its use is system dependent.

<float.h>

The header file <float.h> provides several macros that give various limits and characteristics pertaining to floating-point values, where a normalized floating-point value has been expressed in an exponential notation with a mantissa greater than or equal to 1 and less than 10. The macros and their corresponding definitions are as follows:

```
int FLT_ROUNDS;
```
specifies the rounding mode for floating-point addition
```
int FLT_RADIX;
```
radix of the exponent representation for floating-point values
```
int FLT_MANT_DIG;
int DBL_MANT_DIG;
int LDBL_MANT_DIG;
```
number of radix base digits in the normalized mantissa for a float, double, or
long double value
```
int FLT_DIG;
int DBL_DIG;
int LDBL_DIG;
```
number of decimal digits in the normalized mantissa for a float, double, or
long double value
```
int FLT_MIN_EXP;
int DBL_MIN_EXP;
int LDBL_MIN_EXP;
```
integer used to determine the minimum radix exponent for a normalized value for a
float, double, or long double value

```
int  FLT_MIN_10_EXP;
int  DBL_MIN_10_EXP;
int  LDBL_MIN_10_EXP;
```
integer used to determine the minimum exponent for a base-10 normalized value for a `float`, `double`, or `long double` value
```
int  FLT_MAX_EXP;
int  DBL_MAX_EXP;
int  LDBL_MAX_EXP;
```
integer used to determine the maximum exponent for the radix base for a normalized value for a `float`, `double`, or `long double` value
```
int  FLT_MAX_10_EXP;
int  DBL_MAX_10_EXP;
int  LDBL_MAX_10_EXP;
```
integer used to determine the maximum exponent for a base-10 normalized value for a `float`, `double`, or `long double` value
```
float FLT_MIN;
double DBL_MIN;
long double LDBL_MIN;
```
minimum representable value for a `float`, `double`, or `long double` value
```
float FLT_MAX;
double DBL_MAX;
long double LDBL_MAX;
```
maximum representable value for a `float`, `double`, or `long double` value
```
float FLT_EPSILON;
double DBL_EPSILON;
long double LDBL_EPSILON;
```
difference between 1 and the smallest value greater than 1 for a `float`, `double`, or `long double` value

`<limits.h>`

The header file `<limits.h>` provides several macros that give various limits and characteristics pertaining to integer values. These macros and their definitions are as follows:

```
int  CHAR_BIT;
```
number of bits for the smallest nonbit value
```
int  CHAR_MIN;
int  CHAR_MAX;
```
minimum and maximum values for type `char`
```
int  INT_MIN;
int  INT_MAX;
```
minimum and maximum values for type `int`
```
int  LONG_MIN;
int  LONG_MAX;
```
minimum and maximum values for type `long int`
```
int  MB_LEN_MAX;
```
maximum number of bytes in a multibyte character
```
int  SCHAR_MIN;
int  SCHAR_MAX;
```
minimum and maximum values for type `signed char`

```
int  SHRT_MIN;
int  SHRT_MAX;
```
   minimum and maximum values for type `short int`
```
int  UCHAR_MAX;
```
   maximum values for type `unsigned char`
```
int  UINT_MAX;
```
   maximum value for type `unsigned int`
```
int  ULONG_MAX;
```
   maximum value for type `unsigned long int`
```
int  USHRT_MAX;
```
   maximum value for type `unsigned short int`

`<locale.h>`

The header file `<locale.h>` defines two functions, one type, and several macros relative to the formatting of numeric values. This information allows numeric values to be formatted in ways that address internationalization issues; included is information relating to monetary formatting and times.

`<math.h>`

The header file `<math.h>` defines functions that are often needed to perform engineering calculations. These functions, also described in detail in Chapter 2, are as follows:

```
double acos(double x);
```
   computes the arccosine, or inverse cosine, of x, where x must be in the range $[-1, 1]$; returns an angle in radians in the range $[0, \pi]$
```
double asin(double x);
```
   computes the arcsine, or inverse sine, of x, where x must be in the range $[-1, 1]$; returns an angle in radians in the range $[-\pi/2, \pi/2]$

```
double atan(double x);
```
   computes the arctangent, or inverse tangent, of x; returns an angle in radians in the range $[-\pi/2, \pi/2]$
```
double atan2(double y, double x);
```
   computes the arctangent, or inverse tangent, of the value $\frac{y}{x}$; returns an angle in radians in the range $[-\pi, \pi]$
```
int ceil(double x);
```
   rounds x to the nearest integer toward ∞ (infinity)
```
double cos(double x);
```
   computes the cosine of x, where x is in radians
```
double cosh(double x);
```
   computes the hyperbolic cosine of x, which is equal to $(e^x + e^{-x})/2$
```
double exp(double x);
```
   computes the value of e^x, where e is the base for natural logarithms, or approximately 2.718282
```
double fabs(double x);
```
   computes the absolute value of x
```
int floor(double x);
```
   rounds x to the nearest integer toward $-\infty$ (negative infinity)
```
double log(double x);
```
   computes ln x, the natural logarithm of x (to the base e); errors occur if $x \leq 0$

```
double log10(double x);
```
computes $\log_{10}x$, the common logarithm of x (to the base 10); errors occur if $x \leq 0$
```
double pow(double x, double y);
```
computes the value of x to the yth power, or x^y; errors occur if $x = 0$ and $y \leq 0$, or if $x < 0$ and y is not an integer
```
double sin(double x);
```
computes the sine of x, where x is in radians
```
double sinh(double x);
```
computes the hyperbolic sine of x, which is equal to $(e^x - e^{-x})/2$
```
double sqrt(double x);
```
computes the square root of x, where $x \geq 0$
```
double tan(double x);
```
computes the tangent of x, where x is in radians
```
double tanh(double x);
```
computes the hyperbolic tangent of x, which is equal to (sinh x)/(cosh x)

<setjmp.h>

The header file <setjmp.h> contains a macro, a function, and a type declaration used to by-pass the normal function call and return processes. These operations are not generally recommended and thus are not discussed here.

<signal.h>

The header file <signal.h> contains a type definition, two functions, and several macros for handling various signals, which are conditions that may be reported during the execution of a program and that are generally considered to be fatal errors. Since signal handling is non-portable, we do not discuss the information in this header file here. In general, the default handling provided by a system is sufficient for handling signals.

<stdarg.h>

The header file <stdarg.h> contains a type definition and three macros for working with functions that allow a variable number of arguments. While this capability is very powerful, it is not commonly used in engineering applications.

<stddef.h>

The header file <stddef.h> contains type definitions and macros that are essentially unrelated. These types and macros are not commonly used in engineering applications and thus are not addressed here.

<stdio.h>

The header file <stdio.h> defines the types, macros, and functions required to perform input and output. Of the new types defined, the type FILE is the one most useful in engineering applications, because it is used in conjunction with data files. Many of the input/output functions are discussed in Chapters 2 and 3; those functions and several additional functions are summarized here:

```
void clearerr(FILE *stream);
```
clears the end-of-file and error indicators for the stream pointed to by stream
```
int fclose(FILE *stream);
```
closes the file associated with the file pointer

```
int feof(FILE *stream);
```
tests the end-of-file indicator for the stream pointed to by stream
```
int ferror(FILE *stream);
```
tests the error indicator for the stream pointed to by stream
```
int fflush(FILE *stream);
```
causes unwritten data for the stream to be written to the file
```
int fgetc(FILE *stream);
```
returns the integer equivalent of the next character in the stream
```
int fgetpos(FILE *stream, fpos_t *pos);
```
returns the current value of the file position indicator for the stream pointed to by
stream in the object pointed to by pos
```
char *fgets(char *s, int n, FILE *stream);
```
reads into the array pointed to by s at most one less than the number of characters
specified by n from the stream pointed to by stream
```
FILE *fopen(const char *filename, const char *mode);
```
opens the file whose name is a string pointed to by filename
```
int fprintf(FILE *stream, const char *format, ...);
```
writes the output to the stream pointed to by stream, using the format specified and
the values that follow the format reference; returns the number of characters printed
```
int fputc(int c, FILE *stream);
```
writes the character specified by c to the output stream pointed to by stream
```
int fputs(const char *s, FILE *stream);
```
writes the string pointed to by s to the stream pointed to by stream; does not write
the terminating null character
```
size_t fread(void *ptr, size_t size, size_t nmemb, FILE *stream);
```
reads into the array pointed to by ptr up to nmemb elements whose size is specified by
size from the stream pointed to by stream
```
FILE *freopen(const char *filename, const char *mode, FILE *stream);
```
reopens the file whose name is a string pointed to by filename
```
int fscanf(FILE *stream, const char *format, ...);
```
reads input from the stream pointed to by stream, using the format specified and the
addresses that follow the format reference; returns the number of input values assigned
```
int fseek(FILE *stream, long int offset, int whence);
```
sets the file position indicator for the stream pointed to by stream
```
int fsetpos(FILE *stream, const fpos_t *pos);
```
sets the file position indicator for the stream pointed to by stream according to the
value of the object pointed to by pos
```
long int ftell(FILE *stream);
```
returns the current value of the file position indicator for the stream pointed to by
stream
```
size_t fwrite(const void *ptr, size_t size, size_t nmemb,
      FILE *stream);
```
writes from the array pointed to by ptr up to nmemb elements whose size is specified
by size to the stream pointed to by stream
```
int getc(FILE *stream);
```
returns the integer equivalent of the next character in the stream
```
int getchar(void);
```
returns the integer equivalent of the next character from the standard input stream

```
char *gets(char *s);
```
reads characters from the input stream into the array pointed to by s until a new-line character or end-of-file is encountered; the new-line character is replaced by a null character in the array

```
void perror(const char *s);
```
maps the error number in the integer expression errno to an error message

```
int printf(const char *format, ...);
```
writes the output to the standard output stream, using the format specified and the values that follow the format reference; returns the number of characters printed

```
int putc(int c, FILE *stream);
```
writes the character specified by c to the output stream pointed to by stream

```
int putchar(int c);
```
writes the character specified by c to the standard output stream

```
int puts(const char *s);
```
writes the string pointed to by s to the standard output stream and adds a new-line character to the output in place of the null character

```
int remove(const char *filename);
```
removes accessibility of the file pointed to by filename

```
int rename(const char *old, const char *new);
```
changes the file originally pointed to by old to the one pointed to by new

```
void rewind(FILE *stream);
```
sets the file position indicator for the stream pointed to by stream to the beginning of the file

```
int scanf(const char *format, ...);
```
reads input from the standard input stream, using the format specified and the addresses that follow the format reference; returns the number of input values assigned

```
void setbuf(FILE *stream, char *buf);
```
specifies the type of buffering to be used with the stream pointed to by stream

```
int setvbuf(FILE *stream, char *buf, int mode, size_t size);
```
specifies the type of buffering to be used with the stream pointed to by stream

```
int sprintf(char *s, const char *format, ...);
```
writes the output to the array pointed to by s, using the format specified and the values that follow the format reference; a null character is appended to the end of the characters written; returns the number of characters printed, not including the null character

```
int sscanf(const char *s, const char *format, ...);
```
reads input from the string pointed to by s, using the format specified and the addresses that follow the format reference; returns the number of input values assigned

```
FILE *tmpfile(void);
```
creates a temporary binary file that is automatically removed when it is closed

```
char *tmpnam(char *s);
```
generates a string that is a valid file name and is not the same as the name of any existing file

```
int ungetc(int c, FILE *stream);
```
pushes the character specified by c back onto the input stream pointed to by stream

```
int vfprintf(FILE *stream, const char *format, va_list arg);
```
writes the output to the stream pointed to by stream, using the format specified and the values contained in the variable argument list; returns the number of characters printed

```
int vprintf(const char *format, va_list arg);
```
writes the output to the standard output stream, using the format specified and the values contained in the variable argument list; returns the number of characters printed
```
int vsprintf(char *s, const char *format, va_list arg);
```
writes the output to the array pointed to by s, using the format specified and the values contained in the variable argument list; a null character is appended to the end of the characters written; returns the number of characters printed, not including the null character

\<stdlib.h\>

The header file \<stdlib.h\> defines types, macros, and functions that do not fit into any of the other header files. The types div_t and ldiv_t are structures for storing a quotient and a remainder. The macros are as follows:

NULL
an integer value of binary zero
EXIT_FAILURE
EXIT_SUCCESS
integral expressions used to return unsuccessful or successful termination status, respectively, to the host
RAND_MAX
an integral expression that is the maximum value returned by the RAND function
MB_CUR_MAX
a positive integer expression whose value is the maximum number of bytes in a multibyte character

The functions that are most likely to be used in engineering applications are listed next, with the function prototype statement and a brief description (many of these functions are discussed in Chapters 3 and 6):

```
void abort(void);
```
causes an abnormal termination of the program
```
int abs(int k);
long int labs(long int k);
```
computes the absolute value of the integer k
```
int atexit(void (*func)(void));
```
registers the function pointed to by func, called without arguments at normal program termination
```
double atof(const char *s);
int atoi(const char *s);
long int atol(const char *s);
double strtod(const char *s, char **endptr);
long int strtol(const char *s, char **endptr, int base);
unsigned long int strtoul(const char *s, char **endptr, int base);
```
converts the initial portion of the string pointed to by s to a numerical representation
```
void *bsearch(const void *key, const void *base, size_t n, size_t
             size, int(*compar)(const void *,const void *));
```
searches an array of n objects for the value pointed to by key
```
void *calloc(size_t n, size_t size);
```
allocates space for an array of n objects, each of size size

```
div_t div(int numer, int denom);
ldiv_t ldiv(long int numer, long int denom);
```
   computes the quotient and remainder of the division of numer by denom
```
void exit(int status);
```
   causes normal program termination to occur
```
void free(void *ptr);
```
   deallocates the space pointed to by ptr
```
void *malloc(size_t size);
```
   allocates space for an object of size size
```
void qsort(void *base, size_t nmemb, size_t size, int
          (*compar)(const void*, const void *));
```
   sorts an object of n objects into ascending order
```
int rand(void);
```
   returns a pseudorandom integer in the range from 0 to RAND_MAX
```
void *realloc(void *ptr, size_t size);
```
   changes the size of the object pointed to by ptr
```
void srand(unsigned int seed);
```
   uses the seed seed to initialize a new sequence of values from the RAND function

`<string.h>`

The header file `<string.h>` defines the type `size_t`, which is an unsigned integer, and the macro NULL, which has the value of binary zero. In addition, the header file defines the following functions for handling strings (these functions are also discussed in Chapter 6):

```
void *memchr(const void *s, int c, size_t n);
```
   returns a pointer to the first occurrence of c in the initial n characters of the object
   pointed to by s
```
int memcmp(const void *s, const void *t, size_t n);
```
   returns an integer greater than, equal to, or less than zero, accordingly, as the string
   pointed to by s is, respectively, greater than, equal to, or less than the string pointed to
   by t
```
void *memcpy(void *s, const void *t, size_t n);
```
   copies n characters from the object pointed to by t into the object pointed to by s
```
void *memmove(void *s, const void *t, size_t n);
```
   copies n characters from the object pointed to by t into the object pointed to by s,
   using a temporary area
```
void *memset(void *s, int c, size_t n);
```
   copies the value of c into the first n characters of the object pointed to by s
```
char *strcat(char *s, const char *t);
```
   concatenates the string pointed to by t to the end of the string pointed to by s; returns
   a pointer to the string pointed to by s
```
char *strchr(const char *s, int c);
```
   returns a pointer to the first occurrence of the character c in the string pointed to by s
```
int strcmp(const char *s, const char *t);
```
   compares string s with string t, element by element; returns an integer greater than,
   equal to, or less than zero, accordingly, as the string pointed to by s is, respectively,
   greater than, equal to, or less than the string pointed to by t

```
int strcoll(const char *s, const char *t);
```
returns an integer greater than, equal to, or less than zero, accordingly, as the string pointed to by s is, respectively, greater than, equal to, or less than the string pointed to by t
```
char *strcpy(char *s, const char *t);
```
copies the string pointed to by t to the string pointed to by s; returns a pointer to the string pointed to by s
```
size_t strcspn(const char *s, const char *t);
```
returns the initial number of characters in the string pointed to by s that consists entirely of characters not in the string pointed to by t
```
size_t strlen(const char *s);
```
returns the length of the string pointed to by s
```
char *strncat(char *s, const char *t, size_t n);
```
concatenates at most n characters of string t to string s; returns a pointer to the string pointed to by s
```
int strncmp(const char *s, const char *t, size_t n);
```
compares at most n characters of string s with string t, element by element; returns an integer greater than, equal to, or less than zero, accordingly, as the string pointed to by s is, respectively, greater than, equal to, or less than the string pointed to by t
```
char *strncpy(char *s, const char *t, size_t n);
```
copies at most n characters from the string pointed to by t to the string pointed to by s; if t has fewer characters than s, then s is padded with null characters; returns a pointer to s
```
char *strpbrk(const char *s, const char *t);
```
returns a pointer to the first occurrence in the string pointed to by s of any character of the string pointed to by t
```
char *strrchr(const char *s, int c);
```
returns a pointer to the last occurrence of the character c in the string pointed to by s
```
size_t strspn(const char *s, const char *t);
```
returns the initial number of characters in the string pointed to by s that consists entirely of characters in the string pointed to by t
```
char *strstr(const char *s, const char *t);
```
returns a pointer to the start of the string pointed to by t within the string pointed to by s

`<time.h>`

The header file `<time.h>` defines two macros, four types, and several functions for representing and manipulating calendar time and local time. The types `clock_t` and `time_t` are arithmetic types capable of representing times, and the structure tm contains a calendar time broken into seconds (tm_sec), minutes (tm_min), hours (tm_hour), day of the month (tm_mday), months since January (tm_mon), years since 1900 (tm_year), days since Sunday (tm_wday), days since January 1 (tm_yday), and a daylight saving time flag (tm_isdst); the order of the values in the structure is system dependent. The related macros are the following:

```
CLOCKS_PER_SEC
```
number per second of the value returned by the clock function
```
NULL
```
an integer representing binary zero

Function prototypes and brief descriptions of their related computations are as follows:

```
char *asctime(const struct tm *timeptr);
```
   returns a pointer to the string containing a converted time
```
clock_t clock(void);
```
   returns the current processor time
```
char *ctime(const time_t *timer);
```
   returns a pointer to a string containing a converted time
```
double difftime(time_t time1, time_t time0);
```
   computes the difference between two calendar times
```
struct tm *gmtime(const time_t *timer);
```
   returns a pointer to a time expressed in Coordinated Universal Time
```
struct tm *localtime(const time_t *timer);
```
   returns a pointer to a time converted from calendar time
```
time_t mktime(struct tm *timeptr);
```
   converts the structure values for time to a calendar time value
```
time_t time(time_t *timer)
```
   returns the current calendar time
```
size_t strftime(char *s, size_t maxsize, const char *format,
                const struct tm *timeptr);
```
   converts time into a formatted multibyte character sequence

APPENDIX B

ASCII Character Codes

The table that follows lists the 128 ASCII characters and their equivalent integer values and binary values. The characters that correspond to the integers 1 through 31 have special significance to the computer system. For example, the character BEL is represented by the integer 7 and causes the bell to sound on the keyboard.

The order of the characters from low to high is the collating sequence and has several interesting characteristics. Note that digits are less than uppercase letters, and uppercase letters are less than lowercase letters. Note also that special characters are not grouped together—some are before digits, some are after digits, and some are between uppercase and lowercase characters. Here is the table:

Character	Integer Equivalent	Binary Equivalent
NUL (Binary Zero)	0	0000000
SOH (Start of Header)	1	0000001
STX (Start of Text)	2	0000010
ETX (End of Text)	3	0000011
EOT (End of Transmission)	4	0000100
ENQ (Enquiry)	5	0000101
ACK (Acknowledge)	6	0000110
BEL (Bell)	7	0000111
BS (Backspace)	8	0001000
HT (Horizontal Tab)	9	0001001
LF (Line Feed or New Line)	10	0001010
VT (Vertical Tabulation)	11	0001011
FF (Form Feed)	12	0001100
CR (Carriage Return)	13	0001101
SO (Shift Out)	14	0001110
SI (Shift In)	15	0001111
DLE (Data Link Escape)	16	0010000
DC1 (Device Control 1)	17	0010001
DC2 (Device Control 2)	18	0010010
DC3 (Device Control 3)	19	0010011
DC4 (Device Control 4-Stop)	20	0010100
NAK (Negative Acknowledge)	21	0010101
SYN (Synchronization)	22	0010110
ETB (End of Text Block)	23	0010111
CAN (Cancel)	24	0011000
EM (End of Medium)	25	0011001
SUB (Substitute)	26	0011010
ESC (Escape)	27	0011011
FS (File Separator)	28	0011100
GS (Group Separator)	29	0011101
RS (Record Separator)	30	0011110
US (Unit Separator)	31	0011111

SP (Space)	32	0100000
!	33	0100001
"	34	0100010
#	35	0100011
$	36	0100100
%	37	0100101
&	38	0100110
' (Closing Single Quote)	39	0100111
(	40	0101000
)	41	0101001
*	42	0101010
+	43	0101011
, (Comma)	44	0101100
- (Hyphen)	45	0101101
. (Period)	46	0101110
/	47	0101111
0	48	0110000
1	49	0110001
2	50	0110010
3	51	0110011
4	52	0110100
5	53	0110101
6	54	0110110
7	55	0110111
8	56	0111000
9	57	0111001
:	58	0111010
;	59	0111011
<	60	0111100
=	61	0111101
>	62	0111110
?	63	0111111
@	64	1000000
A	65	1000001
B	66	1000010
C	67	1000011
D	68	1000100
E	69	1000101
F	70	1000110
G	71	1000111
H	72	1001000
I	73	1001001
J	74	1001010
K	75	1001011
L	76	1001100
M	77	1001101
N	78	1001110
O	79	1001111

P	80	1010000
Q	81	1010001
R	82	1010010
S	83	1010011
T	84	1010100
U	85	1010101
V	86	1010110
W	87	1010111
X	88	1011000
Y	89	1011001
Z	90	1011010
[	91	1011011
\	92	1011100
]	93	1011101
^ (Circumflex)	94	1011110
_ (Underscore)	95	1011111
' (Opening Single Quote)	96	1100000
a	97	1100001
b	98	1100010
c	99	1100011
d	100	1100100
e	101	1100101
f	102	1100110
g	103	1100111
h	104	1101000
i	105	1101001
j	106	1101010
k	107	1101011
l	108	1101100
m	109	1101101
n	110	1101110
o	111	1101111
p	112	1110000
q	113	1110001
r	114	1110010
s	115	1110011
t	116	1110100
u	117	1110101
v	118	1110110
w	119	1110111
x	120	1111000
y	121	1111001
z	122	1111010
{	123	1111011
|	124	1111100
}	125	1111101
~	126	1111110
DEL (Delete/Rubout)	127	1111111

APPENDIX C

Using MATLAB to Plot Data from Text Files

To understand engineering problems and engineering solutions to problems, it is important to be able to visualize the numerical information that is involved. Therefore, the ability to easily obtain simple *xy* plots from data files is important in solving engineering problems.

In this appendix, we present a simple C program that generates a data file, and we then show how to use MATLAB to obtain a plot of the data. We chose MATLAB (MATrix LABoratory) to generate the plots in this appendix and also in the text chapters because it is an extremely powerful software environment for interactive numeric computations, data analysis, and graphics.

In the example that follows, we use a C program to generate a text file and MATLAB to plot the information. A text file can also be generated by means of a word processor, and then the same steps can be used to plot the information with MATLAB. If the data file is generated via a word processor, it is important to select the options for saving the file such that it is saved as a text file instead of as a word processor file.

The program shown next generates a data file containing 100 lines of information. Each line contains the corresponding time and function value from the damped sine function

$$f(t) = e^{-t} \sin(2\pi t)$$

where $t = 0.0, 0.1, 0.2, \ldots, 9.9$ seconds. The statements that open the data file, write information to it, and close it are discussed in Chapter 3.

C Program to Generate a Data File

```
/*-----------------------------------------------------------------*/
/*  Program chapterc_1                                             */
/*                                                                 */
/*  This program generates a data file of values from a           */
/*  damped sine function.                                         */

#include <stdio.h>
#include <math.h>
#define PI 3.141593
#define FILENAME "dsine.txt"

int main(void)
{
   /*  Declare variables.  */
   int k;
   double t, f;
   FILE *data_out;

   /*  Generate data file.  */
   data_out = fopen(FILENAME,"w");
   for (k=1; k<=100; k++)
   {
      t = 0.1*(k-1);
      f = exp(-t)*sin(2*pi*t);
      fprintf(data_out,"%.1f %.3f \n",t,f);
   }
```

```
      /*     Close data file and exit program.   */
      fclose (data_out);
      return 0;
}
/*------------------------------------------------------------*/
```

Text Data File Generated by the C Program

The data file generated by the sample program contains two numbers per line. The first few lines of information and the last line of information are as follows:

```
0.0   0.000
0.1   0.532
0.2   0.779
...
9.9   0.000
```

Generating a Plot with MATLAB

To generate a plot of this information with MATLAB, we need only two statements. Before executing the statements below, store the file in the MATLAB work directory (folder). The first statement loads the file into the MATLAB workarea, and the second statement generates the xy plot:

```
>>load dsine.txt
>>plot(dsine(:,1),dsine(:,2))
```

The plot generated is shown in Figure C.1.

Since it is important to label the information in a plot, we could also add statements to give the plot a title, to label the axes, and to add a background grid:

```
>>load dsine.txt
>>plot(dsine(:,1),dsine(:,2)),
>>title('Damped Sine Function'),
>>xlabel('Time, s'),ylabel('f(t)'),grid
```

The plot with these labels is shown in Figure C.2. The statements also assume that the data file is stored in the MATLAB work directory.

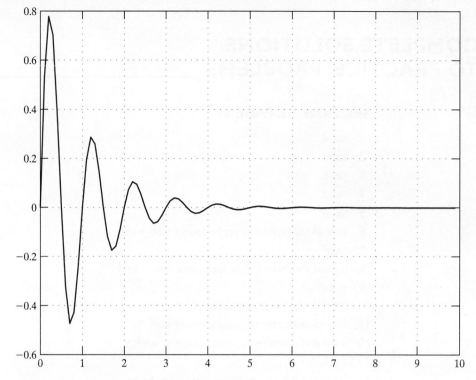

Figure C.1 *Plot of a damped sine function.*

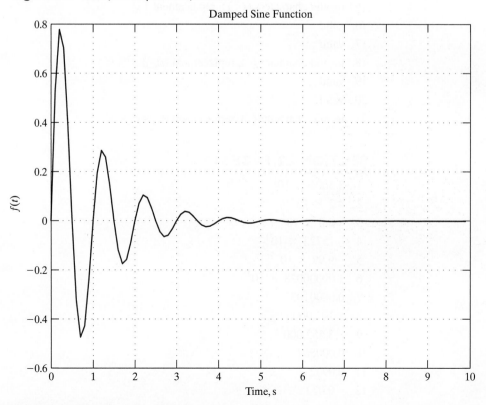

Figure C.2 *Enhanced plot of a damped sine function.*

COMPLETE SOLUTIONS
TO PRACTICE! PROBLEMS _____

SECTION 2.2, PAGE 50

1. valid
2. valid
3. valid
4. valid
5. valid
6. invalid character (-), replacement `tax_rate`
7. valid
8. invalid character (^), replacement `sec_sqrd`
9. valid
10. invalid, keyword, replacement `break_1`
11. invalid character (#), replacement `num_123`
12. invalid character (&), replacement `x_and_y`
13. valid
14. invalid, keyword, replacement `void_term`
15. invalid characters ((,)), replacement `fx`
16. valid
17. valid
18. invalid character (.), replacement `w1_1`
19. valid
20. valid
21. invalid character (/), replacement `m_per_s`

SECTION 2.2, PAGE 51

1. 3.5004×10^{1}
2. 4.2×10^{-4}
3. -5.0×10^{4}
4. 3.15723×10^{0}
5. -9.99×10^{-2}
6. 1.00000028×10^{7}
7. 0.0000103
8. $-105,000$
9. $-3,552,000$
10. 0.000667
11. 0.0902
12. -0.022

SECTION 2.2, PAGE 55

1. #define LIGHT_SPEED 2.99792e08
2. #define CHARGE_E 1.602177e-19
3. #define N_A 6.022e23
4. #define G_MSS 9.8
5. #define G_FTSS 32
6. #define MASS 5.98e24
7. #define MOON_RADIUS 1.74e06
8. #define UNIT_LENGTH 'm'
9. #define UNIT_TIME 's'

SECTION 2.3, PAGE 58

1. 6
2. 4.5
3. 3
4. 3.0

SECTION 2.3, PAGE 60

1. distance = x0 + v0*t + 0.5*a*t*t;
2. tension = (2*m1*m2*g)/(m1 + m2);
3. P2 = P1 + rho*v2*v2*(A2*A2 - A1*A1)/(2*A1*A1);
4. centripetal $= \dfrac{4\pi^2 r}{T^2}$
5. potential energy $= \dfrac{GM_E m}{r}$
6. change $= GM_E m\left(\dfrac{1}{R_E} - \dfrac{1}{R_E + h}\right)$

SECTION 2.3, PAGE 63

1. x `3`   y `4`   z `8`

2. x `3`   y `4`   z `12`

3. x `6`   y `4`

4. x `2`   y `0`

SECTION 2.4, PAGE 67

1. Sum =   65; Average =   12.4
2. Sum =   65
   Average = 12.3680

3. Sum and Average

   65 12.4
4. Character is b; Sum is A
5. Character is 98; Sum is 65
6.   12.37 is the average;
             65 is the sum
7.   12.37 is the average;   65 is the sum

SECTION 2.6, PAGE 75

1. 75.86°   89.35°   111.25°   109.92°
2. 0.67   1.60   1.71   1.87
3. There are five times that correspond to 110°, as can be seen from Figure 2.5. These
   values can be computed to be the following:
       1.98   2.84   3.39   4.42   4.67

SECTION 2.8, PAGE 82

1. -3   2. -2   3. 0.125   4. 3.16
5. 25   6. 11   7. -1   8. 32

SECTION 2.8, PAGE 83

1. `velocity = sqrt(pow(v0,2) + 2*a*(x - x0));`
2. `length = k*sqrt(1 - pow(v/c,2));`
3. `center = 38.1972*(r*r*r - s*s*s)*sin(a)/((r*r - s*s)*a);`
4. frequency $= \dfrac{1}{\sqrt{\dfrac{2\pi c}{L}}}$

5. range $= \dfrac{v_0{}^2}{g}\sin 2\theta$

6. $v = \sqrt{\dfrac{2gh}{1 + \dfrac{I}{mr^2}}}$

SECTION 2.8, PAGE 85

1. `cothx = cosh(x)/sinh(x);`
2. `secx = 1/cos(x);`
3. `cscx = 1/sin(x);`
4. `acothx = 0.5*log((x + 1)/(x - 1));`
5. `acoshx = log(x + sqrt(x*x - 1));`
6. `acscx = asin(1/x);`

SECTION 2.9, PAGE 87

1. x
2. A
3. a
4.
   ac
   C

SECTION 3.2, PAGE 110

1. true	2. true	3. true	4. true
5. true	6. true	7. false	8. false

SECTION 3.3, PAGE 115

1. ```
   if (time > 15)
       time += 1;
   ```
2. ```
   if (sqrt(poly) < 0.5)
       printf("poly = %f \n",poly);
   ```
3. ```
   if (abs(volt_1-volt_2) > 10)
       printf("volt_1: %f, volt_2: %f \n",volt_1,volt_2);
   ```
4. ```
   if (den < 0.05)
       result = 0;
   else
       result = num/den;
   ```
5. ```
   if (log(x) >= 3)
   {
       time = 0;
       count--;
   }
   ```
6. ```
   if (dist<50.0 && time>10)
       time += 2;
   else
       time += 2.5;
   ```
7. ```
   if (dist >= 100)
       time += 2;
   else
       if (dist > 50)
           time += 1;
       else
           time += 0.5;
   ```
   ```

SECTION 3.3, PAGE 117

```
1. switch (rank)
   {
      case 1: case 2:
         printf("Lower division \n");
         break;
      case 3: case 4:
         printf("Upper division \n");
         break;
      case 5:
         printf("Graduate student \n");
         break;
      default:
         printf("Invalid rank \n");
         break;
   }
```

SECTION 3.5, PAGE 127

1. 18	2. 18	3. 17
4. 9	5. infinite loop	6. 15

SECTION 4.2, PAGE 181

1.   Actual Parameters   Formal Parameters

x	25	$\longrightarrow$	a	25
sqrt(x)	5	$\longrightarrow$	b	5
x-30	-5	$\longrightarrow$	c	-5

2.   2

SECTION 4.2, PAGE 183

1. external identifiers: none
2. local variables and scope:
   main function:

```
seed, n, k, component_reliability, a_series,
a_parallel, series_success, parallel_success,
num1, num2, num3, rand_float
```

```
rand_float prototype statement:
    a, b
rand_float function:
    a, b
```

3. external identifiers: none
4. local variables and scope:

```
main function:
    n, k, a0, a1, a2, a3, a, b, step, left, right
check_roots prototype statement:
    left, right, a0, a1, a2, a3
check_roots function:
    left, right, a0, a1, a2, a3, f_left, f_right
poly prototype statement:
    x, a0, a1, a2, a3
poly function:
    x, a0, a1, a2, a3
```

SECTION 4.9, PAGE 218

1. ```
   #define area_sq(side) ((side)*(side))
   printf("area: %f \n",area_sq(side1));
   ```

2. ```
   #define area_rect(side1,side2) ((side1)*(side2))
   sum = area_rect(sidea,sideb) + area_rect(sidec,sided);
   ```

3. ```
   #define area_par(base,height) ((base)*(height))
   area1 = area_par(b,h1);
   ```

4. ```
   #define area_trap(base,height1,height2)
           (0.5*(base)*((height1)+(height2)))
   area += area_trap(base,left,right);
   ```

5. ```
   #define vol_sph(radius) (4.0/3.0*3.141593*pow((radius),3))
   printf("volume: %f \n",vol_sph(5.5));
   ```

6. ```
   #define vol_pyr(area,height) (1.0/3.0*(area)*(height))
   vol1 = vol_pyr(0.5*baseb*baseht,pyrht);
   ```

7. ```
   #define vol_cone(radius,height)
           (1.0/3.0*3.141593*pow((radius),3)*(height))
   vol2 = vol_cone(diameter/2,ht);
   ```

8. ```
   #define vol_cube(side) ((side)*(side)*(side))
   vol3 = vol_cube(sqrt(base_area));
   ```

9. ```
   #define vol_par(length,width,height)
           ((length)*(width)*(height))
   vol4 = vol_par(side1,side1,side1);
   ```
   ```

SECTION 5.1, PAGE 236

1.

-5	4	3	0	0	0	0	0	0	0

2.

'a'	'b'	'c'

3.

?	-5.5	5.5	5.5

4.

-0.4	-0.3	-0.2	-0.1	0	0.1	0.2	0.3	0.4

SECTION 5.1, PAGE 238

1. 3 8   2. 8 30
   15 21
   30 41

SECTION 5.1, PAGE 241

1. 9.8   2. 9.8   3. 3.2   4. 1.5

SECTION 5.4, PAGE 256

1. 9   2. 6   3. 5.36   4. 2.32
5. 2.5   6. 5.75

SECTION 5.8, PAGE 271

1.

1
4
6

2.

5	2
-2	3
0	0
0	0
0	0
0	0

3.

0	0	0	0
0	0	0	0
0	0	0	0
0	0	0	0

4.

1	0	0
0	1	0
0	0	1

5.

0	1	2	3	4
1	2	3	4	5
2	3	4	5	6
3	4	5	6	7
4	5	6	7	8

6.

1	-1	1	-1	1
1	-1	1	-1	1
1	-1	1	-1	1
1	-1	1	-1	1
1	-1	1	-1	1

SECTION 5.8, PAGE 273

1. 9
2. 4
3. −6
4. 3

SECTION 5.8, PAGE 276

1. 13
2. 2
3. 18
4. 22

SECTION 5.10, PAGE 285

1. 5
2. −8
3. $\begin{bmatrix} 5 & -1 & 3 \\ 3 & -3 & 2 \end{bmatrix}$

4. $[-4 \quad 12]$

5. $\begin{bmatrix} -2 & -2 & 4 \\ 7 & -9 & 10 \end{bmatrix}$

6. $\begin{bmatrix} 24 \\ -20 \\ 18 \end{bmatrix}$

SECTION 5.11, PAGE 292

1. $x = 2, y = 1$
2. $x = 3, y = -1, z = 2$

SECTION 6.1, PAGE 314

1.

ptr

a | 1 | b | 2 |

2.

ptr

a | 2 | b | 2 |

3.

ptr

a | 1 | b | 5 | c | 1 |

4.

ptr

a | 2 | b | 2 | c | 2 |

SECTION 6.2, PAGE 318

1. ptr_2   ptr_1

x | 15.6 | y | 31.2 |

2.   ptr_2

w | 10 | x | -8 |

3.   ptr_1
    ptr_3 ptr_2

x | 2 | 4 | 6 | 8 | 3 |

4. first_ptr   last_ptr

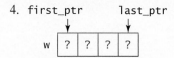

SECTION 6.2, PAGE 320

		offset
g[0]	2	0
g[1]	4	1
g[2]	5	2
g[3]	8	3
g[4]	10	4
g[5]	32	5
g[6]	78	6

1.	2	5.	2
2.	4	6.	8
3.	3	7.	4
4.	32	8.	32

SECTION 6.2, PAGE 321

1.

		offset
d[0][0]	1	0
d[0][1]	6	1
d[1][0]	0	2
d[1][1]	0	3
d[2][0]	0	4
d[2][1]	0	5
d[3][0]	0	6
d[3][1]	0	7

2.

		offset
g[0][0]	5	0
g[0][1]	2	1
g[0][2]	-2	2
g[0][3]	3	3
g[1][0]	1	4
g[1][1]	2	5
g[1][2]	3	6
g[1][3]	4	7
g[2][0]	0	8
g[2][1]	0	9
g[2][2]	0	10
g[2][3]	0	11

3.

		offset
h[0][0]	0	0
h[0][1]	0	1
h[0][2]	0	2
h[1][0]	0	3
h[1][1]	0	4
h[1][2]	0	5
h[2][0]	0	6
h[2][1]	0	7
h[2][2]	0	8

SECTION 6.2, PAGE 321

		offset
x[0][0]	1	0
x[0][1]	8	1
x[0][2]	7	2
x[0][3]	6	3
x[1][0]	2	4
x[1][1]	4	5
x[1][2]	-1	6
x[1][3]	0	7

1. 1   3. 3
2. 7   4. 14

SECTION 6.2, PAGE 322

1.
```
int a[4][6], sum=0, *ptr=&a[1][0];
...
for (k=0; k<=5; k++)
   sum += *(ptr+k);
```

2.
```
int a[4][6], sum=0, *ptr=&a[0][2];
...
for (k=0; k<=3; k++)
   sum += *(ptr+k*6);
```

3.
```
int a[4][6], *ptr=&a[0][0];
...
max = a[0][0];
for (k=0; k<=17; k++)
   if (max < *(ptr+k))
     max = *(ptr+k);
```

```
4. int a[4][6], *ptr=&a[0][0];
   ...
   min = a[0][2];
   for (i=0; i<=3; i++)
      for(j=2; j<=5; j++)
         if (min > *(ptr+i*6+j))
             min = *(ptr+i*6+j);
```

SECTION 6.4, PAGE 328

1. invalid, arguments are not pointers to integers

2.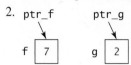

3. invalid, arguments are not pointers

4. invalid, arguments are not pointers to f and g

5.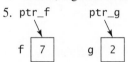

6. invalid, arguments are not pointers

SECTION 7.1, PAGE 357

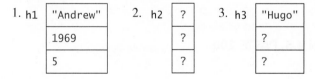

SECTION 7.1, PAGE 358

1. Audrey
   Frederic

2. Category 4 Hurricane: Audrey

SECTION 8.3, PAGE 396

1. 15012.368

2. 15012.368

3. 150
   12.368

4. 150
   12

5. 150,12.4

6. 150,12.368

SELECTED SOLUTIONS
TO MODIFY! PROBLEMS _____

SECTION 2.4, PAGE 72

```
1. /*-----------------------------------------------------------*/
   /*  Program chapter2_mod                                    */
   /*                                                          */
   /*  This program includes a division by zero               */
   /*  to determine the system response.                      */

   #include <stdio.h>

   int main(void)
   {
      /*  Declare and initialize variables.  */
      double a=5, b=0, c;

      /*  Execute a division by zero.  */
      c = a/b;

      /*  Print the value of c.  */
      printf("c = %f \n",c);

      /*  Exit program.  */
      return 0;
   }
   /*-----------------------------------------------------------*/
```

SECTION 4.5, PAGE 200

```
   /*-----------------------------------------------------------*/
   /*  Program chapter4_5mod                                    */
   /*                                                           */
   /*  This program generates and prints ten random            */
   /*  floating-point values between user-entered limits.       */

   #include <stdio.h>
   #include <stdlib.h>

   int main(void)
   {
      /*  Declare variables and function prototypes.  */
      unsigned int seed;
      int k;
      double a, b;
      double rand_float(double a, double b);

      /*  Get seed value and interval limits.  */
      printf("Enter a positive integer seed value: \n");
      scanf("%u",&seed);
```

```
        srand(seed);
        printf("Enter limits a and b (a<b): \n");
        scanf("%f %f",&a,&b);

        /*  Generate and print ten random numbers.  */
            printf("Random Numbers: \n");
        for (k=1; k<=10; k++)

            printf("%f ",rand_float(a,b));

        /* Exit program.  */
        return 0;
    }
    /*-----------------------------------------------------------*/
```
        (No changes in the rand_float *function from page 179.)*
```
    /*-----------------------------------------------------------*/
```

SECTION 5.6, PAGE 264

```
1.  /*-----------------------------------------------------------*/
    /*   Program chapter5_mod                                */
    /*                                                       */
    /*   This program initializes an array and then uses     */
    /*   the selection sort to reorder it.                   */

    #include <stdio.h>

    int main(void)
    {
    /*  Declare variables and function prototypes.  */
        int k;
        double x[10]={4,8,-2,16,19,6,-4,0,20,3};
        void sort(double x[], int n);

        /*  Print original order.  */
        printf("Original Order \n");
        for (k=0; k<=9; k++)
            printf("%.1f ",x[k]);
        printf("\n");

        /*  Sort values.  */
        sort(x,10);

        /*  Print new order.  */
        printf("New Order \n");
        for (k=0; k<=9; k++)
            printf("%.1f ",x[k]);
        printf("\n");

        /*  Exit program.  */
        return 0;
    }
    /*-----------------------------------------------------------*/
```
            (No changes in the sort *function from page 243.)*
```
    /*-----------------------------------------------------------*/
```

COMPLETE SOLUTIONS TO END-OF-CHAPTER SHORT-ANSWER PROBLEMS

CHAPTER 1

1. F
2. T
3. T
4. T
5. F
6. T
7. F
8. F
9. F
10. T
11. F
12. F
13. F
14. T
15. T
16. F
17. F
18. F
19. T
20. T
21. (d)
22. (c)
23. (c)
24. (d)
25. (a)

26. (b)
27. (e)
28. (c)
29. (d)
30. (c)
31. (d)
32. (b)
33. (c)
34. program
35. hardware
36. central processing unit
37. output devices
38. system software
39. algorithm
40. compilation
41. spreadsheet
42. syntax
43. operating system
44. arithmetic logic unit
45. debugging
46. logic errors
47. ANSI C
48. microprocessor
49. supercomputer
50. machine language

CHAPTER 2

1. T
2. T
3. F
4. T
5. F
6. incorrect, `register` is a keyword and cannot be used as an identifier
7. correct
8. incorrect
   ```
   long double d = 12.78909876;
   ```

9. incorrect
   ```
   float a1, a2;
   ```
10. correct
11. (c)
12. (d)
13. (c)
14. (c)
15. (e)

16.
x 27   y 8   z 3

18. f= +8.092
19. value_4= 6.65e01
20. 97 a

17. a 3.8

x 2.8

n 2

CHAPTER 3

1. T
2. F
3. F
4. F
5. T
6. T
7. incorrect; commas should be semicolons
8. No error
9. incorrect: duplicate case in function main(), since ASCII of 'A' is 65

10. (c)
11. (c)
12. (a)
13. (a)
14. (c)
15. (b)
16. (d)
17.

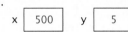

x 500   y 5

CHAPTER 4

1. T
2. F
3. T
4. T
5. (b)
6. (a)
7. (d)
8. x=8 y=9
9. x=9 y=9
10. x=10 y=12
11. No, this function cannot swap two integers passed as arguments. In this function, though the first argument, i, is called by reference, the second argument, j, is called by value. As i contains the address of the first argument, *i has access to the actual argument. However, only a copy of the actual argument is stored in j. So, any changes within the function body, does not change the value of the second argument. Hence, it fails to swap two integers passed as arguments.

CHAPTER 5

1. T
2. F
3. F
4. T

5. (a)
6. (a)
7. (a)
8. (b)

9.

1	2	3	5	8	13	21	34	55	89

11. 6

12. 8

10.

4	4
3	2
2	4
1	2

CHAPTER 6

1. T
2. T
3. F
4. F
5. (c)
6. (b)
7. (b)
8. (b)

9.

name | 20.5

x | 20.5

a | 14

10. Value of `*i` will be `'P'` and that of `*j` will be `'S'`.

11. Value of `str[0]` will be `'P'` and that of `str[8]` will be `'S'`.

12. `Problem`

13. `gnivloS melborP`

CHAPTER 7

1. F
2. T
3. T
4. T
5. T

6. (c)
7. (a)
8. (b)
9. (d)

10. start_date

?	month
?	day
?	year

end_date

?	month
?	day
?	year

11. start_date

9	month
?	day
?	year

end_date

12	month
?	day
?	year

12. start_date

9	month
?	day
2005	year

end_date

12	month
?	day
2008	year

13. start_date

9	month
?	day
2005	year

end_date

12	month
30	day
2008	year

CHAPTER 8

1. T
2. F
3. T
4. F
5. T
6. T
7. F
8. T
9. T
10. F
11. (a)
12. (c)
13. (b)

SELECTED SOLUTIONS TO END-OF-CHAPTER PROGRAMMING PROBLEMS _____

CHAPTER 2

```
/*------------------------------------------------------*/
/*   Program chapter2_prob39                            */
/*                                                      */
/*   This program computes the molecular weight         */
/*   for an amino acid.                                 */

#include <stdio.h>
#define O 15.9994
#define C 12.011
#define N 14.00674
#define S 32.066
#define H 1.00794

int main(void)
{
    /*  Declare variables.  */
    int num_o, num_c, num_n, num_s, num_h;
    double weight;

    /* Prompt the user for numbers of atoms.  */
    printf("Enter number of oxygen atoms: \n");
    scanf("%i",&num_o);
    printf("Enter number of carbon atoms: \n");
    scanf("%i",&num_c);
    printf("Enter number of nitrogen atoms: \n");
    scanf("%i",&num_n);
    printf("Enter number of sulfur atoms: \n");
    scanf("%i",&num_s);
    printf("Enter number of hydrogen atoms: \n");
    scanf("%i",&num_h);

    /*  Compute molecular weight.  */
    weight = O*num_o + C*num_c + N*num_n +
             S*num_s + H*num_h;

    /*  Print the molecular weight.  */
    printf("amino acid molecular weight = %.5f \n",
           weight);

    /*  Exit program.  */
    return 0;
}
/*------------------------------------------------------*/
```

CHAPTER 3

```
/*-----------------------------------------------*/
/*  Program chapter3_prob37                       */
/*                                                */
/*  This program prints a table showing the       */
/*  number of acres of land reforested at the     */
/*  end of each year, for 20 years.               */

#include <stdio.h>
#define UNCUT 2500
#define RATE 0.02

int main(void)
{
   /*  Declare variables.  */
   int year;
   double old_forest=UNCUT, new_forest;

   /*  Print report.  */
   printf("Reforestation Summary \n");
   printf("Year    Total Acres Forested \n");
   for (year=1; year<=20; year++)
   {
      new_forest = old_forest*RATE;
      old_forest += new_forest;
      printf("%4i    %10.2f \n",year,old_forest);
   }

   /*  Exit program.  */
   return 0;
}
/*-----------------------------------------------*/
```

CHAPTER 4

```
/*-----------------------------------------------*/
/*  Program chapter4_prob14                       */
/*                                                */
/*  This program simulates rolling two six-sided  */
/*  dice and computes the percentage of time that */
/*  the sum of the dots on the dice equals 8.     */

#include <stdio.h>
#include <stdlib.h>

int main(void)
{
   /*  Declare variables and function prototypes.      */
   unsigned int seed;
   int rolls, k, die_1, die_2, sum=0;
   int rand_int(int a, int b);
```

```
   /*  Get seed value.                             */
   printf("Enter a positive integer seed value: \n");
   scanf("%u",&seed);
   srand(seed);

   /*  Prompt user for number of rolls.            */
   printf("Enter number of rolls of dice: \n");
   scanf("%i",&rolls);

   /*  Simulate rolls of the dice.                 */
   for (k=1; k<=rolls; k++)
   {
      die_1 = rand_int(1,6);
      die_2 = rand_int(1,6);
      if (die_1+die_2 == 8)
         sum++;
   }

   /*  Compute and print percentage               */
   /*  of rolls with a sum of 8.                   */
   printf("Number of rolls: %i \n",rolls);
   printf("Percent with sum of eight: %.2f \n",
          sum*100.0/rolls)

   /*  Exit program.  */
   return 0;
}
/*------------------------------------------------------*/
        (Include function rand_int from page 177.)
/*------------------------------------------------------*/
```

CHAPTER 6

```
/*-----------------------------------------------------*/
/*   Function chapter6_prob18                          */
/*                                                     */
/*   This function determines the number of positive,  */
/*   negative, and zero values in an array.            */

void signs(int x[], int npts, int *npos,
           int *nzero, int *nneg)
{
   /*  Declare variables.  */
   int k;

   /*  Set sums to zero.  */
   *npos = *nzero = *nneg = 0;

   /*  Update corresponding sums.  */
   for (k=0; k<=npts-1; k++)
   {
      if (x[k] < 0)
         *nneg += 1;
      else
         if (x[k] > 0)
            *npos += 1;
         else
            *nzero += 1;
   }

   /*  Void return.  */
   return;
}
/*-----------------------------------------------------*/
```

GLOSSARY

abbreviated assignment an assignment statement that uses a shortened format

abstraction a concept in which a programmer can use modules to accomplish specific tasks without needing to know the details of the steps within the modules

actual parameter a value that corresponds to a formal parameter when a function is invoked

address a positive integer that uniquely defines a memory location

address operator a unary operator that determines the memory address of an identifier

algorithm a step-by-step outline of a solution to a problem

ANSI C an American National Standards Institute standard that provides a system-independent and unambiguous definition of the language C

argument an input to a function

arithmetic logic unit (ALU) the part of the computer that performs the arithmetic and logical operations

array a data structure that allows a group of values to be represented with the use of a common name and to be distinguished via subscripts

ASCII American Standard Code for Information Interchange

assembler a program that converts an assembly language program to binary

assembly language a language written in English-like statements and that is specific to a particular type of CPU

assignment statement a statement that assigns a value to an identifier

associativity the property according to which the operations in an expression are grouped

automatic class a class used to represent local variables

binary two states, usually represented by 0 and 1

binary code a code composed of 0's and 1's

binary operator an operator, such as addition, that operates on two values

binary search a search algorithm that reduces the number of values to search in half with each comparison

binary tree a linked list with nodes that have a left branch and a right branch

bit a binary digit, 0 or 1

bug an error in a program

byte a unit of memory that contains 8 bits or binary digits

call-by-address a function reference in which the address of the actual parameter is used as the address of the corresponding formal parameter

call-by-reference a function reference in which the address of the actual parameter is used as the address of the corresponding formal parameter

call-by-value a function reference in which the value of the actual parameter is passed to the corresponding formal parameter

case label the expression used to control a case structure

case sensitive having the property such that lowercase and uppercase letters are perceived as different characters

case structure a structure in which groups of statements are performed in accordance with the value of a controlling expression

cast operator a unary operator that specifies a type change in the value before the next computation

central processing unit (CPU) the combination of the processor and the ALU

character a data type which represents information that is not restricted to being numeric

character function a function that has character arguments or that returns a character value

character string a character array that ends with a null character

circularly linked list a linked list in which the link in the last node points to the first node in the list

class an abstract data type that consists of both data and functions

class declaration C++ statements that declare the name, data members, and function members of a class

class implementation C++ statements that provide complete function definitions for all function members of a class

cloud computing a technique for accessing large amounts of information remotely

coercion of arguments the process of converting the type of a value to another type before using the value in a computation

collating sequence an ordering of characters for a specific code, from low to high

comment a statement in a program that is not an instruction, but that is used to document the steps in the program

compiler a program that translates a program in a high-level language into machine language

compiler error an error identified during the compilation of a program

composition nesting of functions

compound statement a set of statements that are enclosed in braces

computer a machine designed to perform operations that are specified with a set of instructions called a program

computer simulation a computer program that often uses random numbers to model an event

concatenate to place end-to-end

condition an expression that can be evaluated as either true or false

conditional operator a ternary operator that has three arguments: a condition, a statement to perform if the condition is true, and a statement to perform if the condition is false

constant a value, such as 3.141593, that does not change during the execution of a program

constructor function member function of a class that provides for the automatic initialization of class objects

control character one of the following characters: FF (form feed), NL (new line), CR (carriage return), HT (horizontal tab), VT (vertical tab), BEL (bell), BS (backspace)

control string a string in an output statement that specifies the format to use for an output line

controlling expression the expression used in a `switch` statement

conversion specifier a specifier that describes the format to be used in printing a value

data file a file that contains data that can accessed or generated by a program

data member a variable associated with a structure

database management tool a software tool for manipulating and retrieving information from large amounts of data

debug to identify and remove bugs or errors from a program

debugger a program that assists in identifying and removing bugs or errors from a program

declaration a statement that defines variables to be stored in memory

decomposition outline an outline of the general steps necessary to solve a problem

default constructor a constructor function that is called when an object is defined but not initialized

default label a label in a `switch` statement that is used to indicate statements to execute if none of the other statements are executed

dereference an operation that references the value contained in an address that is stored in a pointer

desktop publishing the production of professional-looking documents by a powerful word processor with a high-quality printer

determinant a specific value computed from the entries in a matrix

divide and conquer strategy for solving a large problem by breaking it into smaller problems

dot operator an operator used when objects call member functions

dot product the sum of the products of the values in corresponding positions in two vectors

doubly linked list a linked list in which each node contains a forward link and a backward link

driver program a program whose purpose is to provide a simple interface for testing a function

dynamic data structure a data structure that grows in size by using dynamic memory allocation as data is added

dynamic memory allocation a technique that allows a programmer to specify a memory allocation during the execution of a program

EBCDIC Extended Binary Coded Decimal Interchange Code

electronic copy information that is stored in a computer or in a form that the computer can read, such as on a CD

element a value in an array

empty list a list in which the pointer to the first node contains a NULL character

empty statement a semicolon that is used within an if structure to represent no action

end-of-file indicator a special character at the end of a file to indicate that the end of the file has been reached

EOF character a special character that indicates the end of a text stream

error condition a condition that should not occur in the desired execution of a program

escape character a backslash (\) used in a control string

execution the process of executing the steps described by a program

exponential notation notation that uses the letter e to separate the mantissa from the exponent in scientific notation, as in 3.1e02

expression a group of terms composed of constants, variables, and operators that can be evaluated as a single value

external class a class used to represent global variables that have the entire program as their scope

extraction operator the characters $<<$ used with the cout object

factorial a function of a positive integer that is the product of the integer and all integers between it and 1

Fibonacci sequence a sequence of values that begins with the values 1,1 and continues with each succeeding value being the sum of the two previous values

field width the value specified that controls the minimum number of positions used to print a value

FIFO structure another term for a queue, or a first-in–first-out structure

file open mode a character that indicates the status of a data file

file pointer a pointer variable that is associated with a data file

floating-point value a value that can represent both integer and noninteger values

flowchart a diagram used to describe the steps in an algorithm

for loop a loop that is executed a specified number of times

formal parameter an identifier used in the definition of a function to represent an input value

format flag a flag used in C++ to format output

function a module that returns at most one value to the invoking statement

function prototype a statement that identifies the information necessary to invoke a function

garbage value a value from a previous program that is in a memory location until it is initialized

Gauss elimination a numerical technique for finding the solution of a set of simultaneous equations

global variable a variable defined outside the `main` function or other programmer-defined functions

hardware the computer equipment, such as the keyboard, the mouse, and the hard disk

head a pointer that points to the first node in a linked list

high-level language a language with English-like commands that is not specific to a particular type of CPU

hyperbolic function a function of the natural logarithm function or of the natural exponential function

hyperplane the space represented by an equation with more than three variables

I/O diagram a simple block diagram that defines the input and the output information for a program

identifier name used to reference the value stored in a memory location

ill conditioned lacking a unique solution (said of a system of equations)

incremental search a numerical technique for estimating the roots of a function

indirection an operation that references the value contained in an address that is stored in a pointer

inheritance the ability for a class to inherit attributes from an existing class

initial value the value first given to a variable (often included in the declaration statement)

inner product the dot product

insertion operator the characters $>>$ used with the `cin` object

invoke to call or reference a function (or module)

iteration one pass through a loop

kernel component of the operating system that manages the interface between the hardware and software applications

keyword word with special meaning to the C compiler

least squares a technique that minimizes the square of the difference between a model and a given function or a given set of data points

left justify an alignment in which there are no spaces to the left of a value

library function a function that is included in the files that accompany a compiler

LIFO structure another term for a stack, or a last-in–first-out structure

linear interpolation a numerical technique for estimating the value of a function by assuming that it falls between two points on a straight line

linear modeling modeling a set of data values with a straight line

linear regression a numerical technique for determining the equation of a straight line that best fits a set of data values

linked list a data structure in which each data member includes information that links to the next data member

linking/loading the process of preparing an object program for execution

local variable a variable whose scope is the function in which it is defined

logic error an error in the logic of the steps used to solve a problem

logical operator an operator that is used to compare conditions

loop a set of statements that are repeated

loop control variable a variable used to control a for loop

machine language a language in which instructions are written as binary strings

macro a preprocessing directive that can be used to define a simple function

magnitude absolute value

math function a function that computes the value of a common function, such as the square root of x

matrix a set of numbers arranged in a rectangular grid with rows and columns

matrix multiplication an operation between two matrices that determines a new matrix

mean average value of a list of values

median the middle value in a group of sorted values if there is an odd number of values; otherwise, the average of the two middle values

member function function associated with a class

memory the part of a computer that stores information

memory snapshot a diagram that shows the contents of a memory location at a specified point in the execution of a program

microprocessor a CPU that is contained in a single integrated circuit chip that is smaller than a postage stamp

mixed operation an operation between values with different types

modularity the result of the process of separating the solution to a problem into a group of modules

module a set of statements that perform an operation or that compute a value which can be considered to be a unit in terms of functionality

module chart a diagram that shows the modular structure of a program

modulus an operation that computes the remainder upon division of one integer by another

multiple assignment a statement that allows multiple variables to be assigned values

network an interconnection of computers such that they can share resources and information

node a structure that consists of data plus a pointer to another node

nonsingular a characteristic of a set of equations that have a unique solution

NULL character a constant with the value of binary zero

object an instance of a class

object program a program in machine language

object-oriented programming an approach to programming characterized by the use of classes

offset an integer value that gives the number of positions from the first element in the memory allocation for an array

one-dimensional array a data structure that can be visualized as a list of values arranged in either a row or a column

operating system software that provides an interface between the user and the hardware

overflow an error caused when the result of an arithmetic operation is too large to store in the memory assigned to it

overload to give an operator different meanings, depending on the data type with which it is used

parameter the input to a function; also called an argument

parsing examining the individual characters in an array or a string of characters

personal computer (PC) a small, inexpensive computer that is designed around a microprocessor chip

pointer a variable that contains the memory address of another variable

pointer operator an operator used instead of a structure member operator when data members are accessed by pointing to a structure

polymorphism the ability to assign many meanings to the same name

postfix a position after an identifier

power average squared value of a set of values

precedence the order in which operations are evaluated in an expression

precision the number of decimal digits specified by the mantissa of a value in scientific notation

prefix a position before an identifier

preprocessor directive a statement that gives an instruction to the compiler

private member a member function of a class that may be referenced only by other member functions

problem-solving process a methodology for approaching new problems

processor the part of the computer that controls all the other parts

program the set of instructions that describe the operations to be performed by a computer

program walkthrough a technique in which an algorithm or a program designed to solve a complicated problem is presented in detail to a new group of people in order to get their feedback and suggestions

programmer-defined function a function written by a programmer

prompt a message printed by a program to indicate that information should be entered

pseudocode a set of English-like statements used to describe the steps in an algorithm

public member a member function of a class that can be referenced anywhere in the user program

queue a data structure in which nodes are added at one end and removed from the other end; also called a first-in–first-out (FIFO) structure

random number a number that is defined by statistical properties rather than an equation

random number seed a value that is used to initialize a random sequence

real-time program a program usually written in assembly language so that it executes very fast

recursion a methodology that implements the solution to a problem using a process that invokes itself

register class a class used to represent variables that need to be accessed frequently

relational operator an operator that is used to compare two expressions

repetition a control structure that contains a set of steps that are repeated as long as a condition is true

reusability the result of a process in which software is developed in modules that can be used in a variety of solutions to problems

right justify an alignment in which there are no spaces to the right of a value

root a value of x for which $f(x)$ is equal to zero

scientific notation notation that expresses a value as a mantissa times a power of 10, as in 3.1×10^2

scope the portion of a program in which it is valid to reference a function or a variable

scope resolution operator the symbols :: used between a class name and a function name to specify that the function is a member of the class

selection a control structure that contains one set of steps to perform if a condition is true and another set of steps to perform if the condition is false

selection sort algorithm a sort algorithm that performs several passes through an array, exchanging minimum values with values in specified positions

sequence a control structure composed of steps performed one after another

sequential search a search algorithm that begins with the first value in a list, and searches sequentially for a specified value

sentinel signal a value included at the end of a data file to indicate that the end of the file has been reached

simultaneous linear equations a set of linear equations with a common solution

software the set of programs that describe the steps we want the computer to perform

software life cycle the stages in the development of a large software project

software maintenance the work necessary to add enhancements to existing software, to fix errors identified in the software, and to adapt the software so that it works with new hardware and software

software prototype a software package that does not have all the functions of the final system, but that has much of the user interface so that it can be evaluated by the user

software tool a program written to perform common useful operations, such as generating a report or a graph

sorting a technique in which a group of values is put into ascending or descending order

source program a program in a high-level language

spreadsheet a software tool that works with information that can be displayed in a grid of rows and columns

square matrix a matrix with the same number of rows as columns

stack a linked list in which the last item added to the list is the first item removed from the list; also called a last-in–first-out (LIFO) structure

Standard C library a library of constants and functions that can be accessed from a C program

standard deviation square root of the variance

statement a comment or an instruction in a program

static class a designation which specifies that the memory allocated for a variable should be retained during the entire execution of a program

stepwise refinement the process of breaking the solution to a problem into a sequence of smaller and smaller steps

storage class a designation that determines the scope of a variable

stream a sequence of characters

structure a collection of variables that can be of the same data type or that can be of different data types

structure chart a diagram that shows the modular structure of a program

structure member operator the period used to separate a structure variable name and the data member name

structured program a program written with simple control structures to organize the solution to a problem

subscripts integers used to distinguish elements in an array

summation notation a mathematical notation used to describe the sum of a set of values

symbolic constant a constant that is assigned an identifier by a preprocessor directive

syntax the grammatical rules of a language

system dependent the property whereby a feature may not be available on all computer systems

system of equations a set of equations with a common solution

tag the name associated with a defined structure

test data data designed to test the correctness of a program

top-down design a design methodology that starts with a general "big picture" description of the solution to a problem and then refines the solution

trailer signal a value included at the end of a data file to indicate that the end of the file has been reached

transpose a matrix generated from another matrix such that the rows of the original matrix form the columns of the new matrix

trigonometric function a function that computes a value from a trigonometric or inverse trigonometric function

truncate to drop the fractional portion of a value

two-dimensional array a data structure that can be visualized as a table or grid of values displayed in rows and columns

type specifier a term that distinguishes the various types of forms in which C can store numeric values

unary operator an operator (e.g., negation) that operates on a single value

underflow an error caused when the result of an arithmetic operation is too small to store in the memory assigned to it

utility a program for performing a common function, such as copying a file from a hard disk to a CD

validation and verification two processes aimed at verifying that a program is correctly performing its objectives and that those objectives solve the problem at hand

variable a memory location that is given a name and whose contents may or may not change during the execution of a program

variance average squared deviation of a group of values from their mean

vector a matrix composed of one row or of one column

void pointer a pointer returned by a function that does not specify the type of variable to which the pointer is to point

while loop a loop that is executed as long as a condition is true

white space the space character or one of the following characters: FF (form feed), NL (new line), CR (carriage return), HT (horizontal tab), VT (vertical tab)

word processor a software tool for entering and formatting text that may be used in reports or in a computer program

zero crossing a point at which a function crosses the x-axis

INDEX